Internet E-mail:
Protocols, Standards, and
Implementation

For a complete listing of the *Artech House Telecommunications Library,*
turn to the back of this book.

Internet E-mail:
Protocols, Standards, and
Implementation

Lawrence Hughes

Artech House
Boston • London

Library of Congress Cataloging-in-Publication Data

Hughes, Lawrence E.
    Internet E-mail : protocols, standards, and implementation /
Lawrence Hughes.
        p.   cm. — (Artech House telecommunications library)
    Includes bibliographical references and index.
    ISBN 0-89006-939-5 (alk. paper)
    1. Electronic mail systems.   2. Internet (Computer network)
I. Title.   II. Series
TK5105.73.H84   1998
004.692—dc21   98-20082
                                                CIP

British Library Cataloguing in Publication Data

Hughes, Lawrence
    Internet E-mail : protocols, standards, and implementation.
    —(Artech House telecommunications library)
    1. Electronic mail systems   2. Internet (Computer network)
    I. Title
    004.6'92

    ISBN   0-89006-939-5

Cover and text design by Darrell Judd

© 1998 ARTECH HOUSE, INC.
685 Canton Street
Norwood, MA 02062

International Standard Book Number: 0-89006-939-5
Library of Congress Catalog Card Number: 98-20082

10 9 8 7 6 5 4 3 2 1

To Dylan Edwin Hughes, who was born during the writing of this book. My wife is from the Philippines, and she returned there for Dylan's birth. My first glimpse of my new son was via a scanned-in JPEG image attached to an e-mail message and sent halfway around the world in a matter of seconds. Dylan will grow up in a world of instant global communication that I can scarcely imagine. I hope this book will contribute in some small way to making his world the best possible one.

Contents

Preface

I have worked in the area of computer communications for many years, most recently with Internet messaging and computer security. Amazed at the dearth of available books on Internet e-mail (especially on the technical aspects, such as the protocols), I decided to write one myself.

This book is intended for several audiences:

▶ Anyone who is developing or planning to develop Internet messaging software or even software that is "messaging enabled." An example of the former would be an e-mail client program or even an Internet e-mail server or gateway (say, between SMTP and some proprietary e-mail system, such as Banyan mail). An example of the latter would be an accounting program that sends out invoices or accepts purchase orders in the form of Internet e-mail messages (possibly compatible with Electronic Document Interchange, or EDI).

▶ Network administrators responsible for installing, managing, or troubleshooting Internet e-mail server software. Doing that is difficult or impossible without a good understanding of the protocols and message flow. This book also explains how to use tools like Telnet to help debug such systems.

▶ Anyone evaluating messaging systems for use in any size organization, especially if there is a need to exchange messages with users outside the organization. An in-depth understanding of the technologies (e.g., the recent extensions to the SMTP protocol) can help tremendously in comparing products. An understanding of the emerging global standards-based messaging infrastructures also will help these people avoid possible costly errors by choosing technologies and products that are more likely to work with this emerging infrastructure.

▶ Anyone currently using e-mail who would like to peek under the hood. This book will allow them to make more effective use of their tools. It will increase their ability to exchange messages and attachments with other users. It also will make them aware of recent advances, especially in the area of secure messaging. An educated end-user community is more likely to create a demand for high-quality, standardized, globally interoperable messaging systems.

▶ Students in advanced undergraduate or graduate courses who would like to understand Internet messaging.

The book is divided into three parts. A serious reader should cover all three parts, because later chapters build on earlier ones. Advanced readers already familiar with the subject area might find individual chapters useful out of context, such as the ones on IMAP or S/MIME. Ideally, between the text in the book and the text of the RFCs included on the accompanying CD-ROM, most readers should be able to obtain a complete understanding of Internet e-mail.

Part I is an overview of the two basic messaging architectures, shared file and client/server, and discusses the relative merits of each (with an admitted bias toward the latter). It also covers the technical details of secure messaging.

Part II covers the underlying infrastructure on which messaging is built, including TCP/IP, DNS, and directory services. There are other books that cover each of these subjects in considerably greater detail, but if you understand the concepts as they are covered here, you should have no trouble understanding the concepts in the rest of the book.

Part III examines the various protocols and standards used in Internet e-mail, including SMTP, ESMTP, MIME, DSN, POP, IMAP, and NNTP. If you are not already familiar with the field, once you have read through this section, this mess of alphabet soup should become second nature to you.

Be aware that e-mail even now is undergoing dramatic and rapid evolution. Existing standards are being revised and new standards introduced (possibly replacing old standbys). I have reconciled myself to the fact that this book will be a lifelong project, with revisions every few years. I only hope that the current version is not hopelessly outdated in the short time that the publisher takes to transform my rough scribbles into a presentable book, print it, and distribute it to bookshelves (someday electronic publishing will drastically reduce even that short amount of time). The World Wide Web (WWW) has given us a tantalizing taste of such immediate distribution of information from writer to reader. I hope that someday future versions of this book will be available through "instant publishing." Perhaps that will be the only way to keep up with the breakneck pace of innovation in this subject area.

So why should you want to read this book? Your time is valuable, and you probably already have a giant stack of "must-read" books on your shelves.

Simply put, Internet e-mail is one of the most important topics in today's online world. E-mail in general is already handling some 30% of the traffic formerly handled by the U.S. Postal Service (and probably a much higher percentage of the *useful* traffic). Much of that traffic goes through the Internet either directly or via gateways. E-mail is easily the most important application on the Internet (in survey after survey, the Web, while flashier and far more hyped, ranks far lower in terms of business-critical applications). It will become only more important in years to come, especially concerning how to make it work globally and securely.

It is also critical that you understand some of the social aspects of Internet e-mail (again, especially with respect to cryptography), because government policies are being written and put into place that, in the decades to come, will profoundly affect your freedom of speech and especially your right to privacy. Most people are not even aware that these issues exist, let alone who is on what side. Many of the big-brother types in government are hoping they can put controls in place before the general populace is even aware of the issues. I hope this book will help wake up at least some people to the issues involved.

Unfortunately, there is a real dearth of information on how this very important technology really works (quick, what is the RFC that defines the ESMTP protocol, and where can you find a copy?). I have tried to pull together most of the details on how it all works and explain it in terms that are, I hope, more approachable than "RFC-ese" (a curious dialect of computer-speak).

Contents

System architecture

E-mail (short for "electronic mail") is fast becoming one of the most commonly used channels for communication in both office and home environments. Most experts agree that e-mail is the single most important and widely used application on the Internet, if not the most hyped (the World Wide Web holds that distinction). Some estimates indicate that as much as 30% of all correspondence that previously would have gone via traditional channels, such as the U.S. Postal Service, is now being exchanged via e-mail. While various e-mail systems have been around for decades, the real growth has occurred in the last few years as a part of the explosion of the Internet. Many people are using the Internet directly, but a large number of e-mail users still participate in this global connectivity only indirectly, via proprietary e-mail systems connected to the Internet with gateways. It is my somewhat controversial prediction that over the next few years most of those proprietary systems will be phased out and replaced with native Internet e-mail technology and protocols. In effect, the core Internet technology will be pushed all the way to the user's desktop.

This book attempts to bring together all the important information related to Internet e-mail technology. Where it is necessary to be specific about operating systems and products, examples are drawn from the Microsoft Windows NT world rather than from the birthplace of most of this technology, UNIX. That is because Windows NT is the most widely deployed platform capable of supporting the technology; it is clearly the future, and in many companies it already has become the present. Creation and sales of Windows NT versions of e-mail and related products have already significantly outpaced those for the aggregate UNIX market. That is due in part to the marketing prowess of Microsoft, in part to the continuing failure of the UNIX market to unify (it is highly fragmented into numerous incompatible submarkets), and in part to the fact that Windows NT is such a fine platform for developing commercial-grade systems of this kind. Only the very high end of the Internet market, characterized by the AT&T Worldnets of the industry, requires more horsepower than is available with Windows NT–based systems; even that final refuge of UNIX should fall within a couple of years (at least in terms of performance).

To describe the technology and the issues in this important and rapidly evolving field, it is necessary to use quite a bit of jargon, words that have specific meanings to professionals in the field. To help the reader deal with such terminology, many of the chapters in this book start with what appears to be a glossary. The entries go into more detail than a typical glossary and are listed in semantic rather than alphabetic order (the terms build on the ideas or definitions covered in preceding entries).

Even professionals who have been working in the field for many years should read the jargon sections carefully, so we all are talking the same language. Some of the terms appear to be common English words and indeed have technical meanings related in some way to the common usage (e.g., *gateway*). Some of the terms have meanings quite different from the common English usage or are based on some infrequently used meaning of a common word (e.g., *store*, which in e-mail terms is a place you temporarily put things, not a place you buy things). Professionals from different backgrounds may use some terms in different ways or entirely different terms for the same thing. For example, people from an

Internet background would call the e-mail program on a personal computer a mail client, while people from an International Standards Organization (ISO) background would call it a user agent.

Jargon

Bit   A bit (short for binary digit) is the smallest possible quantum of information. At any given time, a bit can have either of two possible values, off (also interpreted as 0 or FALSE) or on (also interpreted as 1 or TRUE). A bit can be represented physically with one of two possible voltages (0 VDC for off and 5 VDC for on) or various other physical quantities, either inside a computer or on physical communications channels, such as one of two audio frequencies on a telephone line, as in a modem.

Small groups of bits, say, 8 to 32 of them, can be used to represent items of information: an alphanumeric character (e.g., A), a numeric value (e.g., 255 or $6.02 \cdot 10^{23}$), or a computer instruction (e.g., "add 1 to the accumulator register"). A group of bits also can represent the color of a pixel in an image (e.g., cyan), the network address of a particular computer connected to the Internet (e.g., 123.45.67.89), or an approximation of the analog voltage of an audio signal at a given instant (one sample on an audio compact disc).

Collections of such groups of bits can be used to represent complex objects, such as the entire contents of this book (including formatting and pictures), a photographic-quality picture of your mother, the network addresses of all computers on the Internet, or a Beethoven symphony. Technically, anything that can be represented in bits can be communicated between two computer users via e-mail, provided the sheer number of bits does not exceed the limitations of the technology used. It is also necessary that both the sender and the receiver have compatible e-mail tools that know how to interpret the bits.

Byte   A byte is an ordered sequence of (typically) 8 bits. In most cases, a byte is the smallest unit of information that can be fetched, stored, or operated on by a computer. It can represent an integer numeric value in the range of 0 to 255 or a single character from a character set that has up to 256 distinct members. Some older systems used other numbers of bits (e.g., 6 to 9) per byte. A more precise term for an 8-bit byte is an *octet*, which is used in many requests for comment (RFCs).

Double byte   A double byte is an ordered sequence of 16 bits, or equivalently 2 bytes. Character sets with more than 256 distinct members (e.g., Unicode) may require a double byte to represent each character. Windows NT happens to use the double-byte Unicode character set internally and throughout the NTFS file system (which makes it ideal for use in countries like Japan, China, and Thailand).

Word   A word is an ordered sequence of 16 to 64 bits, depending on the central processor unit (CPU) architecture, and is the amount of information most typically fetched, stored, or operated upon by a computer in a single operation. It is also typically the width of the accumulator or working registers in the CPU. Most popular microcomputers today have a word size of 32 bits (DEC's Alpha AXP processor can work in either 32- or 64-bit modes). A 16-bit word can contain an integer numeric value from 0 to 65,535, or two 8-bit characters. A 32-bit word can contain an integer numeric value from 0 to 4,294,967,295, or four 8-bit characters. A 64-bit word can contain an integer numeric value from 0 to approximately $1.84467 \cdot 10^{19}$ (a truly large number), or eight 8-bit characters. The Intel 486 and Pentium series have a word size of 32 bits, although they have special instructions that can store, fetch, and operate on 16-bit or 8-bit quantities (and some floating-point instructions that can store, fetch, and operate on 80-bit quantities).

ASCII   The American Standard Code for Information Interchange, or ASCII (pronounced "AS-key"), is the American National Standards Institute (ANSI) standard that defines a specific set of 128 commonly used characters, including uppercase and lowercase alphabetic characters, the digits 0 through 9, many common punctuation symbols, and some control codes such as carriage returns. The standard also specifies the binary representation of each of those characters so that everyone will interpret a given bit pattern as the same character. (Appendix A gives a complete list of the characters and the 7-bit binary codes used to represent them.) For example, all ASCII-compliant computers use the numeric value 65 to represent an uppercase A. To be precise, that is 65 decimal (base 10), which is a numeric value that also can be written as 0x41 hexadecimal (base 16), or 0100 0001 binary (base 2). Only a few legacy computers (holdovers from an earlier epoch of computing) use different basic character sets (and corresponding binary values for those characters), for example, IBM's extended binary coded decimal information code (EBCDIC). The advantages or disadvantages of any given scheme for representing the basic working characters are minor. However, the advantages of everyone agreeing to use the *same* standard are over-whelming. Fortunately for users of e-mail systems, ASCII is close to being a true lingua franca. (The character set also is referred to, especially outside the United States, as US-ASCII.)

ISO 8859 ISO 8859 is an international standard that is a slight general-ization of US-ASCII using 8-bit numeric values for each character (which allows up to 256 distinct members). ISO 8859 actually defines a family of several character sets (8859-1 through 8859-10). Each of those 8859 character sets includes US-ASCII as the first 128 members and various extended characters as the second 128, as appropriate for various lan-guage families (e.g., the 8859-1, or Latin-1, character set is for English, French, German, and Italian, and the 8859-5 character set for Cyrillic lan-guages). E-mail tools that support ISO 8859 can interoperate fully with ASCII-based systems if they stick to the first 128 characters, but they also can support the extended characters required by most European lan-guages among users who have tools that support the various ISO 8859 character sets. To sympathize with the plight of a German user with only ASCII support (who must cope with no umlaut characters, the vowels with the two dots above them), imagine that your own e-mail tools sup-ported only consonants and no vowels.

Unicode Unicode is a recent international standard that is a major gen-eralization of US-ASCII using 16-bit (double byte) numeric values for each character, which allows up to 65,536 distinct members. Unicode includes ISO 8859-1 as the first 256 members, then both traditional and simplified Chinese ideographs, Japanese and Thai characters, Korean Hangul syllables, and various other glyphs that cover most of the world's major languages. Few e-mail tools support Unicode today, but we can expect to see more and more as the worldwide Internet phenomenon forces those of us in the West to accommodate users in countries with radically different languages. This is one of the areas in which Windows NT will have a major advantage over competing operating systems (it uses Unicode as its native character set).

Rich text format   Rich text format (RTF) is ASCII text with some additions to allow specification of formatting information, such as typeface, font size, and color. Rich text format allows sending of e-mail messages that use such information to have a more professional appearance or to include highlighting and emphasis difficult to achieve with simple ASCII text. ASCII was sufficient in the days of minicomputers, teletypes, and cathode ray tubes (CRTs). However, with today's graphics user interface (GUI)–based systems, people want to be able to use the same kinds of formatting in their e-mail messages as is available in their word processors. Unfortunately, there are two competing rich text standards: Microsoft's RTF (as used in Word and Exchange) and the one defined in Multipurpose Internet Mail Extension (MIME) (see Chapter 14). Also, many e-mail clients do not support any rich text standard, and messages composed using rich text format may be difficult or impossible to read with such clients.

In general, it is good practice to avoid using rich text format content in mail messages unless you specifically know that your recipient supports it (and in a manner compatible with the rich text format features of your own e-mail tools). Good systems can keep track of which recipients support rich text format and automatically convert rich text format messages to simple ASCII for anyone else. Some systems send both simple ASCII and rich text format versions of a message, but most currently available e-mail tools do not know what to do with the rich text format version (it will appear as a spurious attachment). Because of the many possible problems currently associated with rich text format, considerable thought should be given before using it. In the future, ideally, all e-mail tools will support a single standard.

ASCII text file   An ASCII text file is a collection of data on a computer disk whose bytes contain only printable ASCII characters, plus a limited set of control codes. The characters are organized into one or more "lines," each of which consists of a short sequence of up to 80 or so printable characters followed by one carriage return (CR) and one line feed (LF) control character. A simple ASCII file can be created by a program such as Windows Notepad or the old DOS-style *edit* program. The command-line *type* command displays a file's contents correctly only for simple ASCII text files. If you create a file with a word processor or Wordpad, it is probably more complex than a simple ASCII text file. Note that the internal representation of ASCII text files on UNIX computers uses just an LF character (0x0A) as a line terminator (presumably to save disk space before computers had gigabytes of disk space). Most Internet protocols, for example Simple Mail Transport Protocol (SMTP), require conversion to and from the standard CR/LF pair (0x0D, 0x0A) when such systems are communicating with other computers (even if both computers are UNIX based).

Most e-mail tools work primarily with simple ASCII content (at least for the primary body part of the message). If you save a typical simple e-mail message on disk, it usually will be an ASCII text file. Some e-mail tools have the ability to load or import a file into the main message body, possibly via cutting and pasting in a Windows GUI-style program. Any such files imported into the main body part of an e-mail message must be simple ASCII files. Many (but not all) ASCII text files in DOS and Windows use the filename extension .txt, for example, names.txt.

Binary file A binary file is a collection of data on a computer disk, the contents of whose bytes are not limited to ASCII characters but may have any value from 0 to 255 (in decimal or base 10). A binary file need not be organized into lines. Examples include executable programs (winfile.exe), spreadsheets (accounts.xls), and word processing files (chap01.doc). The output of a compression utility such as pkzip or encryption programs such as pgp typically are binary files (although pgp has the option of writing the encrypted data in an ASCII form if desired). Most audio or image files are binary files (whistle.wav or mom.jpg). To test whether a given file is an ASCII file or a binary file, you can try opening it with Notepad. If you see readable text, it probably is simple ASCII text. If you see apparently random gibberish, you probably are dealing with a binary file. Note that actually all ASCII text files really are just a special case of a binary file with more restrictive rules on content. That means that anywhere we discuss binary files, you also can use any ASCII text file. The reverse, of course, is not true.

E-mail message An e-mail message is ASCII data composed or read on a computer system and exchanged between users via terminals on a single time-sharing system or between users of different computers connected via a network or dialup modems. E-mail messages typically are sent from one person to one recipient (or to a list of individual recipients or members of a mail group). Messages sent in reply to e-mail messages are intended to be read only by the original sender or, in some cases, by the sender and the other specified recipients. In comparison, in a newsgroup or bulletin board system (BBS), both the original message and the responses to it are posted for any interested party to see. E-mail messages typically are delivered into a mailbox and are brought to the recipient's attention (a push technology). By comparison, in a newsgroup or BBS system, you must check for new messages in each topic group and specifically retrieve any of interest to you (a pull technology).

E-mail attachment   An e-mail attachment is a binary file (which could be an ASCII file as a special case) that is not in the main body part of the message but tacked onto the main body part. An attachment tags along with the main body part as it goes through the mail distribution system. On the recipient's end, there is some indication that attachments have been received. For example, the e-mail program may insert a summary line for each such attachment indicating the name of the file on disk where the attachment was stored on receipt. Some more sophisticated clients, for example, Windows GUI programs, present an icon representing the attachment, which when clicked upon bring up the associated application (e.g., Microsoft Word for a .doc file) and feed the attached file into it (so the document appears in Word when it starts). Such programs use a Windows technology called object linking and embedding (OLE).

Because most mail transport systems (e.g., the Internet e-mail servers that the message flows through during delivery) support only ASCII content, a binary attachment must be converted into an ASCII-only form during transport, and then back into the original binary form on receipt. Such a conversion increases the size of the attached file by 30% to 100% while it is in ASCII form. (Chapter 3 gives further details on this process.)

Shared-file system   Most network operating systems have some way to make files on a server computer available via the network to programs running on client computers. The network files appear to client programs as if they actually were located on a local physical hard disk drive, but in reality they physically are located on a hard disk on the server computer. The remote access is accomplished via a scheme called file input/output (I/O) redirection. When the program running on the client computer tries to read from or write to any disk file, a special part of the network operating system that runs on the client computer (called a redirector) intercepts the disk I/O request. The redirector either passes the request down to the local operating system, in the case of a local file, or redirects it out the network adapter to the network operating system (NOS) running on the server computer in the case of a network file. In that case, the network OS actually does the real disk I/O, using the data sent from the client computer in the case of a write operation or returning the data to the client in the case of a read operation.

One interesting aspect of the shared-file system scheme is that a single network file can be accessed from multiple client computers simultaneously. Some form of cooperation between such programs (called file and record locking) must be enforced to prevent corruption of the network files. Older, simpler e-mail systems, such as MS Mail, were developed using the shared-file mechanism.

Client/server system A way to design network software that allows programs running on different computers to establish a direct connection between them via the network (as opposed to making the network connections look like disk file I/O, as in a shared-file system). A server program runs on one computer and processes requests from a client program running on another computer. The requests and the responses to them are sent over direct network connections, for example, named pipes or sockets. The client program sends requests to the server program when it needs something from it. Typically much greater control can be imposed on such a connection and greater efficiency (fewer actual data being sent via the network) can be achieved compared to shared-file systems. Most modern e-mail systems (Internet e-mail, in particular) use the client/server approach.

Multiuser computer-based e-mail systems

It is possible to implement an e-mail system for multiple users on a single multiuser computer (such as a mainframe or UNIX-based computer). Such a system need not involve a network at all. It is kind of a special case of a shared-file e-mail architecture in which all the clients run on the same computer as the server, and they read and write the message store using disk I/O (ideally with record and file locking to prevent data loss). It is also possible to implement a socket-based client/server architecture, even Internet e-mail, completely on one multiuser computer, with or without networking. For the purposes of this book, such systems are just degenerate cases of the more general network based systems described.

Shared-file versus client/server e-mail architecture

We will now compare the two basic approaches to designing e-mail systems. The coverage of shared-file systems is primarily for historical purposes, to help you understand what is new and better about Internet e-mail architecture. It also will help you understand some of the issues involved in connecting older e-mail systems to the Internet via gateways.

Shared-file e-mail architecture

The first generation of personal computer–based e-mail systems, such as MS Mail and cc:Mail, have all their functionality (intelligence) in the client programs. There really is no active server component as there is with client/server designs. All that exists on the server is a shared collection (or subtree) of directories and network files called the message store. Each client computer has to mount that shared subtree of files using some existing network shared-file scheme, such as Novell NetWare or even Microsoft workgroup file sharing. The actual e-mail software is independent of the network file sharing scheme used, which means it can work equally well with NetWare or other file sharing schemes. Figure 1.1 shows a typical shared-file architecture mail system.

Compared to a client/server architecture, a shared-file architecture has three major disadvantages.

> ▶ Exposed message store. If the mail client program can read and write the files, so can any other program on the client computer. In fact, the user on the client computer typically can view the entire mail subtree with tools such as File Manager. They may be able to view, delete, or even alter files in the subtree, which makes possible privacy violations, fraud, and vandalism. Clever users may discover

they can write their own files in that network share, thereby circumventing disk space limits to which they may otherwise be subject.

▶ Unnecessary network I/O. Because access to the shared message store is accomplished only via the underlying network OS file sharing scheme, more data have to be moved back and forth over the network than with a client/server design. For example, if the client program wanted to search through all messages in the store for a particular word, the entire message store would have to be transferred over the network for the client to search through.

▶ Limited ability to scale to a large number of users. Because of the relatively inefficient mechanisms involved in making network I/O look like disk file I/O, shared-file e-mail systems typically are limited to a few hundred users. In addition, a shared-file e-mail server typically must be located on the same local area network (LAN) as the client program. That is because shared-file access via wide area network (WAN) links would be too slow and insecure.

While there is no active server component per se in a shared-file e-mail system (a shared-file message store is sometimes called a passive server), most of these systems provide some way to link multiple LAN-based e-mail systems into an enterprise mail system using message

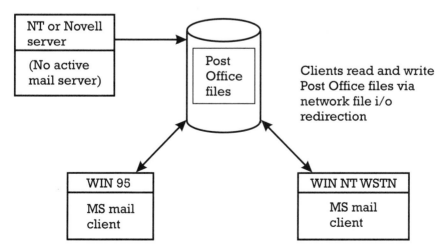

Figure 1.1   Shared-file architecture mail system.

transfer agents (MTAs). MTAs are specialized programs that can relay inter-LAN mail between the component LAN-based e-mail systems. They typically provide some kind of directory synchronization so a new user added to any site will appear in the address books of all sites within a short time.

MTAs tend to be difficult to install and maintain, due in large part to the fact that most were implemented on a single-tasking operating system (MSDOS). One of the main innovations of MS Mail v3.5 over its predecessor (MS Mail v3.2) is a multitasking MTA (MMTA) that can run on Windows NT as a service (no dedicated computer is required). In reality, it is the old OS/2 MMTA from v3.2 ported to run on Windows NT.

Client/server e-mail architecture

A client/server e-mail system has an active server component with which the client programs communicate via direct network connections, such as named pipes, sockets, or remote procedure calls (RPCs). A client/server e-mail system is independent of any network shared-file system; in fact, you do not even need to have such a system in place. However, both client and server computers must have the appropriate network transport layer and network application program interface (API) in place, for example, TCP/IP and WinSock. In comparison, shared-file e-mail systems will work with any network OS that provides basic file sharing, and they do not require any network API to be available. A programmer creating a shared-file e-mail system need think only in terms of reading and writing disk files. A programmer creating a client/server e-mail system needs to understand how to use the network API and how to implement the protocols.

In practice, the skills required for creating good clients and good servers are so different that most companies specialize in one or the other. Creating a really good server component is particularly difficult (I know because I once wrote my own SMTP/POP3 server for Windows NT). Fortunately, the existence of open standards makes it possible for companies to create the components without access to each other's source code and still have them work together.

A common mistake is to assume that the term *client/server* implies use of a database. Many of the early client/server products were based on distributed database systems, like Oracle or SQL Server, but e-mail does not require such technology. Any software system that has components running on different computers (e.g., e-mail client and e-mail server) connected via interprocess communication (IPC) schemes, like named pipes, sockets, or RPCs, is a distributed system. Distributed systems have a central component that provides some service to the other components are client/server designs. Data storage and retrieval in the context of a relational database is only one kind of service that such a component might provide. In practice, some companies do use database systems to implement certain parts of the functionality of an e-mail system internally (e.g., managing the set of user accounts or even implementing the message store). However, the network I/O aspects of the design are never done with the database system's client/server mechanisms. In fact, in general, the database needs of an e-mail system are much simpler than those provided by a general-purpose relational database system, and the additional cost, complexity, and performance overhead of such tools cannot be justified. Those needs can be met with simpler and less expensive tools, such as "b-tree" packages.

E-mail client software typically has a sophisticated GUI but usually is single threaded, which means it does only one thing at a time. It does not need to be running at all times: the user need run it only to compose new outgoing mail or check for incoming mail. An e-mail client program also needs to maintain only a single network connection at any given time to a single e-mail server.

In comparison, an e-mail server program must be running at all times; hence, it usually is implemented as a UNIX daemon or NT service. That means the server software starts automatically when the computer is booted and runs continuously until the computer shuts down. It must be able to handle numerous open network connections simultaneously (not just quickly one after another) from e-mail clients or other servers. Such a design must be at least multitasking and possibly even multithreaded if it is to take advantage of a symmetric multiprocessor (SMP) server computer. On the other hand, e-mail servers have little or no user interface, GUI or otherwise.

It may help you to think of the e-mail client as the equivalent of a telephone handset, the instrument in your home or office with push buttons, a speaker, and a microphone. That makes the e-mail server the equivalent of a telephone company central office, the building downtown where thousands of wires from handsets in the homes and offices terminate and the magic happens that links the pieces into a communication system. Without the central office, all those handsets would be so much useless metal and plastic. Likewise, without the central e-mail server, all those e-mail clients on individual users' computers are of little value.

To take the analogy one step further, in much the same way that all the central offices in the world are linked with high-capacity trunk lines, e-mail servers are linked to each other to exchange interserver messages. For example, if I call my next-door neighbor, the call goes into my central office and right back out directly to my neighbor. If I call my mother in the next state, the call goes into my central office, takes a leap over a long distance line to her central office, then over her local line to the handset in her home. Client/server e-mail systems work in much the same way. Figure 1.2 shows a typical system.

The three main weaknesses of the shared-file architecture previously described are addressed in the client/server architecture as follows:

▶ Protected message store. With client/server designs, the message store is accessible only via the server program, which would allow appropriate accesses only from authenticated users. The message store itself is completely invisible to programs such as File Manager (unless the message store has been shared accidentally and unnecessarily via a shared-file scheme).

▶ Less network I/O compared to a shared-file design. With a client/server scheme, if the client wants to search for a given word in all messages in the message store, it can send a short request to the

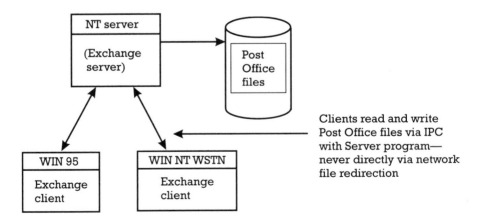

Figure 1.2   Client/server e-mail system.

server to search for that word. The server then reads through the messages locally and returns only the locations where the word was found, if at all. The total network traffic is only a few bytes, regardless of how large the message store is.

▶ Scales to a large number of users. A client/server system can be designed to take advantage of high-performance symmetric multiprocessor server computers (using a multithreaded concurrent design). Good client/server protocols minimize the amount of network I/O compared to shared-file systems. A single server computer can handle tens of thousands of users, and a distributed architecture design can scale to literally millions of users. A correctly designed client/server system can work well and securely even with a WAN link between the client and the server.

Microsoft's Exchange Server is an example of a client/server design. It uses Microsoft's messaging API (MAPI) as the native protocol between client and server, which is implemented using RPCs over various network transports. Native Internet e-mail servers are another example of true client/server designs. They are implemented using SMTP, Post Office Protocol v3 (POP3), or Internet Messaging Access Protocol (IMAP) over TCP/IP sockets.

Scalability of e-mail systems

One of the main differences between shared-file and client/server e-mail systems is in their scalability. Scalability refers to their ability to work equally well with any number of users, from a handful to a few thousand or even millions. There are two aspects to this issue: reliability and performance. In general, a given e-mail system on a particular server computer will work well and have reasonable response time and throughput up to some number of users. Above that number of users, it may continue to work reliably, but response time and throughput will degrade dramatically with additional users. Past a certain number of users, the system will lose e-mail messages, refuse connections from clients or other servers, or even crash.

Response time refers to the time it takes for the e-mail system to complete a given action, such as accepting a connection from a client and retrieving a message. A lightly loaded e-mail server might allow retrieval to begin immediately and take only seconds per kilobyte of mail message. A moderately loaded e-mail server might take 5–10 seconds before it accepts a connection and starts the retrieval, then take many times as long to transfer each message as would a lightly loaded server. An overloaded e-mail server might not accept the connection at all before the client times out (gives up). If it does accept the connection in time, there may be long, mysterious delays during the message transfers, or it may even stop mid-transfer. In general, users will become impatient if either the time to accept the connection exceeds a few seconds (especially if their client times out) or messages appear to be taking much longer than usual to transfer. Although much of this is highly subjective, users are fairly sensitive to response time increasing much above what they are used to.

Throughput refers to the amount of information that the e-mail system can process in a given time, in bytes per second. It is possible to measure the throughput of a specific client/server connection or of the e-mail system as a whole.

Good e-mail systems will provide ways to measure response time and throughput of the system to determine whether the resources allocated for the server are sufficient to handle the peak load acceptably. However, such performance may be affected by other activity on the server computer (e.g., if it is also running SQL Server) and general network load (e.g., if a large network backup job is running). A really good system might run periodic checks on performance and alert the administrator via e-mail or pager if it degrades unacceptably.

Several factors limit the scalability of shared-file systems.

▶ Unacceptable behavior over WAN links. File sharing systems typically are limited to use only within a given LAN. The network file sharing software has no knowledge of the e-mail application or its specific information transfer needs; hence, it must behave in the most general manner. Without the ability to implement "server-side intelligence," such as searching or rules processing, shared-file systems typically must exchange too much data over the network to work well over a slower WAN link. Also, making general shared-file systems available outside the LAN may present unacceptable security risks. Typically, shared-file e-mail systems require one e-mail server per LAN and use schemes other than network file sharing to exchange mail between servers. A client/server system can be designed so that a single e-mail server can service an entire WAN.

▶ Coarse granularity of shared-file locking. A simple shared-file system may be able to lock only entire files within the message store, as opposed to only those parts of the files actually being used by a given client. That can lead to worse contention problems (one user waiting for another to finish) as the number of users increases. Because the message store in a client/server system is not accessed directly by the clients (it can be accessed only indirectly via the server component), the designer can implement a finer degree of granularity in the file and record locking schemes, resulting in fewer contention problems.

▶ Shared-file e-mail servers typically are limited to running on a single server. It is difficult or impossible to distribute the functionality of a shared-file e-mail server over several computers and still make it appear to the client as if it really were all a single computer. Perhaps one day with distributed-file systems, that will become feasible, but at the current state of the art it essentially is impossible. In comparison, it is fairly easy to accomplish this with a client/server design (at least using Internet technology). A distributed architecture e-mail server can scale to any number of users, just by adding additional computers. A typical system might have several computers running a message store service, others running SMTP, and still others running POP3 or IMAP4 access servers. This is accomplished by using a trick called a domain name service (DNS) round robin scheme to make multiple servers appear to have the same nodename (e.g., smtpserver.megacorp.com) and maintaining a central directory that keeps track of which message store server a given user's message store is on. The new Microsoft Commercial Internet Services (MCIS, formerly known as "Normandy") is just such a distributed design. No matter how much memory and CPUs you add to a single server computer, eventually you will reach a limit of how many users you can support. A distributed architecture design can scale up to essentially any number of users (given a sufficient number of computers).

Standards and interoperability

One of the main things that drives the e-mail industry is open standards. The term *open* has been misused by commercial companies that want to have the appearance but not the reality of being open. A good working definition is that a standard is open if it has been created or at least adopted by a recognized standards body, such as the Internet Engineering Task Force (IETF), ISO, ANSI, or International Electrical and Electronics Engineers (IEEE). The standard should be readily available (and, ideally, free or inexpensive) to any implementor. Preferably, no (or very small) royalties should be involved in using the technology. It should not be controlled by a single company (as is Microsoft's MAPI), even if it originated with one. For example, the IBM PC's bus was a proprietary design that was copied by many companies and later refined and adopted by an industry coalition as the industry standard architecture (ISA) bus. In comparison, the IBM Micro Channel architecture (MCA) bus introduced in their PS/2 computers failed as a product because it remained a captive, proprietary standard with high royalties for use.

Good open standards are specified in enough detail so it is not only possible but likely that if two companies independently develop products based on those standards they will interoperate (work with each other). The standards also need to be available in time to be of use. For example, the IEEE 696 standard based on the S100 bus came along just about the time that the market for that kind of computer disappeared. Unfortunately, ISO standards (such as X.400) have a four-year approval cycle (sometimes multiple four-year cycles are required). Also, the high level of politicization often leads to late, overly complex standards that do not ensure interoperability.

Open standards are the lifeblood of network software. The field depends more than most on a large number of products from different vendors being able to work together. Many companies have tried to force proprietary standards on the industry, usually with limited or no success.

Even the ISO standards, which have been thoroughly engineered, have not met with overwhelming success. Today virtually every company, even Microsoft, recognizes the importance of everyone using the same network protocols. The winner by acclaim is the Internet protocol suite. The basic network protocols will be described in detail later in this book: TCP/IP in Chapter 7 and DNS in Chapter 8.

In the more specific area of e-mail, early vendors tried to create proprietary systems and impose them on the industry. That worked fairly well as long as most e-mail systems were isolated in their own little islands (LANs or workgroups). However, as the scope of people's interactions widened to the entire enterprise, a series of Band-aid fixes evolved, in the form of gateways coupling otherwise incompatible systems. In the last few years, the further expansion of the scope to worldwide and the increasing sophistication of e-mail clients, not to mention the sheer volume of e-mail, are making such schemes collapse under the strain. Most vendors already have discovered that rather than have each system support gateways to each of the others, it is better to have all such systems couple into a single backbone e-mail transport, specifically Internet Mail (SMTP). Forward-thinking institutions, such as Harvard University, already have progressed to the next, inevitable step in the evolution, which is to push the Internet Mail standards all the way out to the desktop, eliminating proprietary systems and gateways altogether. The Internet Mail standards are rapidly evolving to incorporate all the functionality of the best proprietary systems and beyond. The latest generation of e-mail vendors is creating standards-based e-mail servers and clients that incorporate those features, with commercial-grade reliability and functionality, that allow worldwide interoperation.

Summary

In this chapter, I have defined some of the jargon used in this highly technical field. I have aslo explained and compared the two basic architectures of e-mail systems: shared-file and client-server. I have hopefully made a case for the superiority of the latter. In short, I have tried to set the stage for the rather more detailed sections to follow.

Contents

Internet e-mail architecture

The basic architecture of Internet e-mail systems is not overly complicated. It is a client/server, store-and-forward system that employs two primary protocols, SMTP and POP3 (with IMAP4 coming on strong as an alternative to POP3). It is highly dependent on the underlying Internet network architecture. It is also scalable to enormous systems with literally millions of users located anywhere on earth, which no other currently existing system can claim.

Again, we first will define a few technical terms to simplify the material in this chapter and to make sure we all are on the same wavelength. Many of the terms are equally applicable to discussing any client/server e-mail system, such as X.400.

Jargon

User agent   A user agent (UA) (also called an e-mail client) is what most people think of as their e-mail program: the primary entry and exit points to an e-mail system, the program that an e-mail user uses to compose or read messages, the e-mail equivalent of the common telephone deskset. A UA also can be a mail-enabled application program. A UA program need not be running at all times, only when its user wants to compose or read mail or check for new mail. A given UA typically establishes a single network connection to the same e-mail host or server (the one on which the user has an account) only during the time that it is sending mail to or retrieving mail from the e-mail server. Specifically in Internet e-mail, it is possible for the incoming mail to be handled by one computer (the one running a POP or IMAP server) and the outgoing mail to be handled by a different computer (the one running the SMTP server).

In smaller systems, the server programs often are run on a single computer. However, all Internet Mail UAs (that I have had experience with) have the option of specifying different computers for those two functions. Most proprietary e-mail systems limit all users to a single e-mail client, from the same vendor that supplies the server component. In the case of Internet e-mail (because of the open standards that define all interaction between UA and server components), there are a large number of possible UAs from which the user can choose. Furthermore, most real-world Internet Mail systems have quite a mix of UAs from different vendors and with different capabilities. In a few cases (e.g., in areas where the standards still are evolving, such as security), it may be important that the sender and the receiver in a given message exchange use the same UA.

Message transfer agent   A message transfer agent (MTA)—also known as an e-mail server—is the internal component of an e-mail delivery system that is responsible for collecting mail from and distributing mail to UAs, in addition to relaying mail between e-mail postoffices (for mail exchanged between users of different mail servers). An MTA is the e-mail equivalent to a telephone company central office switch (kind of a glorified, computerized operator switchboard that can handle about 10,000 telephone lines and one big trunk line into the long distance carrier). Because an MTA should be running (and be ready to receive connections from any number of UAs and other MTAs) at all times (even many simultaneously), it must be a multitasking design and should be installed on a server computer (or at least some machine that is always running). In the case of Internet Mail, an MTA typically consists of two processes (or services): one to handle the SMTP protocol (used to transfer messages between MTAs and from UA to MTA) and another to handle the POP3 protocol (used by a UA to retrieve messages from the MTA). The two processes can be running on the same computer or on different computers. Each UA that uses a particular MTA needs to know the node-names of the computer(s) on which the MTA is running.

Central message store   A central message store (MS) is a specialized free-form textual database used by an MTA to store messages received from UAs or other MTAs and from which a UA retrieves messages. Some designs actually use real database managers, such as Oracle, in their design. Others implement the database functionality internally. It is possible for messages in such a store to be either deleted from the MS immediately on their being retrieved by a UA or retained in the MS for future retrieval (e.g., by another UA at the user's home).

Typically it is up to the UA to keep track of which messages have been read, but the MTA can provide a unique message identifier for each message to help the UA to keep track. There may be a separate process (service or daemon) that implements the central MS, or the MS may be implemented as a part of the MTA process. A central MS holds mail for all users who have mail accounts on that MTA. The MS itself may be implemented on the same machine as the SMTP or POP3/IMAP servers or on yet a third computer. It is even possible on a large system for an MS to be split among several machines. The UA does not need to have any knowledge of how the MS is implemented or even on what machine(s) it is physically located, because the UA will access the MS only indirectly via the SMTP or POP3/IMAP servers.

Message queue   A message queue is an ordered sequence of messages, a temporary holding area. There are two basic kinds of message queues: FIFO (first in, first out), which is like a pipe open at both ends through which messages flow, and LIFO (last in, first out), which is like a pipe open at only one end so that messages come out in reverse order from how they went in. An Internet postoffice may have several FIFO message queues that it maintains internally, for example, to temporarily store messages that it could not deliver (until it either succeeds or gives up). A well-designed Internet postoffice will allow an administrator to manage (view, check the length of, process, empty) any queue(s) that exist.

Outgoing message queue   An outgoing message queue is a FIFO message queue used by an MTA that holds messages prior to retransmission (either when an earlier attempt to transmit has failed or when the message is being relayed to another MTA for some reason, e.g., because the recipient's account is on that MTA).

Local message store   A local MS is similar to a central MS and is used by a UA program to hold messages for the user(s) of that UA. A local MS usually is organized into folders based on some criteria such as subject, sender, or time period in which it was received (managed by the user). Typically located physically on the same computer on which the UA is running (usually a different computer from the one on which the MTA is running). Which MS something is in becomes important when you disconnect a notebook computer from its LAN and take it on the road.

Network protocol   A network protocol is a set of rules that defines the syntax and semantic content of commands and possible responses exchanged between two or more parties (in this case, computers on a network), plus the order in which such commands can be specified. A given protocol typically implements the functionality for a specific function, like sending a message.

SMTP   SMTP (Simple Mail Transfer Protocol) is the primary network protocol used in Internet-style mail and is defined in RFCs 821 and 822. SMTP is the protocol used to exchange messages between MTAs and also by a UA to send messages to its host MTA. An SMTP server (or daemon) is the software program that implements the protocol. The classic freeware UNIX SMTP server is called Sendmail. Commercial Internet Mail postoffices usually include both an SMTP server and a POP3 server and possibly even other servers (IMAP, LDAP, Security, etc.).

POP3   POP3 (Post Office Protocol version 3) is the most common protocol used by UAs to retrieve messages from the MS of the host postoffices. The protocol is defined in RFCs 1082 and 1460. (There is also a POP2 protocol that is similar to POP3 in nature, but quite different in details, and totally non-interoperable. POP2 is not just a subset of POP3.) A POP3 server (or daemon) is the software program that implements the protocol. The classic freeware UNIX POP3 server is called Popper. Most commercial Internet Mail postoffice products include a POP3 server.

IMAP4 IMAP4 (Internet Message Access Protocol version 4) is a later-generation protocol (more recent than POP3) that also can be used to retrieve messages from an Internet postoffice MS (typically in parallel to POP3, so that either type of client can be used). An IMAP server is the software program that implements this protocol. IMAP4 has superior capabilities for doing server-based searches, retrieval of just message headers, and selective retrieval of messages (compared to POP3, which can retrieve only complete messages). It is particularly well suited to low-bandwidth dialup connections, for example, over Point-to-Point Protocol (PPP). The protocol, originally defined in RFCs 1730 to 1733, affects only transactions between a given UA and its host postoffice (both must support the protocol). The sender of a message does not need to support IMAP4 in any way for it to be fully functional for the receiver (i.e., it is not an end-to-end protocol).

MIME MIME (Multipurpose Internet Mail Extension) is a set of extensions to the original RFC 822 SMTP message syntax and was originally defined in RFCs 1521–1523. It allows attaching various kinds of files (ASCII and binary) to text mail and supports various message organizations (multipart, parallel). MIME defines support for a rich text format (not to be confused with Microsoft's RTF) for messages that contain different typefaces, type sizes, even color. Servers can ignore the extensions altogether, since the result of doing a MIME attachment is simply an ASCII text message that complies with the original RFC 822 syntax (even if one or more binary files are attached). However, it is possible for a server to optimize its performance in certain ways if it does process some of the MIME header information. Most contemporary UAs implement only a small part of the full MIME standard (e.g., those aspects dealing with attaching files but not those dealing with parallel message organization or rich text format). Also, few UAs currently support the MIME standards for audio or image attachments, even if those kinds of attachments can be included.

uuencode   uuencode is an alternative (and much simpler) standard used by older mail clients to attach binary files to mail messages. A good UA should be able to process either MIME or uuencode style attachments since a number of older UAs still are in use on the Internet that support only uuencode. A very good UA would keep track of what kind of attachment scheme to use with each user (as an attribute associated with each user in the address book) and automatically handle either kind of attachment on incoming messages.

Node   A node is a device connected to a network. It has a numeric network address (a 32-bit number called an Internet protocol, or IP, address) and typically also a symbolic name (e.g., "smtp_mail" or "dragon") associated with it.

Domain   A domain is a group of nodes managed by a single DNS server. The domain itself has a symbolic hierarchical name (e.g., guys.com). Hierarchical means that there is a treelike structure to the domains, and, specifically with Internet domains, each succeeding field of the name (e.g., guys and com) represents a level in the hierarchy. Thus, the domain guys.com is subsidiary to the domain com. The set of all Internet domains spans the entire Internet, such that every node connected to the Internet is in one domain or another. Every complete domain name (including the path from the root of the hierarchy to the domain in question (e.g., guys.com, not just guys) is unique within the Internet. Therefore, every node in the Internet can be uniquely specified by adding the full domain name after the nodename (e.g., mail.guys.com). *Note:* Do not confuse Internet (IP) domains with Windows NT administrative (NetBIOS) domains, which is a similar but unrelated concept.

DNS   DNS (domain name service) is a distributed, hierarchical database used to map domain-qualified nodenames to IP addresses and also to keep track of other network information for each domain, such as the preferred nodes to handle incoming SMTP mail, via mail exchange (MX) records.

TCP network connection   A Transmission Control Protocol (TCP) network connection is a connection between two computers on an internet (using the TCP of TCP/IP) that is initiated, stays in effect for some length of time (called a session), and then is terminated (dropped). While the connection is established, information can be sent back and forth between the two computers with the following specific "guarantees":

▶ Neither end can overwhelm the other with too much data too fast (flow control).

▶ If any errors occur, they will be corrected, possibly by retransmission (error correction).

▶ Information will be received in the same order in which it was sent (packet sequencing).

Underlying architecture (infrastructure)

Network applications tend to be far removed from the physical transmission medium (voltage levels on a cable). Between the applications and the physical transmission medium lie several layers of functionality, each of which builds on the layers beneath it. What most developers think of as the e-mail applications (UAs or MTAs) together with the WinSock system library they use to do network I/O are at the top (the process/application) layer. The WinSock library makes use of a lower (host-to-host) layer that implements the TCP. That layer in turn makes use of an even lower (Internet) layer, which implements the IP and Internet Control Message Protocol (ICMP) protocols. Finally, that layer makes use of the lowest (network) layer, which consists of the Ethernet protocol implementation and the driver software for a particular network interface card. The only thing below the software layer is hardware, in the form of a network interface card and the physical wires or coax cable linking the interface card to other computers in the network.

Everything below the level of the actual application (from WinSock on down) is really the same for any Internet application, so it cannot be considered to be strictly a part of the Internet Mail architecture. However, Internet Mail could not exist without it, and it must be installed and configured correctly before any e-mail application can be put at the top of the stack. For that reason, this book covers the underlying architecture in some detail, along with another critical piece of Internet infrastructure that is especially important to e-mail, the DNS.

Scalability issues

This section covers an advanced topic, and the casual reader may wish to skip it. Anyone interested in supporting tens of thousands or even millions of users on a single Internet postoffice may find it interesting.

Scalability (being able to handle a wide range of user community sizes) has a major impact on the system architecture of an Internet Mail server. A simple system that is intended for fairly small user communities might be single-threaded and require all components, that is, MTA, POP/IMAP servers, and MS server (MSS), to be on a single computer or at best the MTA on one computer and the POP server on another. Such a design (e.g., Sendmail and Qpopper) cannot scale to a very large user community (perhaps 10,000 on a very powerful computer, such as a 300-MHz DEC Alpha AXP with lots of memory). Making each component multithreaded and deploying it on an SMP (multi-CPU) server would help considerably, but no single box design can scale to millions of users.

Supporting very large user communities (millions of users) requires a somewhat different design. The Internet Mail server part of MCIS is an example of that kind of design. With such a distributed architecture, it is possible to have one or more computers running each server (MTA/SMTP, POP, IMAP, directory, message store, etc.). Being able to distribute the MS over several machines allows you to add new machines as needed to support the very large storage needs of such a monster mail system (with reasonable access time to any particular user's messages). Being able to have multiple-user access servers and MTAs (POP, IMAP, SMTP) allows you to add machines as required to ensure that sufficient connections will be available to support the large number of users. Furthermore, it ensures that queued outgoing mail will not overflow or take too long to process.

There are two tricks to implementing such a distributed architecture. The first trick involves setting up a DNS round robin scheme, in which a single shared nodename (e.g., smtphost.megacorp.com) can be mapped onto any of n IP addresses (basically, $1/n$th of the clients requesting DNS to resolve that nodename will get one of the possible IP addresses). That can be accomplished with any version of the freeware BIND version of DNS from 4.9.3 on, simply by defining multiple A records for the shared nodename. The second trick involves having a high-speed directory (called the Routing Table in MCIS IMS) that can be used to determine quickly which MS computer contains the messages for a given user.

The basic message flow for a UA sending messages to a distributed mail server (as seen in Figure 2.1) is as follows:

1. UA requests DNS to resolve the shared nodename for the MTA (e.g., smtphost.megacorp.com) and DNS responds with one of the n IP addresses corresponding to the n MTAs in the system (it does not matter which one—call it MTA[i]).

2. The UA connects to MTA[i] on TCP port 25.

3. MTA[i] uses the Routing Table to determine whether the recipient is a valid one, and if so, which MSS contains the messages for that user (call it MSS[j]).

4. Assuming the incoming message(s) are for a known user, MTA[i] accepts the message via SMTP and adds them to the messages for

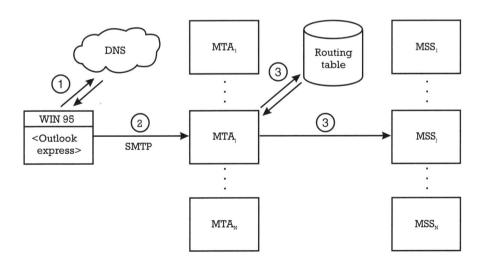

Figure 2.1 Sending messages to Scalable E-mail Server with SNMP.

that user on MSS[j] via an internal IPC connection between MTA[i] and MSS[j].

The basic message flow (as seen in Figure 2.2) for a POP-based UA retrieving messages from a distributed mail server is as follows (an IMAP-based one would work exactly the same way but use the IMAP protocol in place of the POP protocol and port 143 in place of port 110):

1. The UA requests the DNS to resolve the shared nodename for the POP server (e.g., pophost.megacorp.com), and the DNS responds with one of the m IP addresses corresponding to the m POP servers in the system (it does not matter which one—call it POP[k]).

2. The UA connects to POP[k] on TCP port 110.

3. POP[k] uses the Routing Table to determine whether the specified user account is a valid one, and if so, which MSS contains the messages for that user (call it MSS[j]).

4. Assuming it is a valid user, POP[k] makes an internal IPC connection to MSS[j] and allows the UA to retrieve messages from that server via the POP protocol.

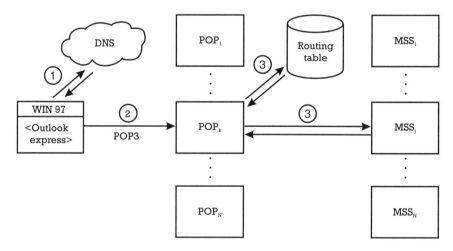

Figure 2.2 Retrieving message from Scalable E-mail Server with POP3.

Basic message exchange functionality

The core functionality of Internet e-mail is concerned with sending and retrieving messages. The overall design is known as a store and forward system. All that means is that the sender and the receiver need not be involved in the message exchange at the same time, as they would with simple voice telephones. It is more like leaving messages for each other on answering machines. However, if a telephone answering machine records an incoming message for someone other than the machine's owner, it generally is not clever enough to realize that and forward it to the correct answering machine (Internet MTAs do exactly that).

We will describe two scenarios involving sending and receiving Internet Mail messages. The first one is the simpler one, because no inter-MTA forwarding is required, while the second one does involve such forwarding.

Scenario 1: sender and receiver on the same postoffice

Say we have a TCP/IP domain called guys.com, with an Internet Mail postoffice (both SMTP and POP3 servers) on node *mail* in that domain, as shown in Figure 2.3. Furthermore, we have two guys named Albert and Barney who have Internet-style UAs on their computers. Both of them specified the nodename mail.guys.com for both smtp host and pop3 host when they configured their UAs. Say that Albert wants to send mail to Barney. The following steps must take place:

S1. Albert starts his UA and composes a new message addressed to Barney (he could even use the more formal style Barney@guys.com, but within a single postoffice that is not necessary). He then clicks his Send button.

S2. In response to the Send button being clicked, Albert's UA makes a network (TCP) connection to the node mail.guys.com on port 25, which is the port used by the SMTP server. The SMTP server there accepts the connection.

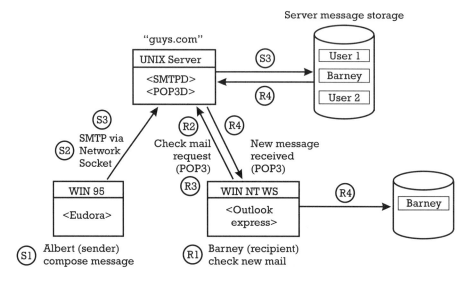

Figure 2.3 Scenario 1: Internet e-mail system: sender and recipient on same server.

S3. As soon as the connection is established, Albert's UA identifies Albert as the sender and Barney as the recipient (which are both acceptable to the SMTP server). Albert's UA then sends the actual message (including the message headers, the message body, and the "signature," if any). Once the message is received, the SMTP server looks at the recipient, realizes that the mail is intended for someone who has a local account in this postoffice (Barney), and hands it to the MS to file in that user's mailbox.

S4. Albert's UA terminates the network connection.

At some point, Barney will complete the mail exchange by retrieving the message Albert just sent. The following steps take place:

R1. Barney starts his UA and clicks the Check for New Mail button (or perhaps his UA has been running all the time and checking for new mail every few minutes).

R2. In response to the Check for New Mail button being clicked, Barney's UA makes a TCP network connection to the node mail.guys.com on port 110, which is the port used by the POP3 server. The POP3 server accepts the connection.

R3. As soon as the connection is established, Barney's UA identifies him as the user and supplies his POP3 password. Barney has a valid POP3 account and specified the correct password for that account, so all this is accepted also.

R4. Barney's UA requests a list of messages in his area of the MS and realizes a new one is there. The UA then retrieves the entire message over the network connection and stores it in Barney's local MS (in his InBox folder). Barney did not select the Leave Messages on Server option in his UA, so his UA tells the POP3 server to delete the new message, which it dutifully marks for deletion.

R5. Barney's UA realizes there are not any other new messages, so it sends a Quit message, which causes the POP3 server to actually delete the marked message, and then terminates the network connection.

R6. A summary line for the new message magically appears in Barney's InBox folder, showing that it was sent by Albert. Barney can view the message, print it, reply to it, forward it on to Charley, delete it, or do whatever he wants with it.

Scenario 2: sender and receiver on different postoffices

Say we have the same TCP/IP domain called guys.com, with an Internet Mail postoffice (both SMTP and POP3 servers) on node mail. Say we also have another TCP/IP domain elsewhere on the Internet, called gals.com. This is shown in Figure 2.4. We have good old Albert, still in the domain guys.com, and Alice in gals.com, both with UAs on their respective computers. Remember that Albert specified his SMTP and POP3 hosts both as mail.guys.com. However, in the gals.com domain, the SMTP server is on node smtp–mail and the POP3 server is on a different server called pop_mail. So Alice specified smtp_mail.gals.com for her SMTP host and pop_mail.gals.com for her POP3 host. Say that Albert wants to send mail to Alice. The following steps must take place:

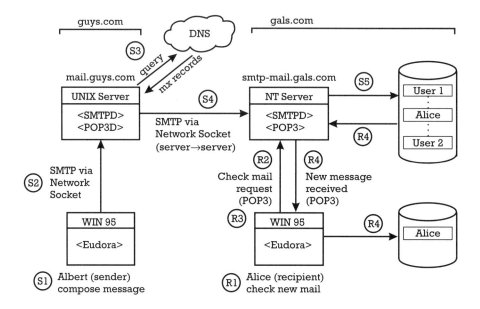

Figure 2.4   Scenario 2: Internet e-mail system sender and recipient on different servers.

S1. Albert starts his UA and composes a new message addressed to Alice@gals.com (he cannot use the short form "Alice" because Alice does not have an account on the local postoffice). He then clicks his Send button.

S2. In response to the Send button being clicked, Albert's UA makes a TCP network connection to the node mail.guys.com on port 25, which is the port used by the SMTP server. His SMTP server (on mail.guys.com) accepts the connection.

S3. Albert's UA identifies him as the sender and Alice over in gals.com as the recipient (which are both acceptable to the SMTP server). His UA sends the actual message. His SMTP server realizes that this mail is addressed to someone in another domain, so it does not store it in its local MS. Instead, it does a query on his local DNS server to see what the preferred node is to accept mail for the domain gals.com. If mail has been sent recently from guys.com to gals.com, his local DNS may have the answer right off the top of its "head" (in cache). If not, his DNS will ask other DNS servers until one (possibly the one for gals.com) admits to having that information and passes it back to Albert's DNS server. Eventually, the answer returns that smtp–mail.gals.com is the preferred node for incoming mail for gals.com.

S4. On finding the name of the node that handles incoming mail for the domain gals.com, Albert's SMTP server establishes a TCP network connection (possibly via WAN links or even halfway around the world) to the SMTP server on that node (smtp–mail.gals.com). To that SMTP server, the connection looks just like one from a local client and is handled in exactly the same way.

S5. Albert's SMTP server identifies Albert (of guys.com) as the sender and Alice (of gals.com) as the recipient (which are both acceptable to the SMTP server). Albert's SMTP server sends the actual message. Alice's SMTP server realizes that the mail is addressed to someone with a local account and hands it to the message store to file in Alice's mailbox.

S6. Albert's SMTP server terminates the network connection.

At some point in the future, Alice will complete the mail exchange by retrieving the message Albert just sent. The following steps take place:

R1. Alice starts her UA and clicks the Check for New Mail button (or perhaps her UA has been running all the time and checking for new mail every few minutes).

R2. Alice's UA makes a connection to node pop–mail.gals.com on port 110, which is the port used by the POP3 server. The POP3 server accepts the connection.

R3. Alice's UA identifies her as the user and supplies her POP3 password. Alice has a valid POP3 account and specified the correct password, so all this is accepted also.

R4. Alice's UA then requests a list of messages in her mailbox on the MS and realizes that there is a new one there. It then retrieves the entire message over the network connection and stores it in Alice's local MS (in her InBox folder). Unlike Barney, Alice did select the Leave Messages on Server option in her UA, so her UA does not tell the POP3 server to delete the new message. However, her UA makes a note of the unique message identifier (UIDL) for that message, so it will not download the message again.

R5. Alice's UA realizes there are not any other new messages, so it sends a Quit message and terminates the network connection.

R6. A summary line for the new message magically appears in Alice's InBox folder, showing that it was sent by Albert, over at guys.com. Alice can view the message, print it, reply to it, forward it on to Barbara, delete it, or do whatever she wants with it.

Basically all other kinds of mail transactions come down to one or the other of those scenarios, as far as the mail servers are concerned. All other functionality (e.g., replying, forwarding mail, etc.) is done by the UA (e.g., by swapping the contents of the From: and To: header lines to do a reply).

It is possible to tell an Internet Mail MTA to always send outgoing mail to a single node rather than doing distribution directly to the destination nodes. That allows setting up gateway or proxy systems for more elaborate routing. One reason for doing that might be to hide the internal locations of the e-mail servers in a large company from the outside world so internal servers could be added or moved without the outside world having to know about such changes. Another reason might be so that all incoming mail would come through a single proxy server, which can be thought of as a carefully controlled hole in an otherwise impermeable firewall protecting your LAN from unauthorized outside snooping or malicious tampering. There are even anonymizing gateways through which you can route mail that will strip any trace of where the mail actually originated.

Contents

Attachments and address books

The concepts of attachments and address books go beyond the basic functions of an e-mail system, but they are important to its usability. While both areas have had good solutions in proprietary mail systems for a number of years, there is considerable variation at present in the level of sophistication, interoperability, and even the existence of these features in current Internet e-mail products. In the case of attachments, an excellent standard (MIME) is already available and is being supported (at least in part) by more and more clients. In the case of address books, several standards exist for shared address books, but none has been widely adopted so far. However, a very good standard is well underway and may finally solve that difficult problem on a global scale.

Jargon

Decimal numbers   Decimal numbers are written in base 10, using the digits 0, 1, 2, 3, 4, 5,6, 7, 8, and 9. The rightmost (least significant) digit is the units, the next digit to the left is the number of tens, the next digit to the left is the number of hundreds (10^2), then the number of thousands (10^3), and so on. For example, the decimal number 314 is actually short-hand for 3 hundreds plus 1 ten plus 4 units.

Hexadecimal numbers   Hexadecimal, or hex, numbers are written in base 16, using the digits 0, 1, 2, 3, 4, 5, 6, 7, 8, 9, A, B, C, D, E, and F. The digits A through F stand for the decimal values 10 through 15. The right-most (least significant) digit is the units, the next digit to the left is the number of 16s, the next digit to the left is the number of 256s (16^2), then the number of 4,096s (16^3), and so on. A common way to denote that a number is in base 16 is to precede it with the characters 0x (the conven-tion used in the C programming language). For example, the hexadeci-mal number 0x1FC is actually shorthand for one 256 plus fifteen 16s plus 12, or 508 decimal. Hexadecimal is frequently used in computing and digital communications due to the ease of converting between it and binary, as compared with converting between decimal and binary.

To convert a hexadecimal number to binary, just convert each of its digits individually to binary, according to Table 3.1.

Hence, the hexadecimal value 0x1FC is 0001 1111 1100 in binary. Converting binary to hex is the reverse of that. Hence, 0111 1111 binary is 0x7F hexadecimal.

UNIX   UNIX is an operating system similar to Windows NT but has been around much longer. Originally developed by Ken Thompson and Dennis

Table 3.1
Conversion of hexadecimal numbers to binary

0x00 = 0000	0x04 = 0100	0x08 = 1000	0x0C = 1100
0x01 = 0001	0x05 = 0101	0x09 = 1001	0x0D = 1101
0x02 = 0010	0x06 = 0110	0x0A = 1010	0x0E = 1110
0x03 = 0011	0x07 = 0111	0x0B = 1011	0x0F = 1111

Ritchie (together with C programming language) at Bell Labs, it has been popular at colleges and universities (due, in part, to its being free for them to use and available for many machines used at such institutions). UNIX is a classic multiuser, multitasking design. It was one of the first platforms to support TCP/IP and the "socket" network programming API. It supports daemon processes (ones that start when the computer is booted and run continuously until the computer is shut down, regardless of whether any user is logged in or not). Its native character set is 7-bit ASCII, which limits it primarily to English.

UNIX was by far the most common operating system on nodes connected to the Internet for many years. Many Internet protocols were first implemented on UNIX-based computers. Even today, many ISPs primarily use UNIX computers to provide their DNS, e-mail, and even Web services. One of the greatest strengths of UNIX (and one of its greatest weaknesses) is that it is not controlled by any one company. That is a strength in that no one company controls its future, but a weakness in that each of the vendors that supports it has made it proprietary in various ways (hence the market is highly fragmented). UNIX is available, albeit in largely incompatible variants, for many different hardware architectures, including the Intel x86.

Windows NT   Windows NT is a more recent operating system similar to UNIX. Like UNIX, it is multitasking and supports TCP/IP and the socket network programming API. It also supports daemon processes (called services). Unlike UNIX, it has native support for threads (lightweight processes), is single user (at a time, per node), and uses 16-bit-per-character UNICODE as its native character set. It has been ported to many other languages, including Far Eastern ones such as Chinese and Thai. It is particularly well suited to distributed computing.

Windows NT already has displaced much of the Novell NetWare market; as it grows in power and sophistication, it also is displacing much of the traditional UNIX market. One of it greatest strengths (and weaknesses) is that it is totally controlled by one company (Microsoft). That is a strength in that the market is completely unified (there is only one variant of Windows NT) but a weakness in that we are all subject to Microsoft's whims and future plans. Windows NT is currently available for the most common hardware architecture (Intel x86) with up to four (or even more) parallel processors. It is also available for computers based on the

Digital Alpha AXP processor (one of the fastest reduced instruction set computer (RISC) architectures).

Apple Macintosh   Apple Computer's Macintosh ("Mac") is an alternative to the WinTel (Windows/Intel) platform. Even more so than Windows NT, both the hardware and the OS are highly proprietary and totally controlled by a single company. Apple has aggressively fought attempts by competitors to create an industry standard around their platform and now are almost out of business because of that. Like essentially everything else about their platform, they have their own incompatible standard for e-mail binary attachments (BINHEX). That is due in part to the fact that their disk files are also incompatible with those of other operating systems (each file has two "forks"). Even the standard Apple networking (AppleTalk) is highly incompatible with TCP/IP, Ethernet, and the Internet. Fortunately, you can replace AppleTalk with Ethernet and TCP/IP on most models today.

Encoding   Encoding is converting the representation of data from one form to another, for example, from binary to ASCII radix-64 or from digital signals to analog tones (as in a modem). No information is lost. Such conversions typically are done to allow data that normally would not be able to be sent through some communications channel (e.g., an e-mail system or a voice-grade telephone line) to assume a form that is compatible with that channel.

Decoding   Decoding is the reverse of encoding and typically is done to recover the original representation of the data after they have traversed some limited communications channel. For example, ASCII radix-64 data would be converted back into binary data, or analog tones would be converted back into digital signals.

MIME   MIME (Multipurpose Internet Mail Extensions) is an elaborate scheme for attaching binary files to e-mail messages, among other things. MIME is covered in greater detail later in the book. For now, consider it as an alternative to uuencode (the "UU" part of the name is short for "UNIX to UNIX"). Unlike uuencode, MIME requires that new lines be added in the e-mail message headers (the lines at the start of the message containing information about who sent the message, to whom it is addressed, when it was sent, etc.).

Address book   Address book is a general term for any kind of facility that allows you to look up an addressee by a "human-friendly" name, to find the computer e-mail address. It is the e-mail equivalent of a telephone book, which allows you to look up telephone numbers by people's names. Such a facility could be private and local, for use by only one person, shared by a workgroup via a network, or even global for use by anyone in the world. A good e-mail client program might support several different levels of address books, including a personal one. You may prefer the term directory service, but this text uses "address book" to refer to the general case and "directory service" to refer to global, distributed address books, such as X.500-based systems.

Attachments

One of the basic limitations of almost all e-mail systems is that the primary messaging channel is limited to transmitting only ASCII text. Specifically, this means relatively short (say up to 100 characters) lines of "printable" characters (those characters that if sent to a printer cause a mark of some kind to be written to the page, such as numeric digits, upper and lower case alphabetic characters, and a few punctuation characters). Each such line is followed by an "end of line" (typically an ASCII Carriage Return hexadecimal 0x0D, followed by an ASCII Line Feed, hexadecimal 0x0A). No provision is made for characters from non-English languages (e.g., the German "umlaut" characters), special codes to indicate font, font size, color or other attributes (sometimes called "rich text"), or raw binary data. In particular, even data such as a word processing document typically could not be sent as the main body of an e-mail message.

Attachments are not an issue for e-mail MTAs (in general), as the attachments look just like ASCII text at the time they pass through the MTA. On the client side, they can be handled (clumsily) by separate pre- and post-processing utilities, or (more conveniently) with embedded facilities in the UA. However, attachments can be a real problem with gateways between dissimilar mail systems (e.g., an MS Mail-to-Internet gateway), as we will see in later chapters.

uuencode/uudecode: the
UNIX attachment standard

Early e-mail UAs offered no solution for people who wanted to send general forms of information. Many years ago, in the UNIX/Internet world, someone came up with the idea of a utility that could encode arbitrary binary data (the most general case of computer files, of which all other forms, including simple ASCII text files, are special cases) into a form that could go through text-only e-mail systems. On receipt, the encoded form could be decoded to recover the original binary data. Think of it as the e-mail equivalent of a modem, which performs similar encoding from a digital bit stream into an audio signal that passes through a voice-grade telephone line and then back into the original digital bit stream at the other end.

A standard way of doing binary-to-text encoding was agreed on, using a scheme that encodes 6 bits of binary data into each ASCII character. Hence, three 8-bit bytes of raw data result in four encoded characters (24 bits of binary data in either representation). The actual encoded character is the 6-bit value plus an offset of 32, with the exception that a resulting "space" (code 32) is mapped onto the back quote (code 96). The characters were chosen to be compatible with the largest number of possible e-mail systems and legacy computers. Each line starts with an encoded count (of original bytes in this line), then the encoded form of the 45 actual data bytes, which is 60 ASCII characters, followed by an end of line (CR, LF). The process is continued until all the binary data have been encoded (the last line usually encodes fewer than 45 bytes). The overhead items allow the decoding program to identify how many characters are on each line and to include "start" and "end" lines (in case there are extraneous lines before or after the encoded data or even multiple files encoded in a single message). Typically, the name of the file encoded also is included. Unfortunately, things like file date and time, ownership, or even the exact size in bytes are not included explicitly.

Table 3.2 lists the mapping from the 64 possible 6-bit values for the standard "UU" encoding.

Table 3.2 can be read as follows: the character used to encode the 6-bit binary value 0x28 (0x20 + 8) is H. You can also use it in reverse, for example, the character W is the encoded form of the 6-bit binary value 0x37 = 0x30 + 7.

Table 3.2
Mapping for standard UU encoding

	0x00	0x10	0x20	0x30
+0	'	0	@	P
+1	!	1	A	Q
+2	''	2	B	R
+3	#	3	C	S
+4	$	4	D	T
+5	%	5	E	U
+6	&	6	F	V
+7	'	7	G	W
+8	(	8	H	X
+9	)	9	I	Y
+A	*	:	J	Z
+B	+	;	K	[
+C	,	<	L	\
+D	-	=	M	]
+E	.	>	N	^
+F	/	?	O	_

Table 3.3 shows the alternative mapping called XX encoding.

If the encoded text contained any lowercase alphabetic characters, the alternative XX mapping was used. If the encoded text contained any punctuation other than + or −, the UU mapping was used. The XX mapping is less common but more likely to go through modern communication channels. A good uuencode program should allow you to force use of either mapping, and a good uudecode program should be able to recognize and decode either mapping automatically.

The first byte of each line is an encoded count of data bytes on that line (typically M, which is the ASCII character with code 77 decimal, which is 45 + 32, since typical lines encode 45 characters). There also may

Table 3.3
Mapping for XX encoding

	0x00	0x10	0x20	0x30
+0	+	E	U	k
+1	-	F	V	l
+2	0	G	W	m
+3	1	H	X	n
+4	2	I	Y	o
+5	3	J	Z	p
+6	4	K	a	q
+7	5	L	b	r
+8	6	M	c	s
+9	7	N	d	t
+A	8	O	e	u
+B	9	P	f	v
+C	A	Q	g	w
+D	B	R	h	x
+E	C	S	I	y
+F	D	T	j	z

be a checksum at the end of each line (the sum of the encoded characters) or a checksum at the end of the entire file, to detect transmission errors.

For example, encoding this file (test.txt, whose size on disk is 139 bytes),

```
Now is the time for all good men to come to the aid of their neighbors.
The quick brown fox jumped over the lazy dog's back. 0123456789.
```

produces the following encoded output (whose size on disk is 228 bytes):

```
begin 777 test.txt
M3F]W(&ES('1H92!T:6UE(&9O<B!A;&P@9V]O9"!M96X@=&\@8V]M92!T;R!T
M:&4@86ED(&]F('1H96ER(&YE:6=H8F]R<RX-"E1H92!Q=6EC:R!B<F]W;B!F
M;W@@:G5M<&5D(&]V97(@=&AE(&QA>GD@9&]G)W,@8F%C:RX@,#$R,S0U-C<X
$.2X-"G5M
end
```

In the above example, each of the first three lines contain 45 bytes of original data (encoded count = M, whose numeric value in ASCII is 77 = 45 + 32). The 45 bytes expand to 60 encoded characters. The UU mapping is used. No line checksums are present. The final line contains 4 bytes of original data (encoded count = $, whose numeric value in ASCII is 36 = 4 + 32). Because the number of original data bytes is encoded in each line, the sum of those yields the exact size of the original file, in bytes (which otherwise might be difficult to recover, given the way data are divided into 6-bit chunks).

To give you an idea of the decoding process, do a reverse mapping of the first few encoded characters, through the UU mapping:

```
3  →  0x13,  F  → 0x26,  ]  → 0x3D,  W  → 0x37
```

Rewrite these hex values into binary:

```
00010011 00100110 00111101 00110111
```

Then discard the top 2 bits of each group (always zero) and regroup into 8-bit bytes:

```
01001110 01101111 01110111
```

Now rewrite those binary values into hex:

```
0x4E 0x6F 0x77
```

Finally, translate the hex codes into ASCII:

```
N o w
```

Running the encoded file through a decoder program recovers the original data. In this case, the original file happened to be ASCII text, but any file could have been fed into the encoder program, including a word processor file, a "zipped" archive, an executable program, or an audio .WAV file. The resulting encoded form still would have been an ASCII text file suitable for going through even the most restrictive communications channel (note that not even lowercase alpha is required).

For a small file like this, the overhead results in an expansion to 164% of the original size. The larger the original file, the lower the impact of the overhead, approaching 140% of the original size as a limit.

Most UNIX systems include simple utility programs called uuencode and uudecode. Similar programs exist for most computers today, including DOS and NT. Public domain versions of uuencode and uudecode for DOS or Windows are available for download from the Internet. Today, most e-mail UAs do support this encoding scheme (often in addition to other schemes, such as BINHEX or MIME), and it is much more convenient to use an embedded facility in a UA than separate programs such as these, so ideally you never will need to use the utilities directly. The utilities are available, however, for those who need them. Occasionally, you will find that the encoded data have not been recognized for some reason, or the automatic decoding was disabled in your UA by mistake, and you receive an attachment still in the encoded form. In those cases, such utilities can be quite useful.

Using separate utilities to do encoding requires that your UA has the ability to read a text file on disk and insert it into the current message you are composing. With a GUI UA, you at least have the option of loading the encoded file into the Notepad utility, cutting (copying) the entire encoded data, and then pasting it into the edit control in your UA. Using a separate utility to do decoding requires that your UA has the ability to save the text (possibly complete with all headers) to a text file, which then can be fed into the decoder utility.

The number of variations in the uuencode scheme (different mapping tables, checksums or no checksums, etc.) are characteristic of UNIX "standards." The situation makes it much more difficult to write software, because something like a uuencode decoder must be flexible enough to handle a large number of possible variations in input.

For a more complete example, I created a binary data file containing the values from 0 to 255 decimal, called demo.dat, which is dumped in hexadecimal, as shown below:

```
000000 - 00 01 02 03 04 05 06 07   08 09 0A 0B 0C 0D 0E 0F ........ ........
000010 - 10 11 12 13 14 15 16 17   18 19 1A 1B 1C 1D 1E 1F ........ ........
000020 - 20 21 22 23 24 25 26 27   28 29 2A 2B 2C 2D 2E 2F  !"#$%&' ()*+,-./
000030 - 30 31 32 33 34 35 36 37   38 39 3A 3B 3C 3D 3E 3F 01234567 89:;<=>?
000040 - 40 41 42 43 44 45 46 47   48 49 4A 4B 4C 4D 4E 4F @ABCDEFG HIJKLMNO
000050 - 50 51 52 53 54 55 56 57   58 59 5A 5B 5C 5D 5E 5F PQRSTUVW XYZ[\]^_
000060 - 60 61 62 63 64 65 66 67   68 69 6A 6B 6C 6D 6E 6F 'abcdefg hijklmno
000070 - 70 71 72 73 74 75 76 77   78 79 7A 7B 7C 7D 7E 7F pqrstuvw xyz{|}~.
```

```
000080 - 80 81 82 83 84 85 86 87   88 89 8A 8B 8C 8D 8E 8F ........ ........
000090 - 90 91 92 93 94 95 96 97   98 99 9A 9B 9C 9D 9E 9F ........ ........
0000A0 - A0 A1 A2 A3 A4 A5 A6 A7   A8 A9 AA AB AC AD AE AF !"#$%&' ()*+,-./
0000B0 - B0 B1 B2 B3 B4 B5 B6 B7   B8 B9 BA BB BC BD BE BF 01234567 89:;?
0000C0 - C0 C1 C2 C3 C4 C5 C6 C7   C8 C9 CA CB CC CD CE CF @ABCDEFG HIJKLMNO
0000D0 - D0 D1 D2 D3 D4 D5 D6 D7   D8 D9 DA DB DC DD DE DF PQRSTUVW XYZ[\]^_
0000E0 - E0 E1 E2 E3 E4 E5 E6 E7   E8 E9 EA EB EC ED EE EF `abcdefg hijklmno
0000F0 - F0 F1 F2 F3 F4 F5 F6 F7   F8 F9 FA FB FC FD FE FF pqrstuvw xyz{|}~.
```

I sent the file to a mail server using Eudora Pro 3.0 (with attachment style set to uuencode), then retrieved it with a POP3 client I wrote that did not process the attachment (to see exactly what was created). The result is the following:

```
Return-Path: <Lawrence.Hughes@bangkok.software.com>
Received: from bangkok ([10.3.101.9]) by bangkok.software.com
         (Post.Office MTA v2.2 ID# 0-0U10) with SMTP id AAA303
         for <lawrence.hughes@bangkok.software.com>;
         Thu, 19 Dec 1996 09:50:26 -0800
Message-Id: <3.0.32.19961219095025.009bd2b0@bangkok.software.com>
X-Sender: LHugh@bangkok.software.com (Unverified)
X-Mailer: Windows Eudora Pro Version 3.0 (32)
Date: Thu, 19 Dec 1996 09:50:26 -0800
To: lawrence.hughes@bangkok.software.com
From: Lawrence.Hughes@bangkok.software.com (Lawrence Hughes)
Subject: binary attachment using UUENCODE
Mime-Version: 1.0
Content-Type: multipart/mixed;
boundary="=====================_851046626==_"
—=====================_851046626==_
Content-Type: text/plain; charset="us-ascii"

This is the message body
—=====================_851046626==_
Content-Type: application/octet-stream; name="demo.dat"
Content-Transfer-Encoding: x-uuencode
Content-Disposition: attachment; filename="demo.dat"

begin 600 demo.dat
M''$"'P0%!@<("0H+#`T.#Q`1$A,4%187&!D:&QP='A\@(2(C)"4F)R@I*BLL+2XOMT'#$%.T.E$#$8#!T"'X9#%$!+$%@E 1'Y'#'$$'*BLL8'4`%4%@
M+2XO,4E$9$5
M6E-5A!91E9N%
MA
M8&9!%!$!L:!$(%!#($
MT=+$H$K
M
end

—=====================_851046626==_
Content-Type: text/plain; charset="us-ascii"


—=====================_851046626==_-
```

Note that Eudora wrapped the attachment in MIME-style headers, but they are ignored by the uudecode program (or any other software that supports uuencode).

BINHEX: the "Mac" attachment standard

BINHEX is another binary-to-ASCII text encoding scheme similar to uuencode, except that it uses a slightly different (and incompatible) encoding scheme. The scheme is used mostly in the Macintosh world, but it is available in the Eudora UA due to the fact that Eudora originated in the Mac world. BINHEX is available in both the Mac and Windows versions of Eudora (the Windows version of Eudora is really just a port of the Mac version, which is why it does not meet many Windows interface standards). I strongly recommend that, unless you are communicating with someone using a Macintosh e-mail UA without MIME support, you avoid using BINHEX. It has no advantages over MIME, and there are already too many standards in this area.

Here is the same attachment sent by Eudora Pro 3.0 using BINHEX attachment style:

```
Return-Path:  <Lawrence.Hughes@bangkok.software.com>
Received: from bangkok ([10.3.101.9]) by bangkok.software.com
          (Post.Office MTA v2.2 ID# 0-0U10) with SMTP id AAA235
          for  <lawrence.hughes@bangkok.software.com>;
          Thu, 19 Dec 1996 09:49:47 -0800
Message-Id: <3.0.32.19961219094947.009be2c0@bangkok.software.com>
X-Sender: LHugh@bangkok.software.com (Unverified)
X-Mailer: Windows Eudora Pro Version 3.0 (32)
Date: Thu, 19 Dec 1996 09:49:47 -0800
To: lawrence.hughes@bangkok.software.com
From: Lawrence.Hughes@bangkok.software.com (Lawrence Hughes)
Subject: binary attachment using BINHEX
Mime-Version: 1.0
Content-Type: multipart/mixed;
boundary="=====================_851046587==_"
--=====================_851046587==_
Content-Type: text/plain; charset="us-ascii"

This is the message body
--=====================_851046587==_
Content-Type: application/mac-binhex40; name="demo.dat"
Content-Disposition: attachment; filename="demo.dat"

(This file must be converted with BinHex 4.0)
```

```
:#'4PE@mZC'&d!%**6N&YC'pc!!!!!!%!!!!!!,Rp!!%#!'3&"JF)#3S,$!d1$a!
4%K-8&4BA'"ND'a'G(KmJ)5)M*#8Q*bJT+LXX,5i[-$%b-c3e0MFi16S12$dq2d"
"3N0%48C(5%P+5da06Np389*69&9@9eKC@PYFA9jIB'&LBf4PCQGSD@TVE'eZE"h
aFR0dGAChH(PkHhapIRq!JB+$K)@'KiL*LSZ-MBk2N!#4NT18PCDAQ*QDQjbGRTq
JSD+MT+@QTkLTUUZXVDk[X,'bX15eYVHiZEUl[,fq[m$"'X2%aFE(b-R+bmc0cXr
3dG,6e0A@epMCfY[FhGlIi1(Liq6PjZISkHVVl1hZlr$am[2dpIEhq2Rkqrcpr[p
q93!!:
```

```
--=====================_851046587==_
Content-Type: text/plain; charset="us-ascii"


--=====================_851046587==_-
```

MIME: the Internet standard

The most popular standard for attachments today is MIME. The standard is considerably better specified, more complex, and more encompassing than either uuencode or BINHEX. In fact, typically only a fraction of the complete MIME specification typically is implemented in current products. The complete specification includes facilities for multimedia (images, audio), rich text (RTF), parallel message organization (for multiple versions of the same message in different formats, e.g., simple ASCII text and rich text), and other advanced features.

Unlike the other two Internet attachment schemes, MIME is a formal standard, found in RFCs 1521, 1522, and 1523 from the IETF. The complete texts of those RFCs are on the CD-ROM accompanying this book, along with many others related to e-mail. There is also a standard for how to handle MIME in a mail gateway (RFC 1344). More recently, further extensions to MIME have been made in RFCs 1891, 1892, and 1893 (concerning error handling and delivery notification). The IETF also has created an extension of MIME (called MIME objects security services (MOSS)) for secure mail, but it has not received very wide acceptance. Recently, a consortium of companies apart from the IETF created the S/MIME specification for secure e-mail, based on MIME, which looks as if it will become the standard for secure e-mail outside the military (where DMS quickly is becoming the standard). By the time this book comes out, there likely will be further extensions to MIME, since this is a rapidly evolving area.

Unlike uuencode and BINHEX, MIME includes some parts in the message header section. Therefore, it is not possible to encode a file using a standalone MIME encoder, import it into a non-MIME UA, and then send it. To send a MIME-encoded message, the encoding software must be integrated into the UA. On the other hand, it is possible to extract binary files encoded with MIME from a received message (a separate MIME decoder must have access to both the message header and the message body). If someone has sent you a mail message using MIME encoding, but your UA does not understand MIME (or the automatic MIME decoding has been disabled in some way), you will need a separate MIME decoder to recover the binary attachment.

A MIME encoder can recognize an attachment that consists only of plain ASCII text and transmit it in the original, unencoded form (but still delimited as an attachment, so it can be saved automatically). If only a few characters go beyond the definition of a simple ASCII text file (say, a few special characters in the range 128 to 255, e.g., German umlaut characters or limited 8-bit word processing codes), then just those characters can be encoded using a "printed quotable" scheme (which is much lower overhead than encoding the entire file into ASCII, as is required with general binary data).

In the case of general binary data, the entire attachment is encoded into ASCII text using a radix-64 encoding scheme (Table 3.4), with 6 bits of the original data encoded into each ASCII character, much like the one used in uuencode (but with a different mapping, of course).

MIME also allows the type of each attachment to be specified (e.g., audio, image, rich text), so a clever UA could do something appropriate with the attachment (play it back, display it as an image, etc.).

MIME is such a complex standard and so important to current Internet e-mail that an entire chapter of this book devoted to it (Chapter 14).

Following is an example of an ASCII file attached using MIME (note that the attachment is not encoded into radix-64 first).

```
Return-Path:  <test.user@bangkok.software.com>
Received: from bangkok ([10.3.101.9]) by bangkok.software.com
          (Post.Office MTA vX.Y ID# 0-0U10) with SMTP id AAA150
          for  <lawrence.hughes@bangkok.software.com>;
          Tue, 31 Dec 1996 15:31:29 -0800
Message-Id: <3.0.32.19961231153128.009b9ec0@bangkok.software.com>
X-Sender: Wildcard@bangkok.software.com (Unverified)
X-Mailer: Windows Eudora Pro Version 3.0 (32)
Date: Tue, 31 Dec 1996 15:31:29 -0800
To: lawrence.hughes@bangkok.software.com
```

Table 3.4
Radix-64 encoding scheme

	0x00	0x10	0x20	0x30
+0	A	Q	g	w
+1	B	R	h	x
+2	C	S	i	y
+3	D	T	j	z
+4	E	U	k	0
+5	F	V	l	1
+6	G	W	m	2
+7	H	X	n	3
+8	I	Y	o	4
+9	J	Z	p	5
+A	K	a	q	6
+B	L	b	r	7
+C	M	c	s	8
+D	N	d	t	9
+E	O	e	u	+
+F	P	f	v	/

```
From: test.user@bangkok.software.com (Test User)
Subject: test with MIME ASCII attachment
Mime-Version: 1.0
Content-Type: multipart/mixed;
boundary="=====================_852103889==_"

--=====================_852103889==_
Content-Type: text/plain; charset="us-ascii"

This is the message body
--=====================_852103889==_
Content-Type: text/plain; charset="us-ascii"

Now is the time for all good men to come to the aid of their neighbor
The quick brown fox jumped over the lazy dog's back. 0123456789

--=====================_852103889==_
Content-Type: text/plain; charset="us-ascii"

--=====================_852103889==_--
```

Next is an example of a binary file attached with MIME.

```
Return-Path:  <Lawrence.Hughes@bangkok.software.com>
Received: from bangkok ([10.3.101.9]) by bangkok.software.com
        (Post.Office MTA v2.2 ID# 0-0U10) with SMTP id AAA318
        for  <lawrence.hughes@bangkok.software.com>;
        Thu, 19 Dec 1996 09:48:57 -0800
Message-Id: <3.0.32.19961219094857.009bcb50@bangkok.software.com>
X-Sender: LHugh@bangkok.software.com (Unverified)
X-Mailer: Windows Eudora Pro Version 3.0 (32)
Date: Thu, 19 Dec 1996 09:48:57 -0800
To: lawrence.hughes@bangkok.software.com
From: Lawrence.Hughes@bangkok.software.com (Lawrence Hughes)
Subject: binary attachment using MIME
Mime-Version: 1.0
Content-Type: multipart/mixed; boundary="=====================_851046537==_"

--=====================_851046537==_
Content-Type: text/plain; charset="us-ascii"

This is the message body

--=====================_851046537==_
Content-Type: application/octet-stream; name="demo.dat"
Content-Transfer-Encoding: base64
Content-Disposition: attachment; filename="demo.dat"

AAECAwQFBgcICQoLDA0ODxAREhMUFRYXGBkaGxwdHh8gISIjJCUmJygpKissLS4vMDEyMzQ1Njc4
OTo7PD0+P0BBQkNERUZHSElKS0xNTk9QUVJTVFVWV1hZWltcXV5fYGFiY2RlZmdoaWprbG1ub3Bx
cnN0dXZ3eHl6e3x9fn+AgYKDhIWGh4iJiouMjY6PkJGSk5SVlpeYmZqbnJ2en6ChoqOkpaanqKmq
q6ytrq+wsbKztLW2t7i5uru8vb6/wMHCw8TFxsfIycrLzM3Oz9DR0tPU1dbX2Nna29zd3t/g4eLj
5OXm5+jp6uvs7e7v8PHy8/T19vf4+fr7/P3+/w==
--=====================_851046537==_
Content-Type: text/plain; charset="us-ascii"


--=====================_851046537==_--
```

Examples of using attachments in e-mail

Adding an attachment with a separate utility (uuencode or BINHEX)

1. Obtain or create the file you want to attach.

2. Using the separate utility, encode the attachment into one text file per attachment.

3. With your UA, create a new outgoing message.

4. At some appropriate point in the message body, insert the encoded attachment. If the UA has an Insert File option, use that. If not, and your UA is a GUI application, load the encoded attachment into Notepad, then select the entire contents of the Notepad, cut (copy) the contents, then paste the result into the new message being created. Repeat for each attachment.

5. Send the message, including the encoded attachment.

Adding an attachment with a UA that has an embedded attachment facility

1. Obtain or create the file you want to attach.

2. Check your UA to be sure that the desired attachment scheme is selected (some UAs support multiple schemes, e.g., Eudora).

3. Use the Attach File option to add the attachment. That will encode the file and insert it at an appropriate place in the message (typically the encoded text will not be visible in the message body).

4. Send the message, including the encoded attachment.

Obtaining an attached file with a separate utility

1. Receive the message with attachment.

2. Save the entire message (including headers) to a text file in a convenient directory.

3. Outside of the UA, change to that directory and feed the received message through the decoder utility. The attachments should be output by the utility into the current directory.

Obtaining an attached file with a
UA that has an embedded
attachment facility

1. Be sure the Process Incoming Attachment option is selected in your UA (if it can be disabled).

2. Be sure you have specified a reasonable location for attachments to be stored and that there is sufficient space on that disk drive to hold the attached file.

3. Receive the message with attachment. That will decode the attachment and save the file produced in the prearranged directory for attachments.

Address books

An address book is exactly what it sounds like: a list of people and their associated (in this case, e-mail) addresses. It also can contain other information. Some of that information might be related to e-mail, for example, what kind of attachment a given user requires, or their public key. It also may contain information not related to e-mail, such as voice phone numbers, fax phone numbers, physical addresses, and so on. The address books are maintained on a computer as a simple database. Although proprietary e-mail systems have had good address book facilities (even shared and enterprise-wide synchronized ones), historically the Internet has been weak in the area of address books, especially shared ones. Most UAs have a way for the user to manage a personal address book (one specific to a single user and available only to that UA). A good UA allows names and e-mail addresses for all incoming mail to be added automatically to the personal address book.

The good thing about personal address books is that no standard is necessary, because each one is used only by a single UA. No other software needs to know anything about its internal architecture. The bad thing about them is that each one typically must be painstakingly built by the individual user, name by name. In a typical office, it is much more convenient to have at least the other e-mail users in that office in a common address book on the network so that those names need be entered or deleted only once, as office personnel change. Typically, there also are some people outside the organization with whom multiple people in the office might need to communicate. Those listings should be in a shared address book. In an ideal world, you would be able to find anyone's e-mail address if you knew some minimum amount of information about that person (company, city, telephone number, etc.). Imagine trying to use a voice telephone system without white pages directories or long distance information.

Some directory services currently are available on the Internet with a large number of listings on them (but, unfortunately, no really comprehensive ones). The output of those systems is intended for humans to read (e.g., via Web browsers). While this is better than nothing, it would be a lot nicer if you could search such a system from inside your UA via network protocols. The ideal is to be able to specify a few criteria (e.g., company name, city, etc.) and get back a list of possible hits, with further details on each. Another possibility would be the ability to browse hierarchical directories until you find the correct listing, much like browsing a network file server for a particular file.

Most proprietary mail systems (e.g., MS Mail) have a centralized address book for a given LAN, usually maintained by the e-mail administrator. When an e-mail user is created, the name and address are entered automatically into the shared address book. In the case of MS Mail, in an enterprise environment, you can browse shared address books for any LAN in the enterprise. Furthermore, there are ways for the administrator to manually enter and maintain lists of "foreign" addresses (e.g., Internet addresses available via a gateway). Such a system allows everyone on the LAN to have a common set of often used addresses without everyone having to duplicate the effort. Doing the same thing for the Internet (with 50 million or so users) is indeed a daunting task (good thing we have computers to help us manage it!).

Another problem with a global e-mail directory is privacy. Some companies do not really want for just anyone to be able to determine exactly how many employees they have in each department or to be able to easily find and contact key individuals (e.g., in R&D). The same can be said for government organizations, especially in sensitive areas like Intelligence or the military. And if you think your conventional mailbox at home is full of junk mail, imagine your e-mail mailbox if every company in the world could easily obtain your e-mail address! Obviously there are more than technical issues to resolve before a global directory can become a reality. Perhaps "maintainers of the global directory" is a worthwhile role for the traditional postal monopolies of the world to take over as more and more of their volume shifts to electronic channels.

Although the Internet is still weak in the area of address books, several schemes have been proposed (NIS, Whois, Whois++), but none really has taken hold. The ISO mail system includes a complex, comprehensive scheme (called X.500) to keep track of e-mail addresses for X.400 mail (it actually allows keeping track of many other things, as well). Alas, X.500 requires an ISO network stack (the ISO equivalent of the TCP/IP trans-port protocols) to work, and the e-mail addresses it keeps track of are ISO X.400 format (e.g., C=US, O=SecureIT, OU=development, CN=Lawrence E. Hughes) as opposed to Internet format (e.g., Lawrence.Hughes@software.com).

The good news is that a new standard, Lightweight Directory Access Protocol (LDAP), is being hammered out even as this book is being written. LDAP does basically the same thing as X.500's Directory Access Protocol (DAP), with two important differences. It works just fine over Internet-style transport protocols (no OSI stack required), and it manages Internet-style e-mail addresses. It is modeled after one of the X.500 protocols, Directory Access Protocol (DAP), and it is even possible to have an X.500 directory server agent (DSA) that implements LDAP in addition to its native DAP and DSP, for inter-DSA exchanges. Within a few years, most e-mail clients will provide support for LDAP, and a global infrastructure of LDAP servers and databases will exist, so that finding someone's e-mail address will be as simple for the worldwide Internet as it is today for a single LAN (or enterprise) using MS Mail. Chapter 9 examines LDAP in some detail, as well as a combined X.500/LDAP DSA product from the ISODE Consortium and a slightly different approach from Microsoft that is still highly interoperable with LDAP and X.500 (Active Directory).

Generalized directory services

In practice, e-mail address books are just a special case of a much more general problem that companies have been trying to solve for some time. One of the early contenders in the area was Banyan Systems, with their

StreetTalk directory services. That scheme allowed keeping track of login accounts for the Banyan network, computers, printers, shared files, and so on. Unfortunately, the scheme was tied closely to Banyan's proprietary Vines IP network protocol (similar to, but not interoperable with, Internet IP). Banyan later made a version that could work in standard IP environments and even independent of the rest of the Banyan network, called Universal StreetTalk. The product is still a long way from being compliant with (or even interoperating with) emerging standards, such as X.500 or LDAP. Such is the price of being an innovator.

Another major player, Novell, introduced a comprehensive directory service (NetWare Directory Services, or NDS) in version 4.0 of their NetWare file and print sharing network product. Again, it has most of the concepts of a generalized directory service but is tied to a proprietary network protocol (in this case, Novell IPX/SPX) and is not compliant or interoperable with the emerging X.500 and LDAP standards.

Microsoft has held off on entering the directory services fray until late in the game, which may have been a good decision in this case. Even as this book is being written, Microsoft is creating an elaborate distributed directory service (called Active Directory) that is based on and may be interoperable with X.500 and LDAP. Microsoft has even created a mapping layer, Active Directory Services Interface (ADSI), that will allow other directory services, such as NDS, to be used as kind of a subdirectory within an overall scheme. It is the directory services equivalent of Microsoft's database mapping scheme called Open Data Base Connectivity (ODBC), which allows applications to be written in a vendor-independent manner (i.e., they can work equally well with Microsoft SQL Server, Informix, or Oracle, as long as the application does all database operations through ODBC calls and uses only the functionality common to all those products).

Contents

Basics of cryptography

As people shift more and more of their interaction from traditional paper mail and voice telephone to computer-based communications schemes (e-mail, newsgroups, chat, net-phone, etc.), they are discovering that the Internet is not quite as secure as they might like. It is, in fact, rather like sending all your conventional mail on postcards rather than in envelopes. It is not all that difficult for anyone along the way to intercept e-mail and read it, modify it as it passes through, or even generate fraudulent e-mail that is indistinguishable from the real thing. Once communications are in digital form, the very power and flexibility of computers make it possible for governments, companies, even individuals to violate people's privacy and commit fraud on a scale never before possible. The good news is that the same technology is capable of adding essentially impregnable armor to digital communications. The magic bullet is encryption technology.

Once the exclusive province of governments and the military, computers have allowed encryption technology to come of age and to be used by the masses. Unfortunately, many people still do not understand what encryption is or what it can do for them. Even more unfortunately, their own governments in many cases are going to great lengths to prevent them from being able to obtain or use this important technology. Another serious problem is that not even the creators of software products today (for the most part) understand the technology well enough to make truly secure systems. Finally, the infrastructure required to deploy encryption technology in such a way that it can be used on a global scale and with little or no additional effort on the part of the end user still has a long way to go before it is a reality.

Ideally, this book will go a long way toward promoting a greater understanding of this important technology among all the players involved. In particular, if end users and voters understand the issues, they will force governments and software vendors to clean up their acts and make possible protection from online vandals and criminals.

Basic encryption

In one form or another, encryption, or secret writing, has been around for thousands of years. An example of a very simple cipher is replacing every character with the previous character in the alphabet; hence, the name IBM would become HAL. To decipher (or decrypt) that simple scheme, you would just do the reverse, that is, replace every character with the following character in the alphabet (HAL would become IBM). To make it a little more difficult to decipher an encrypted message, we could substitute not the next, but the *nth following character* (e.g., if n = 5, A would become F; if $n = 2$, A would become C). The value of n is a simple "key." If you knew that a "shift by n" cipher was being used but not which n, it still would not take you long to discover the value of n for a given message. All you would have to do would be try (up to) all 26 possible values, until a readable message resulted. (In practice, you would have to try only 13 values, on average, before you found the right one.) This is called an "exhaustive" or "brute force" attack. While such an encryption scheme might be adequate protection among fourth grade students, people routinely decipher more complicated schemes in newspaper crypto-grams, based on typical frequencies of letters, for casual entertainment.

Several books listed in the bibliography cover the fascinating history of cryptology, which is the science that covers encryption, steganography (hiding information in another object, like an image, so it is not even obvious that it may contain covert information), and cryptanalysis (code breaking). With the addition of computers into the mix, the stakes have increased in both the ability to protect information (encryption) and the ability to discover it anyway (cryptanalysis), so much so that any scheme (other than so-called one-time pads) used by governments as recently as World War II are child's play to cryptanalyze (break) today. A pivotal event in military intelligence in World War II was the breaking of the German mechanical code machine. The Enigma project was successful only because a group headed by Alan Turing in England invented some of the first digital computers. Today, that same task could be assigned to under-graduate students in a computer class who could solve it with inexpensive home computers.

Even though today's encryption technology is millions of times stronger than those World War II schemes, it still is doing basically the same thing; only the degree of scrambling and the sizes of the keys have changed. Computers have, however, made one amazing breakthrough possible. That breakthrough, the most important advance in privacy in the last 2,000 years, is called public key (or asymmetric key) encryption. The scheme is not that difficult to understand, and (like most really good encryption algorithms) a widespread knowledge of the inner workings of the technology does not compromise its strength in any way.

Jargon

Plaintext   Plaintext is information in its normal form, that is, not encrypted. It is the input to an encryption process and the output of a decryption process. Plaintext is useful (if not actually readable, as in the case of a binary data file) to the possessor of that information.

Ciphertext   Ciphertext is information in the scrambled form, that is, encrypted. It is the output of an encryption process. Because ciphertext appears to be random gibberish, it is totally useless to someone intercept-ing it who does not have access to the appropriate decryption scheme and the specific key required to decrypt it. Encryption does not prevent inter-lopers from intercepting transmissions; it merely makes what they have intercepted useless to them.

Encryption   Encryption is the process of scrambling plaintext informa-tion into ciphertext in such a way that the original information can be recovered only by the intended recipient. There also are one-way schemes to scramble information into a predictable result (a hash of the original data) but without the ability to recover the original information (hence, the term "one-way"). Such schemes are not considered encryp-tion, because it is not possible to recover the original information from a hash.

Decryption   Decryption is the process of unscrambling ciphertext back into the original plaintext, which requires the appropriate decryption algorithm and typically the same key (or the complementary key of a pair, in the case of an asymmetric key algorithm). Imagine, if you will, that someone has used a cake mixer to "scramble" the ingredients of a cake into a batter that bears little resemblance to the original ingredients. Decryption would be equivalent to a cake "unmixer" that could unmix the batter back into the original separate ingredients. The secret, of course, is the way in which the mixer did the scrambling in the first place. If ciphertext were truly random, nothing could recover the original data from it.

Cryptanalysis   Cryptanalysis is the recovery of plaintext from cipher-text without prior knowledge of the correct key, either by trying all possi-ble keys until the correct one is found (an "exhaustive search" or "brute force method") or through analytical methods based on weaknesses in the algorithm used. Discovering the key through weaknesses in the key management scheme (i.e., finding the key written down on some paper in the sender's trash bin or bribing an employee to reveal the key) pro-duces the same result (being able to obtain the plaintext, at least for mes-sages encrypted with that key) but would not technically be considered to be cryptanalysis. A cryptanalysis of an algorithm means that any informa-tion protected by that algorithm can be recovered without access to the correct key, which makes that algorithm totally useless.

Encryption algorithm   An encryption algorithm is a well-defined set of steps (a recipe), implemented with mechanical gears and levers, electronic circuits, or computer instructions, that when performed in the correct sequence will reliably (i.e., repeatably and without loss of information) transform plaintext information into ciphertext. A simple (non-key-driven) algorithm would produce the same ciphertext every time from a given plaintext (in which case the strength of the algorithm depends on keeping the algorithm itself a secret from everyone except the sender and the recipient). A key-driven algorithm will produce a different ciphertext from a given plaintext, depending on what specific key is used for a given encryption. A well-designed key-driven algorithm does not depend on the algorithm itself remaining secret (it can be public knowledge). It transfers the need for secrecy to the key(s). A simple (non-key-driven) encryption algorithm could be considered to be a mathematical function of just the plaintext, which could be written as:

ciphertext = f(plaintext)

A key-driven encryption algorithm could be considered to be a mathematical function of both the plaintext and a specific key:

ciphertext = f(plaintext, key)

Decryption algorithm   A decryption algorithm is similar to an encryption algorithm but reliably recovers plaintext information from ciphertext. With a simple (non-key-driven) algorithm, only the correct decryption algorithm is required to recover the plaintext. You could consider such an algorithm to be a function of the form:

plaintext = f^{-1}(ciphertext)

A key-driven algorithm would be a function of both the ciphertext and the appropriate key:

plaintext = f^{-1}(ciphertext, key)

Cryptographic key   A cryptographic key is some information apart from the cryptographic algorithm itself that modifies the specific way the algorithms work when this particular key is used. In the case of a shift-by-N alphabetic substitution cipher (the fancy name for a Captain Marvel

Secret Decoder Ring), the key would be the number of alphabetic positions each character is shifted, which would be in the range 0 to 25. The key 0 (shift each character by zero positions) would be considered a weak key, since the ciphertext would be the same as the plaintext! It would not take long to do an exhaustive search of the entire keyspace (the set of all possible keys), because there are only 26 possible keys (which can be thought of as a key length of just under 5 bits: 26 is larger than 16, which is 2^4, but smaller than 32, which is 2^5). In general, the larger the keyspace (the longer the key length in bits), the longer it takes to do an exhaustive search to discover a key. Today's computers can search literally millions of possible keys a second, so keyspaces need to be quite large. The U.S. Data Encryption Standard, single DES, has a key length of 56 bits, which results in 2^{56} possible keys, or some 72 quadrillion, which written out looks like this (approximately):

72,057,594,037,930,000

Even with a lot of very fast computers, it would take a while to check out that many keys. A recent study, however, showed that a specialized machine could be built using existing integrated circuits (that implement the DES algorithm in hardware), for roughly $1 million that could recover a DES key in about 3.5 hours. Another study showed that a 40-bit key (the longest currently allowed by U.S. export laws) can be discovered through exhaustive search with a typical supercomputer in about 2 weeks, at a cost of about $10,000.

Keyspace   Keyspace is the set of all possible keys for a given algorithm. For example, with DES, there are 2^{56} possible keys. The size of the keyspace refers to the number of possible keys in it, the larger the better, assuming the algorithm used is cryptographically strong and exhaustive search of the entire keyspace is the most effective known attack. Many implementations of security systems are ineffective because they allow the key to be specified in a way (e.g., entering a short ASCII string) that does not ensure that the entire theoretical keyspace was used. As an example of a situation with a keyspace that is too small, there are fewer distinct automobile ignition keys than there are automobiles produced of a given model and year; if you tried your key in enough identical automobiles, you would find one that your key fit.

Key length   Key length is the number of bits it takes to represent a particular key and happens to be the log base 2 of the number of possible keys. If the size of a given keyspace is 65,536, then the key length is 16 bits (the log base 2 of 65,536 is 16, which is another way of saying that you have to raise 2 to the 16th power to get 65,536). If the size of the keyspace does not happen to be a power of 2, it is possible to talk about key lengths that are "fractional bits" long, as is the case with the Captain Marvel Secret Decoder Ring. It has 26 possible keys, which is a key length of about 4.7 bits (which is the appropriate value of the log base 2 of 26). In practice, you round up to the next integral number of bits (in this case, 5). The greater the key length, the stronger the protection. With symmetric algorithms, today 80 bits is considered to be the minimal "safe" length. With asymmetric algorithms, such as RSA, today 768 bits is considered to be the minimal safe key length. In general, the longer the key length, the more time it takes to generate keys, encrypt, or decrypt. The time can go up rapidly with increasing key length, so you do not want just to automatically use very long keys. As time progresses, the minimal safe key length increases (due to advances in computer speed, cryptographic research, and even number theory). It is particularly difficult to establish a minimal safe key length for protecting information for long periods of time, say, 100 years.

"Secure enough"   "Secure enough" is an important concept in security. With home security systems, designed to foil burglars, a system does not have to be impossible to break, it need only take recognizably more time or money to break into it than would be justified by the worth of the contents in the house. (A more cynical person might point out that it only need be more secure than the neighbors' houses.) A home typically is not threatened by the full might of the U.S. Army but by one or a few larcenous individuals. If a given object is "secure enough," based on the (perceived) value of the thing being protected and the expected threat(s), that is sufficient. For example, a 40-bit key is adequate to protect the contents of my bank account but not that of Bill Gates.

Symmetric key cryptographic algorithm   A symmetric key cryptographic algorithm (sometimes called a private key) is the traditional kind of key-driven algorithm, in which the same key is used to encrypt and to decrypt. A car door lock could be considered a symmetric key scheme:

You unlock your car door with the same key (or at least one with the same pattern of ridges and valleys) that you used to lock it. Symmetric key algorithms can be made to be very fast and are suitable for encrypting large quantities of information. Key management tends to be more difficult, because the recipient must have access to the same key as the sender. There are two implications of that fact:

▶ The single key somehow must be securely given to both the sender and the receiver (or created by one and securely sent to the other). Key distribution typically is the weakest point of a secure system based only on symmetric key technology.

▶ There is no way to tell (at least not with encryption technology) whether a particular use of the key was by the sender or the recipient (the recipient could fraudulently pretend to be the sender), because they both must possess the same key.

Asymmetric key cryptographic algorithm   As simple as it sounds, an asymmetric key cryptographic algorithm (sometimes called a public key) is the most important breakthrough in cryptography in 2,000 years. The basic idea is that keys are created in pairs: one to be kept private and the other to be made public (readily available to any interested party). Imagine a door lock that has two physical keys that are different: You lock the door with one key and unlock it with the other. That would be an example of an asymmetric key pair. Asymmetric key algorithms tend to be very, very slow compared to symmetric key algorithms. There are two primary implications of the way asymmetric key algorithms work:

▶ Each person can generate a unique asymmetric key pair and keep the private half secret; the public half can be safely published in an address book or other scheme. It is not necessary to keep the public half secret; in fact, you want to make it as easy as possible for anyone to obtain it. At no point does anyone's private key need to be transmitted or shared with anyone.

▶ Only the owner of the secret part of an asymmetric key pair can use it. Thus, its use establishes the identity of the user (authentication) or proves that a given message originated from a particular sender (nonrepudiation), assuming the private key has not been compromised (revealed or discovered).

Two important properties for using electronic messaging systems in business or government are difficult to accomplish using symmetric key technology. A number of newer algorithms depend on that simple difference. The two main algorithms are public key encryption/decryption and digital signatures. Another important one is secure-session key exchange (Diffie-Hellman). (Chapter 5 covers those three algorithms in detail.) In a nutshell, to encrypt a message with asymmetric key technology, the sender obtains the recipient's public key and "locks" the message with it. The recipient uses the other half of the key pair (the recipient's private key) to "unlock" it. With a digital signature, the sender uses his or her private key to sign a message, then anyone can obtain and use the sender's public key to verify the signature.

Hybrid cryptographic system   In the real world, asymmetric key algorithms are much too slow to be used to encrypt long messages; it takes a long time to process even 64 to 160 bits with such an algorithm. Fortunately, there are enough bits to contain a symmetric key. A hybrid system uses asymmetric key technology to securely transmit a randomly chosen symmetric session key. A fast symmetric algorithm then uses that session key to encrypt a single transmission, then the session key is destroyed and never used again.

Message digest   A message digest is similar to encryption in that it is an algorithm that takes plaintext and a key as input, but the result is a short, fixed-length "digest" that is a function of the exact contents of the plaintext and the key. Unlike encryption, a message digest is a one-way function, and it is not possible to reconstruct the original plaintext from the resulting digest. The idea is to do the one-way process on two different copies of the same message (i.e., before and after transmission via e-mail or when the file is created and on the allegedly same file years later) and then compare the two results. If they match, the probability is high that the messages are identical. This is somewhat like a simple checksum, also called a cyclic redundancy checksum (CRC), but it is more sensitive to differences. An 8-bit checksum of a message easily can be fooled, and there is one chance in 256 (2^8) that two different messages could produce the same checksum. The CRC algorithm produces 16 bits and is somewhat harder to fool. There is a 1 in 65,536 (2^{16}) chance of two different messages producing the same CRC. A typical message digest produces 64 to

160 bits of digest, using an algorithm that amplifies minor differences (i.e., has good cryptographic dispersion), which results in incredibly low probabilities (less likely than that the sun will go nova tomorrow) of two different messages having the same digest. The bottom line is that if you compute a message digest of two messages using a good algorithm (e.g., the Secure Hash Algorithm, or SHA), and the two digests are equal, you safely can conclude that the two messages are identical. If they do not match, you safely can conclude there is at least some difference (possibly an irrelevant one, such as an extra blank being inserted at the end of each line by a mail gateway, or possibly a critical one, such as the change of a bank account number) between the two files. That property is critical to the process of creating a digital signature, as will be seen in Chapter 5.

Block cipher   A block cipher is an encryption algorithm that works on chunks of data at a time. For example, basic DES encrypts a 64-bit chunk (or block) of data as a monolithic operation (that is the granularity of that algorithm). Such algorithms are well suited to processing large quantities of data, especially in situations where data can be buffered or is packetized (as in a network router).

Stream cipher   A stream cipher is an encryption algorithm that works on streams of data (at either the character or even the bit level). Such algorithms are well suited to being embedded in existing communication environments, where the information already is in bit serial form, such as in a secure modem. It is possible to employ a block cipher algorithm, such as DES, in this kind of operation (as a subprocess in a more complex algorithm), but there are algorithms specifically designed to work in this manner. One common scheme is to create a pseudo-random bit stream that can be reproduced at the receiving end and to EXCLUSIVE OR the data bit stream with the cryptographic bit stream. At the receiving end, the encrypted stream again is made EXCLUSIVE OR with the cryptographic bit stream, which recovers the original data. EXCLUSIVE OR is a boolean algebraic function that has the useful property that EXCLUSIVE OR-ing anything with the same thing twice results in the original thing, that is:

- ▶ Encryption: $ciphertext\_bit_i = plaintext\_bit_i$ XOR $crypto\_stream\_bit_i$
- ▶ Decryption: $plaintext\_bit_i = ciphertext\_bit_i$ XOR $crypto\_stream\_bit_i$

In a stream cipher, the problem reduces to being able to create an apparently random, arbitrarily long stream of bits in such a way that the stream of bits can be regenerated reliably (in synchronization with the sender) by the recipient and only by the recipient. Ideally, the string of bits should be as long as the message; if not, it should be at least many thousands or millions of bits long before it repeats. Such a scheme also can be key driven.

Key escrow   Key escrow is a scheme proposed by the U.S. government (as part of the so-called Clipper proposal) that ensures that any key used to encrypt information in e-mail or secure telephones or other communications devices can be obtained by law enforcement agencies via legal due process (i.e., with a search warrant issued by a court). To escrow something means to store it (or in this case, a copy of it) with a third party trusted by both primary parties to a transaction (in this case, the government and the user of the encryption system). Unfortunately, the government (at all levels) has proved consistently able and more than willing to ignore such safeguards in the past with illegal telephone wiretaps. No foreign company is going to be willing to use an American-based security system if they know the U.S. government has potential access to all the keys needed to break the security. The strength of any secure system is completely dependent on the security with which its keys are managed. With any escrow system, keys could be obtained as easily as bribing a law clerk or an underpaid government worker.

Key recovery system   The latest proposals from the U.S. government have changed the term "key escrow" to "key recovery" and claim it would be a benefit to users of a security system who somehow have lost their key. In practice, both terms mean that the government will have access to everyone's private keys. Any cryptographic system that allows keys to be recovered by any means simpler or faster than by an exhaustive key search or elaborate cryptanalysis is, by definition, insecure. A secure communications system generates public/private key pairs locally (on the user's own computer) and never allows the private key to be sent anywhere else (let alone a communal key storage area). Even on the user's own computer, the private key is kept in strongly encrypted form and is decrypted only on presentation of a pass phrase known only to the user and then only long enough to use. For a fascinating look at "industry expert's" view on key recovery, see the "Key Recovery Study" on the accompanying CD-ROM.

Real-world cryptographic algorithms

Nobody would seriously consider using a Captain Marvel Secret Decoder Ring to protect any real business information. Fortunately, a number of good algorithms are in common use today on digital computers that are entirely adequate to protect even Bill Gates's bank account, if applied correctly. One can safely assume that the National Security Agency (NSA) or military intelligence departments have concocted and use even stronger schemes. It also is likely such groups routinely crack even fairly strong commercial schemes, although with the evolution of public key technology and the rapidly growing sophistication of the communications software industry in the area of encryption, that is no longer a given (the specific regulations concerning export of various technologies and other restrictive laws say a lot about the government's abilities in this area). Unfortunately, the U.S. government in particular has imposed draconian restrictions on the export of secure products that allows only weakened, noncompetitive products that are literally millions or even quadrillions times weaker than those from other countries. In the long term, that will accomplish little if anything, other than the loss of the United States's lead in the world software market. The cat is long since out of the bag (strong crypto is available around the world to pretty much anyone who wants it), and the sooner the U.S. government is willing to admit that, the better it will be in the long run.

Symmetric key block ciphers

DES

DES (Data Encryption Standard) is the classic 64-bit block, symmetric key block cipher (officially adopted as a U.S. federal standard on November 23, 1976). It originally was devised by IBM, based on their Lucifer algorithm, and then weakened dramatically by the NSA (the key length was reduced from 112 bits to 56 bits). The algorithm is now standardized under a number of different names (e.g., Federal Information Processing Standards (FIPS) PUB 46 and ANSI X3.92). Other FIPS standards have since been released concerning the implementation and use of DES.

DES originally was intended to be implemented only in hardware; at the time of its release, a software implementation would have been terribly slow. For comparison, hardware chips are available that can process from 1 MB to as much as 200 MB of plaintext per second. A software implementation running on an Intel 486 at 66 MHz is capable of processing only 350,000 Bps (a factor over 500 times slower than the fastest hardware, but only 3 times slower than the slowest). Today, however, software implementations are readily available for downloading over the Internet, even internationally, and are fast enough for use even in real-time communications systems (especially with today's fast Pentiums).

By 1996 standards, single DES (one 56-bit key) is considered by the industry to be inadequate (in terms of cryptographic strength) for most applications. However, in the latest offer from the State Department to U.S. software exporters, that is as good as the carrot gets and then only for two years, if a company can show that it is implementing a key recovery mechanism (which pretty much defeats the purpose of using encryption). Triple DES is considered adequate for most purposes, but it is essentially impossible to obtain export approval for it. Both DES and triple DES are available for use without the payment of royalties.

Triple DES

One of the simplest ways to improve the cryptographic strength of DES is to apply it three times on the same plaintext, with three distinct 56-bit keys. The recommended scheme is to encrypt with key 1, decrypt with key 2, then encrypt again with key 3. The reason for the middle step being decryption is so that single DES is a special case in which the three keys are identical. Of course, to decrypt triple DES, you must undo those steps in reverse order: decrypt with key 3, encrypt with key 2, then decrypt with key 1. While it may seem that the effective key length would be 3 times 56, or 168 bits, it turns out that it really only doubles the effective key length (hence, triple DES should be considered to be about as strong as a monolithic algorithm with 112 bit keys). There is a powerful meet-in-the-middle attack that makes double DES weak, so the next step up from single DES is triple DES.

The advantage of triple DES is that it is dramatically stronger than single DES (some 72 quadrillion times stronger, on the basis of an "effective key length"). The disadvantage is that it is three times slower (which in a software implementation can be a significant hit, even with today's processors). A common variant of triple DES uses only two keys: encrypt with key 1, decrypt with key 2, then encrypt again with key 1 (it is also known as EDE mode). However, "real" triple DES (with three distinct keys) is decidedly stronger; in general, managing 168 bits worth of key is not much more difficult than managing 112 bits worth. Again, EDE mode with two keys is just a special case of a general three-key implementation (just specify keys 1 and 3 as the same value). There is no physical limitation to going beyond even three keys, but the time to process data increases in direct proportion to the number of keys. Three keys generally are considered to be amply strong by today's standards.

DESX

With software source code of DES readily available, it is simple to create variants of DES (although, of course, variants will not interoperate with "real" DES). One such scheme has been implemented by RSA that involves use of an additional 64-bit "whitening" key (in addition to the basic 56-bit DES key) that modifies the input data before feeding it into DES and then doing a complementary step to reverse the effect on the output. Not only does this scheme make DES more resistant to both "known plaintext" and differential linear cryptanalysis, any change to DES (that does not weaken it) is likely to wreak havoc with specialized DES-cracking machines that governments likely possess. This variant has been available in RSA's BSAFE encryption toolkit for some time. It is not much slower than standard DES and is considerably stronger. It also is much faster than triple DES.

RC2

RC2 is a proprietary 64-bit block cipher with a variable length key (the longer the key, the stronger the security) created and owned by RSA. Because details on the algorithm are not disclosed to the general public, not much peer review has been done. However, it likely is as cryptographically competent as its inventor (Ron Rivest, who is the "R" of RSA, one of the most respected companies in the industry). It is faster (3 times, it is claimed) than DES in software and is fairly easy to obtain export authorization if key length is limited to 40 bits. One interesting aspect of the algorithm is that encryption speed is independent of key length (and there is no maximum key length). Implementors using this algorithm must pay royalties to RSA (at least in the United States).

IDEA

The International Data Encryption Algorithm (IDEA) was created in 1990 to 1992 by Xuejia Lai and James Massey. This is a 64-bit block cipher, with a 128-bit fixed-length key. It is about twice as fast as DES in software. The algorithm is very strong but has not been subjected to as much peer review and analysis as DES. It is patented, and implementors who use it must pay royalties. One of its chief claims to fame is its use as the only symmetric algorithm in PGP (Pretty Good Privacy), the most widely used shareware encryption product.

CAST

CAST is a Canadian algorithm designed by Carlisle Adams and Stafford Tavares (whose initials form the acronym). It is cryptographically strong and is used by Northern Telecom (Nortel) in their Entrust package (which was used by Microsoft in implementing Exchange Server's security aspects: encryption, digital signatures, and key management). CAST is patent-pending, so its use in products requires payment of royalties.

SKIPJACK

SKIPJACK is a newer algorithm devised by the NSA as a replacement for DES. It is likely much stronger (assuming use of strong keys that are not escrowed) but was designed from the start to support key escrow, so that any government agency (especially law enforcement) can obtain any keys used through due legal process. One way the NSA plans to enforce this key escrow scheme is interesting from a technical standpoint. It seems that only the government will have the information necessary to generate "strong keys," and of course it will escrow copies of any keys it generates. Any keys you try to generate for SKIPJACK yourself are supposed to be guaranteed to be "weak." (The number of "strong keys" is a vanishingly small percentage of the possible keys; without the secret of how to find the few grains of wheat among the mountains of chaff, the odds of finding a strong key are remote.)

Unlike DES, the implementation details of SKIPJACK are highly classified, and companies must meet stringent requirements to be able to work with it. Implementation of SKIPJACK will be allowed only in VSLI hardware chips. While the government has not yet managed to criminalize the use of other algorithms (although Louis Freeh, the current head of the FBI, has proposed exactly that), it is planning to require all government agencies (except for military and intelligence branches, which have their own schemes) to use only SKIPJACK-based security, which will force anyone wanting to communicate securely with them also to adopt it (although not exclusively, fortunately). This scheme has received some press under the name Clipper, the name of the VLSI chip that implements the SKIPJACK algorithm. The government has since tried to sound more reasonable by using the term key recovery rather than key escrow, but it amounts to the same thing.

BLOWFISH

BLOWFISH is a new 64-bit block, symmetric key encryption algorithm devised and promoted by Bruce Schneier (author of *Applied Cryptography*, the bible of computer cryptography implementors). Its main attractions are its variable-length key (up to 448 bits) and almost 4X speed advantage over DES in software implementations (and even greater on newer large-cache processors like Pentium and PowerPC). Although it has not been subjected to anything like the intense scrutiny of DES, it appears to be cryptographically competent and has the advantage of a variable-length key, plus there likely are no machines currently in existence designed to crack it. The algorithm and the author's own C code implementation are in the public domain, and details are readily available. No royalties are due for its use.

Comparison of some symmetric key block ciphers

Relative performance of some symmetric key block ciphers (normalized to 1.0 for DES; higher numbers mean more bytes per second) are listed in Table 4.1.

For a rough idea on absolute performance, DES running on an Intel 486/33 CPU with good 32-bit code can process about 35 KBps.

Table 4.1
Relative performance of some symmetric key block ciphers

Cipher	Relative performance
DES	1.0
Triple DES	0.34
IDEA	2
BLOWFISH	3.86

Asymmetric key block ciphers

RSA

Named after its inventors (Ron Rivest, Adi Shamir, and Leonard Adelman), RSA is an asymmetric key algorithm based on the difficulty of factoring the product of two very large prime numbers (a classic difficult problem in mathematics that the world's best mathematicians have been trying to solve for hundreds of years). RSA is about 100 times slower than DES when both are implemented in software (but about 1,000 times slower compared with the best current hardware implementations). RSA is patented, but the U.S. patent will expire in September 2000. Implementors currently must pay royalties to use the algorithm in their products.

El Gamal

Created by Taher El Gamal (now a senior scientist at Netscape), El Gamal is an asymmetric key algorithm based on the difficulty of calculating discrete logarithms in a finite field (another classic difficult problem in mathematics). The algorithm is covered by the patent on the Diffie-Hellman key exchange algorithm (due to have expired April 29, 1997).

Message digest algorithms

MD5

Message Digest 5 (MD5) is from RSA and is a stronger, later version of their earlier MD4 algorithm. It produces a 128-bit hash from any size input. It is used in several Digital Signature schemes, for example, in PEM (for Privacy Enhanced Mail). Unfortunately, MD5 has been broken (cryptanalyzed), so its use is not recommended.

SHA

SHA (Secure Hash Algorithm) was created by the NSA and the National Institute of Standards and Technology (NIST) and is used in the Digital Signature Standard (DSS). It produces a 160 bit hash value from any size input. SHA is similar to MD5 but with sufficient differences to make it significantly more difficult to break. As an open standard, there is no royalty required for using SHA. On an Intel 486/33, using good 32-bit code, it is possible to process about 75 MB of text per second.

Contents

Applications of cryptography in messaging systems

Now that you understand the basic concepts of security, we can talk about how that technology can be applied in e-mail products. To follow the ideas in this chapter, it is necessary that you grasp the information presented in Chapter 4.

Jargon

Public key/ private key   With asymmetric cryptography, there are two keys for a given user. One key is kept secret (the private key), and the other is published for anyone to see or use (the public key). If one of the two keys (say, the public key) is used to encrypt something, only the other key of that pair (in this case, the corresponding private key) will be able to decrypt it.

Key management   The aim of key management is to provide a reliable, trusted way to generate public/private key pairs and to make the public keys available to anyone who needs access to them.

Digital envelope   A digital envelope is a method of achieving privacy by encrypting the contents of a message, equivalent to putting a paper letter inside a sealed envelope. It is also referred to as "sealing a message." A message is sealed using the recipient's public key, so only the recipient can unseal (decrypt) it.

Digital signature   A digital signature is the appending of some additional information to a message that is a function of both the entire contents of the message (including the order of the characters) and the sender's private key. That allows the recipient (given access to the sender's public key) to verify that the message really came from the purported sender, and furthermore that it has not been tampered with along the way. A digital signature does not provide any degree of privacy. A given message can be digitally signed, digitally sealed, both, or neither (the two processes are independent of each other).

How cryptography is used in messaging systems

Some systems (e.g., Microsoft Exchange Server) already have implemented the facilities described in this chapter, at least for a limited group (say, a single company). Even as this book is being written, standards are being refined and adopted to allow this technology to be extended to the largest possible scale, the worldwide Internet. The technology is absolutely crucial to the ability of the Internet to serve as an acceptable messaging system for all kinds of business and commerce.

With traditional mail, we depend on a number of important cues to help us detect fraud of various kinds, such as someone pretending to be someone else or alterations made to the original mail. Such cues include letterhead stationery, postmarks applied by a trusted third party (the U.S. Postal Service), and handwritten signatures. Erasures and whiteout usually are readily detected. In the olden days, when couriers were less trusted than today's Postal Service, it was common to seal letters with melted wax and a signet ring. If the letter arrived with the wax seal intact, the recipient had a high level of confidence that it had not been read and was really from the purported sender (the assumption being that no one else owned a signet ring with the same distinctive pattern).

Unfortunately, e-mail gives none of the above cues. It arrives as a set of totally depersonalized bits, which we view on video screens or print with our own printers. Even the font in which it is displayed or printed may have no relation to what the sender used. If it is printed at all, it is on our own paper (no letterhead). Any changes made along the way are totally invisible (no telltale whiteout or smudges). Some e-mail may include a rough equivalent to postmarks (nodenames of computers it passed through along the way, complete with times and dates), but those can be faked. Even the From: address is easy to fake with certain widely used packages (e.g., Eudora). There is no handwriting, not even a signature, we can recognize. Even a bitmap image of a handwritten signature can be scanned in easily. It is amazingly easy for anyone along the way to intercept an e-mail message, read it, and even alter it. Certainly there is no way to drip molten wax onto an outgoing e-mail message, let alone impress it with a distinctive signet ring.

It is remarkably easy to commit a variety of kinds of fraud with e-mail. Already criminals are figuring out how to obtain credit card numbers (complete with full names and expiration dates) from e-mail messages (or other online services, such as the WWW). With the right tools (which I could create in a few days), I could generate e-mail that appears to be from the President of the United States to the U.S. Attorney General, confessing to all sorts of fascinating crimes. It would be essentially impossible to prove that the e-mail did not come directly from the Oval Office. It could even appear to have been originally posted to the mail server at whitehouse.gov (which I was able to determine from home, in about two minutes, to be storm.eop.gov, at IP address 198.137.241.51, using a widely available DNS debugging tool called "nslookup"). If you are amazed at the magic that can be done with digitally retouched photographs and motion pictures these days, you should see what a clever hacker could do with (to) your e-mail. Companies and government agencies routinely scan all mail messages going through their servers for key words or phrases. Or perhaps they archive everything on writable CD-ROMs. They may even be required to do so by law.

So what is one to do? Eschew the speed and convenience of e-mail for anything other than letters to Mom? Use your e-mail software to compose and print out letters locally, sign them by hand, seal them in paper envelopes, and drop them in the corner mail box? Fortunately, some very clever folks have figured out how to use encryption technology to provide levels of privacy, authentication (guarantee of identity), and tamper resistance that is actually millions of times stronger than is possible with traditional mail and, ideally, without losing any of the convenience or speed of e-mail.

The two primary applications of security in e-mail are digital envelopes and digital signatures. A digital envelope is analogous to putting a traditional letter inside an envelope to keep the contents private from handlers along the way and to prevent others from making changes in the letter or even replacing the letter with their own. The digital signature concept is analogous to adding your handwritten signature at the bottom of your letter. It serves to authenticate your letter as having really come from you. In this case, however, the electronic version has benefits that go well beyond those of your handwritten signature. It can provide a guarantee that no one else in the world could have generated that particular digital signature. Furthermore, it can detect even a single character changing in a message the size of *War and Peace*. The technology used to do that easily can supply every living human with a "digital signet ring" that is readily distinguished from all the rest (that alone would be a real trick with physical signet rings).

How is this magic trick done? It is elementary, *if* you have access to public key encryption.

The most obvious use of security technology in e-mail is for privacy. The basic idea is similar to the secure telephone line popular in spy movies. The user creates an e-mail message "in the clear" (plaintext), then scrambles it (into ciphertext). The ciphertext is transmitted via the insecure channel (e.g., Internet e-mail). On receiving the ciphertext, the recipient unscrambles the message and reads it. Note that encryption does not prevent someone from intercepting the message, but what they obtain is useless to them. A side effect is that they cannot tamper with the message, because they not only would have to unscramble the message before changing it but also rescramble it afterward in exactly the same way the real sender scrambled the original message. Since that process has much the same effect as putting a traditional letter inside a paper envelope and sealing it, it often is referred to as using a digital envelope (by the same analogy, today's nonsecure e-mail is equivalent to using post cards).

The other primary use for security technology in e-mail is for the sender of a message to authenticate it, based on the exact contents of that message and the sender's private key (using asymmetric key technology). When the message is received, the recipient can validate the signature (using the sender's public key), which lets the recipient know for certain whether the message has been altered in any way and that it could have come only from someone possessing the sender's private key. Unlike encryption, this method does not prevent a third party from seeing or modifying a message, but it can detect any modification with a high degree of certainty. The process is the electronic analog of your handwritten signature (as well as the old fashioned sealing of a message with wax and a signet ring), which is why it is called a digital signature.

Digital envelopes

It would be possible to encrypt mail using only symmetric key technology (e.g., DES). However, in such a system, key management is difficult and usually the weak point of the system. The recipient (or all the recipients of a mailing list) would have to possess exactly the same key used to encrypt the message in the first place. If the sender generates the key, it must be securely relayed to the recipient(s). If a third party generates the key, it must be securely relayed to both the sender and the receiver. If the sender chooses to change the key, all recipients must be updated. If anyone's copy of the key is compromised, the key is compromised for everyone. Typically, due to the complexity and overhead of key management, a single key is used for some period of time (a given number of messages or time interval), which further increases the odds of its discovery.

With asymmetric key technology, the sender obtains the public key of the recipient. We will cover later how that is done and how the sender can be sure that it is the correct key and not some clever imitation published by a third party. The sender then encrypts the message using the public key, creating a ciphertext for each recipient (if the message is going to more than one receiver, the sender must use the appropriate public key for each). The ciphertext then can be sent safely through any channel (even one as insecure as the Internet). The recipients decrypt the cipher-text using their private keys. At no point has anyone had to share a private key with *anyone*, not even someone who is receiving the encrypted mail. And since only recipients should have access to their own private keys, no one else would be able to decrypt the ciphertext (assuming a strong algorithm and a sufficiently large key are used).

In practice, you could implement a real system in exactly that way, but it would take a very long time for senders to encrypt their mail and for recipients to decrypt their mail (even for fairly short messages of a hundred words or so). For large files, the time required would be prohibitive, even with today's inexpensive desktop mainframe microcomputers. Asymmetric key algorithms are very, *very* slow (thousands of times slower) compared to symmetric key algorithms. In fact, they are useful only for encrypting a few bytes (perhaps 32 to 64 bytes). Conveniently enough, that is plenty of bytes to hold a symmetric key. Real systems use asymmetric technology only to encrypt a randomly generated (used once and thrown away) session key for a symmetric key algorithm. The bulk of encryption work is done using such a session key and a high-speed sym-metric key algorithm. Such a scheme is called a hybrid system. The exact steps involved in sending encrypted mail in a hybrid system are as follows:

S1. Compose the message in plaintext.

S2. Generate a randomly chosen session key for the chosen symmet-ric key algorithm.

S3. Obtain the public key of the recipient.

S4. Encrypt the session key using an asymmetric key algorithm and the recipient's public key and write the result to the outgoing message.

S5. Encrypt the message using the (unencrypted) session key and the chosen symmetric key algorithm and write the result to the out-going message.

S6. Destroy all traces of the session key. It never will be needed or used again by the sender (possibly like the original message, depending on the security of your system and your level of paranoia).

S7. Send the results of the steps S1–S6 (the outgoing message) to the recipient via e-mail.

The steps the recipient goes through to read the message are as follows:

R1. Receive the message and extract the encrypted session key and encrypted message text.

R2. Use his own (the recipients's) private key and the public key algorithm used by the sender to decrypt the session key.

R3. Use the recovered session key and the symmetric key algorithm used by the sender to decrypt the message text.

R4. Destroy all traces of the decrypted session key.

R5. Read, print, archive, reply to, or forward the received (decrypted) message.

Some subtle details are involved in performing many of those steps in a truly secure way. Let us cover those now.

Choosing a session key

Step S2 is a critical one and is often botched even by competent developers not familiar with the subtleties of security work. Netscape recently botched that very step, and some hackers easily were able to crack the supposedly secure HTTP over SSL (Hyper Text Transfer Protocol (HTTP) over the secure socket layer (SSL)). The algorithms and the key lengths were sufficient for a fairly high level of security, but the session key was chosen in such a way that there were far fewer possible keys to try than the symmetric algorithm supported (only a tiny fraction of the symmetric algorithm's keyspace was used). The hackers simply reverse engineered the way the Netscape programmers had generated a session key (by decompiling the object code of the browser). Then they could quickly search that fairly small keyspace. Netscape has since corrected this glaring security hole after much embarrassment. To choose a session key correctly requires the use of quite a bit of random information. Truly random would be best, perhaps obtained by reading the output of a fast analog-to-digital converter being fed from a white noise source, like a reverse-biased diode. Merely using bytes from a pseudo-random number genera-tor (like the C rand function) is inadequate. Short of a source of true ran-dom data, it is possible to make a fair approximation by compositing information together from all over the computer, such as the following:

- ❱ Current date and time to the nearest millisecond;

- ❱ User's login or e-mail account name;

- ❱ User's full name;

- ❱ Computer's node name and IP address;

- ❱ Number of open files on the whole computer;

- ❱ Lengths in bytes of the first hundred items in the NT registry.

Ideally, several hundred bytes worth of such information should be collected. In many cases, that may require detailed knowledge (and result in dependencies on the operating system) to do a good job.

Collecting the random seed is only the start of picking a good session key. Another critical aspect is that the key should be chosen at random from all possible keys in the entire keyspace for the chosen algorithm. For example, with DES there are 2^{56} possible keys. With a good scheme for choosing a session key for DES, there would be no possible keys out of all 72 quadrillion that *never* could be chosen. It is not necessary that each pos-sible key have exactly the same probability of being chosen, just that no part of the keyspace is excluded from *ever* being chosen. In practice, when there is a problem with that, only a tiny portion (perhaps a millionth or less) of the total keyspace may be utilized. For example, many products allow you to specify a short ASCII string, which either is used directly or is processed in some way (e.g., encrypting it with a private key encrypting key) to become the actual DES key. People are likely to choose common English words (of which there are only about 30,000 or so). It takes at least 12 alphabetic characters chosen *completely* at random (26 to the 12th is greater than 2 to the 56th) to ensure that the entire keyspace is utilized. Even using thoroughly mixed uppercase and lowercase plus digits (62 characters) requires about 9 characters. However, humans are notori-ously poor at choosing characters truly at random, so you probably should double those theoretical minimum lengths to account for that.

Obtaining the recipient's public key

Step S3 (obtaining the recipient's public key) also can be tricky. What if a malevolent third party tricks you into using its own public key when you encrypt the message, instead of the desired recipient's public key? Then only that third party would be able to decrypt the message. That would prevent the intended recipient from being able to decrypt the message. Of course, the interloper also would have to intercept a copy of the message (which is depressingly easy on the Internet) to actually be able to read the message.

So how do you get the recipient's public key in the first place? Perhaps the recipient includes it as part of his or her e-mail signature, or you could ask the recipient to mail it to you. That approach, obviously, is not convenient, efficient, or secure. There also is no way to determine if the key has been compromised. A key server can solve several of those problems nicely. If all the people who want to communicate securely with each other deposit their public keys on a key server, then any of them can at any time retrieve any of the other necessary keys from one source (without having to locate or bother the key owner). In fact, a UA can obtain not only a person's e-mail address but also his or her public key from such a server automatically (as part of the addressing process), via an address service protocol such as X.500 DAP or, more recently, LDAP.

It also is possible to maintain on the same key server a list of public keys known to have been compromised, a certificate revocation list (CRL). As soon as you discover that your public key has been compromised, you publish that fact on a key server. If everyone's UA checks the CRL before each use of a public key, exposure from compromised keys will be held to a minimum.

Of course, now you have the problem of who can submit or update keys on a key server or publish the fact that one has been compromised. These and other fascinating aspects of key servers are covered in more detail later.

Regardless of how you obtain a public key, how do you know it is the correct public key for the person with whom you want to communicate? Can you trust the public key? The solution is to not just publish your public key by itself but embed it in a public key certificate. A public key certificate is a file that includes the following information:

▶ The public key itself (usually in ASCII form);

▶ The public key owner's name (and possibly e-mail address);

▶ The time and date the key was embedded in the certificate;

▶ An expiration date and time for the certificate.

The file then is digitally signed, but not by the public key's owner. Typically it is signed by someone both the sender and the receiver trust (perhaps the chief operating officer of the corporation they both work for). The general idea is that a mutually trusted third party digitally signs the information. In effect, that third party vouches for the authenticity of the public key. By digitally signing the certificate, the third party is saying, "I certify that this key was presented to me by the person named in the certificate and that the embedded public key is really that person's." By signing the entire certificate, the recipient can obtain the *signer's* public key and verify that the certificate is intact and that it was really signed by that person.

There is even a widely adopted standard that describes the content and syntax of such a public key certificate, known as X.509 v3. That standard is part of the ISO X.500 Directory Services standard (specifically, the part concerned with public keys and authentication, called X.509, and from the third revision of X.509). Actually X.509 covers far more than just the content and the syntax of public key certificates (it also covers how you go about using them), but the term is used today to describe the certificates regardless of how they are used or the protocols by which they are exchanged. The standard figures prominently in several recent standards, such as Microsoft's "authenticode" and the scheme Visa and MasterCard have agreed on for securing credit card information on the Internet (neither of which uses much of anything else from ISO X.509 or X.500).

But what if there is no common third party that both the sender and the receiver know and trust? There are now certificate authorities that anyone can go to and prove his or her identity to some level of certainty (by presenting a driver's license, original birth certificate, passport, etc.). In practice, such authorities may provide different levels of certificates, depending on what kind of proof is offered. One such company is Verisign (www.verisign.com). Verisign will provide you with several different levels of certificate based on what kind of proof you offer and how much you are willing to pay for its services. The lowest trust level is established just by returning the certificate to the e-mail address included in the certificate (as of the writing of this book, they will issue that from their website for free). Certificates with very high levels of trust require submission of documents to prove your (or your company's) identity and cost a hundred or so dollars a year.

What if you want to exchange secure mail with people in other countries that have their own certificate authorities? There are ways to create authentication chains, wherein the UA can obtain the public key certificate of the person or company that signed the sender's certificate and then the public key certificate of the person or company that signed *that* certificate, and so on, until some lord high protector of public key certificates (the U.N. Secretary General? the Pope?) is reached. Obviously, the asymmetric key used by a major certificate authority is going to be verified and protected very carefully; that one being compromised could compromise thousands or millions of other certificates.

The whole scheme can get very complicated indeed and depends on your trusting some third parties (possibly government agencies) somewhere along the way. One of the most fascinating aspects of the PGP product is an alternative called circle of trust, which depends on the fact that in a given country, generally there is a chain of seven or fewer friendship links between any two people ("I personally know Mike, who went to school with Jill, who is Bill Gates's aunt"). The circle of trust scheme will be covered in considerable detail in later chapters. For now, understand that the difference is one of the main ones between PGP and other secure mail systems, like secure MIME (S/MIME) or Privacy Enhanced Mail (PEM), which depend on the hierarchical trust model just described.

Holistic security

Step S6 and R5 often are overlooked by users (and designers) of secure systems. Many people make the mistake of using very strong algorithms and large keys (sufficient for NATO headquarters), then blow it all by leaving the files they protected in plaintext on their system, which is wide open on the Internet. A security system is only as strong as its weakest point, which is usually the person using the system or the key management scheme. It easily can be the person implementing the system (you should deal only with companies in which you have a high level of trust when it comes to security products, especially if they happen to be competitors).

In particular, both physical and network access to a system can be primary points of attack. If you really do not want someone to see the message you sent securely to someone else, you should either destroy *all* traces of it (which can be more difficult than you realize), or keep it locally only in encrypted form (with good protection of the keys used to encrypt it). The field abounds with rumors of obscure government agencies' ability to read data on a hard disk that has been overwritten up to nine times, pick up the radio frequency (RF) emissions from your CRT and reconstruct the image you are looking at on another CRT many blocks away, and so on, and so on. You can get a good idea of some of the threats in this area by looking at the specifications for the creation of products for defense companies (like TEMPEST-hardened computers and terminals) or for the disposal of old computer equipment (like hard disks) that once may have held secret information. In particular, the Department of Defense (DoD) *Orange Book* is fascinating reading. It defines terms like C2 and B1, two common levels of trusted computer systems that Windows NT can be configured to meet.

Some cynics hold that the only truly secure computer is one that is shut down, disconnected from any outside system (including power), wrapped thoroughly in metal foil and chains, embedded in tons of hardened cement, then dropped into the deepest ocean. Be that as it may, we certainly need to adopt a system view concerning security and not be deluded into thinking a system is strong just because parts of it are. It definitely is difficult to ensure that at least pieces of a sensitive message have not been written out to a swap file (even if the software explicitly never writes it to disk itself) and that deleting a file in most operating systems sets only a single flag somewhere in the directory (otherwise, unerase programs would be impossible). A reasonable security system should include some means to scrub at least those areas of a disk that held the pieces of a file clean by overwriting them (many times) with random data, then deleting all pointers to where those pieces were (in the directory and/or the file allocation table). Creating such a utility requires a high level of knowledge of the file systems a computer might use (e.g., file allocation table (FAT) and new technology file system (NTFS).

At the very least, you should be very careful what you do with sensitive information after you have sent a copy of it and once the recipient has decrypted it at the other end. A good system provides a way for the sender to specify "eyes only" to prevent the recipient from saving, printing, or forwarding secure information (you even may have to be able to defeat Windows-style cutting and pasting). Ideally, received messages are stored in encrypted form and redecrypted each time you want to view them.

A secure e-mail system also should be developed in conjunction with a secure perimeter, such as a proxy server or a firewall. Those technologies are addressed in Chapter 11.

Digital signatures

The other major application of security technology in e-mail is the digital signature, which is independent of using a digital envelope (i.e., encrypting the message). You can do neither, either, or both. Digital envelopes allow you to hide the content of the message from anyone but the intended recipient (and, incidentally, keep anyone from tampering with it). Digital signatures allow you to detect tampering and establish for certain the identity of the sender. It is entirely possible you might not care if people are able to read a message (e.g., a purchase order from a shop supervisor for 10,000 widgets), but you want to make certain that no important information is changed from the time the message leaves the sender and furthermore that it really is from the sender, not from some pretender. Otherwise, won't the supervisor be surprised when an 18-wheeler backs up to the shop door to unload 10,000 widgets!

Governments also tend to be much less concerned by the use of digital signatures than by use of digital envelopes (digital signatures do not keep the government from eavesdropping). For that reason, you may have a situation in which a digital signature can be used legally, but a digital envelope cannot (e.g., e-mail crossing international boundaries). Also, digital signatures do not affect the ability to read messages if the recipient does not happen to have a secure e-mail UA available or is not able to obtain the sender's public key for some reason. Of course, while not able to verify the sender's signature, the recipient would be much worse off with a message in a digital envelope and without the correct secure UA or access to his or her own private key.

The basic idea behind a digital signature is that of a checksum. For a simple example, say you publish a software package and use a seven-digit serial number on every copy. You could add an eighth digit to the serial number so the sum of all the digits (including the added one) is an even multiple of 10 (which is another way of saying that the sum of the digits is equal to 0, modulo 10). Given any seven digits, it is simple enough to come up with an appropriate eighth digit so that will be the case. Just add the seven digits and subtract the sum from the next highest multiple of 10, as shown in Table 5.1.

In Table 5.1, the resulting eight-digit serial numbers have the useful property that it is easy to check for the valid serial numbers. You just add up all eight digits; if the sum is a multiple of 10, then the serial number must be valid (of course, there is a 10% probability in this case that any sequence of 8 digits would be valid).

Because an ASCII text message is represented in the computer as a sequence of binary values between 0 and 127, you also could create a checksum for an entire message. The larger the modulo, the less likely it is for a random message to pass the test for validity. A simple checksum for an ASCII message might be 128 minus the sum of the binary values of all the characters modulo 128. In that case, you add all the binary values for all characters, including the checksum; if the result is an even multiple of 128, then the message is valid. That checksum is not a very good one for several reasons:

▶ The resulting checksum might not be a printable character (it might be 13, which is the ASCII code for Carriage Return).

▶ There is a chance of 1 in 128 (probability = 0.78%) that any random message will validate (the last digit will just happen to be the right one, so that the sum of the binary values of all the characters will be zero modulo 128).

Table 5.1
Example of a checksum

Serial number	Sum	Check digit	Checked serial number
3141592	25	$30 - 25 = 5$	31415925
1234567	28	$30 - 28 = 2$	12345672
1111111	7	$10 - 7 = 3$	11111113

▶ This scheme cannot detect even simple tampering like swapping the position of two characters (e.g., changing an order quantity from 15,000 to 51,000).

There are better schemes (e.g., CRC) that result in a checksum with more bits (the CRC scheme uses 16), so that the probability of a randomly chosen checksum being the right one is only 1 in 2^{16} (1 in 65,536 probability = 0.0015%). It also is more difficult to fool with simple character transposition. The first issue could be handled by converting the 16-bit binary check value into a sequence of four hexadecimal digits (which themselves can be any digits from 0 to 9 or alpha characters from A to F, inclusive).

In the real world, we use an algorithm like MD5 or SHA to produce a checksum of a message. Such algorithms produce much larger residues (MD5 produces 128 bits and SHA 160 bits, which correspond to probabilities of 2.939×10^{-37} and 6.842×10^{-47}, respectively, of two different messages producing the same checksum). They also exhibit good cryptographic dispersion, which means that every bit of input affects every bit of the resulting checksum. Essentially, any modification to the message will result in a radically different checksum. We encode this fancy checksum with 32 (or 40 for SHA) hex digits.

Even that simple checksum would be useful, although not particularly secure (the algorithms are public knowledge, and source code for calculating them are readily available). Therefore, a bad guy simply could strip off the original checksum, modify the message, and generate a new one; the modified message would validate just fine.

To make a checksum useful for security, you need to encrypt it with a public key algorithm (such as RSA or El Gamal) and the sender's private key. The encrypted checksum then is appended to the message, and the result is sent. The recipient decrypts the attached signature using the same public key algorithm and the sender's public key. The result of the decryption is compared with the newly generated checksum of the message itself (minus the checksum). If the decrypted checksum matches the newly regenerated one, the recipient knows three important things (given that only the sender possesses the private key):

▶ The attached checksum was the one originally generated and attached by the sender (no one else could generate a new checksum for a modified message and then encrypt it so that the sender's public key would decrypt it to a checksum that would match).

▶ The message really originated from the purported sender, not some pretender.

▶ Nothing has changed in the message since it was signed by the sender, either by malicious tampering or transmission error.

If the decrypted checksum does not match the newly regenerated one, we know only that at least one of the following is true:

▶ The message (but not the encrypted checksum) was tampered with between the sender and the receiver, possibly in an innocuous way (e.g., by the e-mail system adding a blank to the end of each line); hence, the regenerated checksum is incorrect.

▶ The message was corrupted by a transmission error while in transit (either in the message or in the encrypted checksum).

▶ It was signed and sent by someone other than the purported sender (the wrong private key was used).

For a signature to work, the sender and the recipient have to agree on the following things (which must be addressed in any digital signature standard):

▶ The hash algorithm used to produce the checksum (e.g., MD5, SHA);

▶ The public key algorithm used to encrypt the checksum;

▶ The syntax used to append the encrypted checksum to the message, so it can be recognized and cleanly removed to regenerate the checksum on the original message.

Note that the requirements for the public and private keys are different than for the use of digital envelopes:

When creating a digital envelope, the sender must obtain the public key of the recipient (which anyone can do). The recipient uses his or her own private key to open the envelope (and *only* the recipient can do this).

When creating a digital signature, the sender uses his or her own private key (and is the only person who could do that). The recipient must obtain the sender's public key (which anyone can do), to validate the signature.

To summarize, the steps involved in digitally signing a message are as follows:

S1. Compose the message in plaintext and write it to the outgoing message.

S2. Generate a message checksum using the appropriate hash algorithm.

S3. Encrypt the message checksum using the sender's own private key.

S4. Encode the encrypted checksum into ASCII text form and append it to the outgoing message.

S5. Send the results of the above steps (the outgoing message) to the recipient via e-mail.

The steps the recipient goes through to validate the signature are as follows:

R1. Receive the message and extract the appended encrypted checksum.

R2. Obtain the public key of the sender.

R3. Decrypt the encrypted checksum using the sender's public key.

R3. Regenerate the checksum from the remainder of the message (minus the encrypted checksum).

R4. Compare the decrypted checksum with the regenerated one. If they match, the sender's identity and the message integrity are validated. If not, the sender is bogus or the message's integrity has been compromised in some way.

Again, although public key algorithms are slow, with digital signatures, the total amount of information that must be encrypted or decrypted with them is very small (16 to 20 bytes).

As with digital envelopes, the same issues arise concerning obtaining public keys and holistic security. It is possible to use the same public/private key pair for digital envelopes and digital signatures or a distinct pair for each purpose (there is little technical advantage to either scheme). However, with either scheme, at some time or other, you will need access to the public key of the other party, and the same key server can provide that access in either case (and facilities to trace certificate chains, search CRLs, etc.).

Even in the case of primitive products such as the public domain PGP, some of the steps are taken care of automatically. With PGP, however, you still need to do the encryption/signing and message sending in two distinct steps and the message receiving and decrypting/validating in two distinct steps. Obtaining public keys can be another whole process. For people to make widespread use of the technology, it is absolutely essential that it be no more difficult to send and receive secure mail than it is to send and receive nonsecure mail.

Ideally then, a secure e-mail client would automatically retrieve any required public keys at the appropriate time (from a key server) and provide the option of automatically encrypting or signing mail when you hit the send button and automatically decrypting or validating incoming messages when you read them. It also should be able to keep track (in the address book or directory service) of which users with whom you want to use digital envelopes and, if multiple standards are supported, which standard to use (e.g., PGP to Fred, S/MIME to George, but only plaintext to Sue).

Key servers

One of the areas in which certain proprietary e-mail systems (notably Microsoft's Exchange Server) are considerably more advanced than Internet equivalents is public key management. Microsoft incorporated in Exchange Server some encryption and key management technology from Nortel called Entrust. That technology allows a user to use digital envelopes and digital signatures. These systems are fully general implementations of the concepts but, unfortunately, not in a manner that is compatible with any of the emerging Internet secure e-mail standards. Of most interest here is an optional server component of Exchange Server that you can install (along with the many other NT services that make up Exchange Server), called the KM Server (KM stands for "key manager").

The KM Server has the following functionality, which is typical of a full-service key server:

▶ Generating an asymmetric key public/private key pair, download the private key securely to the end user, and embed the public key in a certificate;

▶ Managing a library of public key certificates;

▶ Allowing UAs to retrieve public key certificates via a network protocol;

▶ Managing a list of revoked public key certificates (a CRL);

▶ Allowing the CRL to be searched by a UA via a network protocol.

If you are interested in the specific details of Exchange Server's security features and its KM Server, refer to my article in *Windows NT Magazine*, "Exchange Mail: Signed, Sealed and Delivered." (See www.winntmag.com or look in the October 1996 issue.) Unfortunately, this technology is of little use in Internet e-mail due to the fact that all parts of it are highly proprietary. It is possible for the security features (and even the KM Server) to be used over an Internet link but only by components of the Exchange Server system operating in their native mode (using the Internet purely as a transport). If you use UAs with standard Internet protocols (SMTP/POP3), none of those features are available (nor can they use the KM Server). If you use the Exchange Client in its native mode, it will not interoperate (securely) with PGP- or S/MIME-based UAs. Ideally, Internet directory servers will evolve to support all the above features of the Exchange Server KM Server, using the LDAP protocol.

Key servers for the Internet still are pretty much in their infancy. A few are available for exchanging PGP keys manually (as opposed to via a protocol that a smart UA could use to retrieve public keys automatically). One key server is at the Massachusetts Institute of Technology (MIT). As of January 1, 1997, it contained 35,683 active keys and 302 disabled (revoked) keys. It is available via both Web forms and e-mail (but currently no network protocol, such as LDAP). The Web version is somewhat easier to use. Connect your browser to either of the following sites and follow the directions you find there:

```
http://bs.mit.edu:8001/pks-toplev.html
http://www-swiss.ai.mit.edu/~bal/keyserver.html
```

To use the PGP key server via e-mail, send a message via Internet e-mail to:

```
pgp-public-keys@pgp.mit.edu
```

with a command in the message subject (the body is ignored and may be left empty). For example, when I sent the following request:

```
To: pgp-public-keys@pgp.mit.edu
Subject: help
```

I got the following response with the (then) current usage information:

```
Date: Wed, 1 Jan 97 21:38:31 +0000

To: "Lawrence E. Hughes" <lehughes@worldnet.att.net>
From: PGP Key Server Administrator
<jis@BIG-SCREW.MIT.EDU>
Subject: Your command, help

Key server software written by Marc Horowitz

For questions or comments regarding this key server site,
contact PGP Key Server Administrator <jis@bs.mit.edu>
Current version: 0.9.1

NOTE!

It is the responsibility of users of public-key
cryptography to ensure that their activities conform to
legal requirements.
MIT does not vouch for the authenticity of the informa-
tion available here.
It is up to each user to determine whether or not a given
PGP key can be trusted to represent the user it claims
to be from!
```

PGP Public Email Keyservers
———————————--

There are PGP public email key servers which allow one to
exchange public keys running using the Internet and UUCP
mail systems. Those capable of accessing the WWW might
prefer to use the WWW interface available via http://
bs.mit.edu:8001/pks-toplev.html and http://www.pgp.net/
pgp/www-key.html and managers of sites which may want to
make frequent lookups may care to copy the full keyring
from the FTP server at ftp.pgp.net:pub/pgp/

This service exists only to help transfer keys between
PGP users. It does NOT attempt to guarantee that a key
is a valid key; use the signatures on a key for that
kind of security.

Each keyserver processes requests in the form of mail
messages.
The commands for the server are entered on the Subject:
line.
——————————————— ======== ——-
Note that they should NOT be included in the body of the
message.
—————--- === ———————————--

        To: pgp-public-keys@keys.pgp.net
        From: johndoe@some.site.edu
        Subject: help

Sending your key to ONE server is enough. After it
processes your key, it will forward your add request to
other servers automatically.

For example, to add your key to the keyserver, or to
update your key if it is already there, send a message
similar to the following to any server:

        To: pgp-public-keys@keys.pgp.net
        From: johndoe@some.site.edu

        Subject: add

```
—-BEGIN PGP PUBLIC KEY BLOCK—-
Version: 2.6

<blah blah blah>
—-END PGP PUBLIC KEY BLOCK—-
```

COMPROMISED KEYS: Create a Key Revocation Certificate
(read the PGP docs on how to do that) and mail your key
to the server once again, with the ADD command.

Valid commands are:

Command	Message body contains
ADD body of msg)	Your PGP public key (key to add is
*INDEX	List all PGP keys the server knows about (-kv)
INDEX userid	List all PGP keys that match the userid (-kv)
*VERBOSE INDEX (-kvv)	List all PGP keys, verbose format
*VERBOSE INDEX userid	List all PGP keys, that match the userid, verbose (-kvv)
*GET (split)	Get the whole public key ring
GET userid	Get just that one key
*MGET regexp	Get all keys which match /regexp/ regexp must be at least two char-
acters long	
LAST days 'days' days	Get the keys updated in the last

* Commands preceeded by asterisks are not implemented on
all keyservers.

In particular the Marc Horowitz "C" based keyservers do
not implement these commands. This includes pgp-public-
keys@pgp.mit.edu where you received this file from.

Examples for the MGET command:

```
    MGET michael    Gets all keys which have "michael"
in them
```

```
MGET iastate      All keys which contain "iastate"

MGET E8F605A5|5F3E38F5 Those two keyid's
```

One word about regexps: These are not the same as the wildcards Unix shells and MSDOS uses. A * isn't ''match anything'' it means ''match zero or more of the previous character'' like:

 a.* matches anything beginning with an a

 ab*c matches ac, abc, abbc, etc.

Just try not to use ''MGET .*'' — use ''GET'' instead.

Note on the ''GET'' command: If at all possible, ftp the keyring from a server such as ftp.pgp.net:pub/pgp/keys rather than using the ''GET'' command to return the whole ring. Currently, this ring comes out to be over 50 files of 52k each. This is a lot of files, and a lot of bother to get in the right order to run through PGP.

Users should normally use the email address 'pgp-public--keys@keys.pgp.net' or your national servers using one of:

 pgp-public-keys@keys.de.pgp.net

 pgp-public-keys@keys.nl.pgp.net

 pgp-public-keys@keys.no.pgp.net

 pgp-public-keys@keys.uk.pgp.net

 pgp-public-keys@keys.us.pgp.net

for the email interface, and 'ftp.pgp.net:pub/pgp/' for FTP access.

Users are recommended to use the "*.pgp.net" addresses above as these are stable and reliable.

Here is a sample request to retrieve my PGP public key from the MIT key server:

To: pgp-public-keys@pgp.mit.edu

```
Subject: get lehughes@worldnet.att.net
Here is the message obtained in the reply to that
request:
Date: Wed, 1 Jan 97 21:41:42 +0000
To: "Lawrence E. Hughes" <lehughes@worldnet.att.net>
From: PGP Key Server Administrator BIG-SCREW.MIT.EDU
Subject: Your command, get lehughes@worldnet.att.net
Key server software written by Marc Horowitz mit.edu
For questions or comments regarding this key server site,
contact PGP Key Server Administrator bs.mit.edu
Current version: 0.9.1
NOTE!
It is the responsibility of users of public-key
cryptography to ensure that their activities conform to
legal requirements.
MIT does not vouch for the authenticity of the
information available here.
It is up to each user to determine whether or not a given
PGP key can be trusted to represent the user it claims
to be from!
—-BEGIN PGP PUBLIC KEY BLOCK—-
Version: 2.6.2
Comment: PGP Key Server 0.9.1

mQBtAzJ07bcAAAEDAMZV2zLv+usZlxndoP/wPmeUNGPT2JugE98xcdRtC5jI28EM
tXR4UjaWM8cC1XNpb1z5AKW87wu6zG3HMchxsXWd+hQe8lfbetQUSM+dfc57/i8k
llwfyQglPxLNvDR7bQAFE7QuTGF3cmVuY2UgRS4gSHVnaGVzIDxsZWh1Z2hlc0B3
b3JsZG5ldC5hdHQubmV0Pg==
Gjyo
END PGP PUBLIC KEY BLOCK
```

Key servers are one area in which quite a bit of additional work needs to be done. Unfortunately, secure e-mail will be somewhat difficult to use until such time as the technology has matured.

Session authentication

Another widely used application of security technology is for a network client to authenticate itself at the beginning of a protocol session (e.g., at the start of a POP3 session to retrieve mail from a server). Many protocols (e.g., POP3, IMAP4) have defined optional cryptographic authentication schemes (in addition to simpler schemes, such as user name and password). Such protocols have two major advantages over simpler schemes:

▶ It is much more difficult for a hacker to gain unauthorized access (for one thing, they cannot just snoop on network transmissions and obtain valid user names and passwords).

▶ The authentication mechanism can exchange a session key and enable a protection mechanism (encryption of the remaining data exchanged in the session) using that key.

The disadvantages are that they are much more complicated to implement and usually require additional servers to be present on the network.

Cryptographic authentication schemes

Several cryptographic authentication schemes currently are defined.

Kerberos v4

Kerberos v4 was developed at MIT and used as the basis of a number of authentication systems. Kerberos v4 is based entirely on DES, which is a symmetric key cryptographic algorithm with a 56-bit key and moderate strength (but still heavily restricted for export). Both Kerberos v4 and the underlying DES are public domain technologies, so no royalties are due for using it in a more comprehensive system (e.g., e-mail). The definition of Kerberos v4 and a reference implementation (for UNIX, not NT) are available free from MIT via the Internet (see http://web.mit.edu/network/kerberos-form.html). Kerberos v4 never has been officially released as an IETF standard.

Kerberos v5

Kerberos v5, a later version of Kerberos, also was developed at MIT and based on DES. This version has been released as an IETF standard (RFC 1510). Versions 4 and 5 are not compatible (the contents of the messages between clients and servers are different). Version 5 fixes several subtle weaknesses in the original system. A client (e.g., IMAP4 UA) could be designed to support both versions. It should check first to see if version 5 is supported; if not, then try to use version 4. If no cryptographic authentication scheme is supported, the client could fall back to plaintext authentication (e.g., LOGIN username password). A given server probably would support version 4 or version 5, but not both. Any new work should use version 5, except specifically for backward compatibility. Implementations of Kerberos v5 are commercially available (e.g., see Cygnus Solutions at www.cygnus.com, for its KerbNet product for UNIX and NT).

S/KEY

S/KEY is a one-time password authentication scheme. It was developed at Bellcore, and information on it (as well as evaluation copies of their client and server for various platforms, including NT) is available at www.bellcore.com. Just go to that homepage and do a search on S/KEY.

ISO X.509 authentication framework

Various authentication schemes are based on asymmetric key (public key) algorithms (e.g., RSA). Such schemes depend on the availability of a system of certificate authorities, trusted companies or agencies that can issue X.509 signed public key certificates. They also require a system (e.g., X.500) for those certificates to be distributed and to determine if a given certificate has been revoked (i.e., a way to maintain and search a CRL).

GSSAPI

Generic Security Service Application Program Interface (GSSAPI) is an IETF standard (RFC 1508). GSSAPI is not really a cryptographic authentication mechanism itself; rather, it is a general framework that allows an application program (either client or server) to make use of various authentication mechanisms, such as Kerberos v5 or the ISO X.509 authentication framework. The commercial product KerbNet includes GSSAPI, the recommended way an application would include support for Kerberos using that product.

Authentication mechanism algorithms

Kerberos

Kerberos requires a trusted third party that is a new server called an authentication server (AS). The server must be running on a physically secure host computer. The AS maintains a database of Kerberos principals (including users and application servers, e.g., an IMAP4 server). The AS keeps a DES key for each principal. All normal Kerberos operations require only read-only access to that database (the only time read/write access is required is to add or delete principals or to change their Kerberos keys). Kerberos also requires that significant new code be added into both client and server components (typically done by calling a few functions in a Kerberos library, which could be implemented using GSSAPI or some proprietary scheme).

There are several kinds of message exchanges in Kerberos, as detailed below:

▶ Client to authentication server: Client sends plaintext request to the AS for credentials, that is, a ticket for the desired service (includes the principal's name and the name of the application server for which access is desired).

▶ Authentication server to client: AS replies to client's request with a ticket granting ticket (TGT) for later use with a ticket granting service (TGS). This is encrypted with the client's private key.

▶ Client to ticket granting service: Client sends the TGT it received from the AS to the TGS, as part of a request for credentials (ticket) to use the desired application server.

▶ TGS to client: The TGS replies with a ticket to use the desired application server, encrypted with the session key from the TGT.

▶ Client to application server: Finally, the client sends its credentials (server ticket), just obtained from the TGS, to the application server. The principals now share a session key, which they have obtained in a secure, authenticated manner.

Once a shared session key has been obtained, it can be used to allow the principals in a transaction to verify their identities (authenticate) to each other, to ensure the integrity of further messages exchanged between them or to ensure the privacy of the messages.

To verify the identities of the principals in a transaction (e.g., IMAP4 client and IMAP4 server), the client transmits the server ticket to the application server. A ticket contains the name of the server and several items encrypted with the server's private key: the client's name, the client's IP address, the time interval during which the ticket is valid, and the session key. The client does not know the server's private key with which to encrypt that information, but the AS does. (The AS actually collects the information, encrypts it with the server's private key, then returns the result to the client. The client cannot decrypt the information, because it does not possess the server's private key, but it can pass the message along still encrypted with the server's private key to the server). The included timestamp prevents a malicious third party from obtaining fraudulent access via a replay attack. The server can obtain all those items from the ticket because it knows its own private key (but no one else does, except for the AS).

The integrity of any future messages exchanged between the principals is accomplished by the sender (either client or server) creating a digest of the message (e.g., with MD5), then encrypting the digest with the session key. The encrypted message digest is sent, along with the message as a fancy checksum. The receiver (the other party in an exchange) also generates a digest of the message (not including the checksum) and encrypts it with the session key. It then compares the result with the checksum the client sent. If they match, then the message has not been tampered with while in transit, since only the client and the server know the session key (and anyone changing the message would have had to know that key to generate the correct cryptographic checksum).

Privacy can be obtained by the sender encrypting the data with the shared session key and the receiver decrypting with the same key.

S/KEY

S/KEY can be implemented with additional code in existing client and server applications (no new servers are required, as they are with Kerberos). The basic algorithm is as follows:

1. Server transmits a challenge string consisting of three tokens (separated by one space, terminated by a space or a new line): the string S/KEY; an integer sequence number; and a seed string, in plaintext (this challenge need not be kept private from eavesdroppers) e.g.,

   S/KEY 110 3268329492

2. Client appends its secret pass phrase (password) to the server's seed string, producing a longer string, e.g.,

3268329492swordfish

3. Client runs the resulting string through the MD5 digest algorithm to obtain a 128-bit hash. MD4 may be supported for backward compatibility with older implementations (originally, only MD4 was supported). However, MD4 has been cryptanalyzed, so it is not recommended. SHA1 can be used for even stronger security, which produces a 160-bit hash. The resulting hash is folded into a 64-bit value (e.g., the first 64 bits are made EXCLUSIVE OR with the second 64 bits).

4. Client repeats step 3 n times (where n = the sequence number from the server's challenge string), using the 64-bit result of the previous iteration as input (rather than the concatenated seed and pass phrase).

5. Client sends the 64-bit result of the final iteration to the server as its response to the challenge. The pass phrase itself is never sent in plaintext over the Net. The 64-bit value sent to the server is a one-time password and will never be used again.

The server validates a password by looking up the user's pass phrase and appending it to the seed string sent to the client, then doing the same calculation that the client did. If that produces the same result as the 64-bit value sent by the client, the client must have known the correct pass phrase. When a successful authentication occurs, the server decrements the sequence number for that client by 1 (for the next authentication). When the sequence count reaches zero, the system must be reinitialized by changing the seed string, then resetting the sequence number (e.g., to something in the 500–1,000 range).

For each client, the server must keep track of a seed string, the client's pass phrase, and the current sequence number. For greater security, the server could keep the result of the first iteration (the seed and the pass phrase run through MD5 once) instead of actually keeping the pass phrase itself on the server.

ISO X.509 authentication framework

The preceding authentication schemes are based solely on symmetric key algorithms (e.g., DES). Asymmetric key (public key) algorithms can be used to accomplish the same thing with much simpler protocols. ISO X.500 includes a complete subsection (X.509) on security and authentication, based on asymmetric key algorithms.

An important part of X.509 is the definition of a signed public key certificate (commonly called an X.509 certificate). Basically, that certificate contains the following items (among others):

▶ The user's name;

▶ The user's public key;

▶ A field that identifies the certificate format;

▶ Fields identifying the issuing certificate authority;

▶ A field that identifies the algorithm used to sign the certificate, together with any necessary parameters;

▶ The time period during which the certificate is valid.

The certificate authority that issues the certificate digitally signs it (with its own private key), by creating a message digest of the information in the preceding list and encrypting that digest with the authority's private key. Anyone can obtain the certificate authority's public key and validate the certificate (verify that it came from the claimed certificate authority and has not been tampered with).

Three authentication schemes are based on those X.509 certificates: one-way, two-way, and three-way.

▶ The one-way scheme involves sending a single message from client to server. Because it depends on a timestamp, the one-way scheme requires a synchronized, authenticated time reference that is available to both client and server.

▶ The two-way scheme involves one message from client to server and a second message from server back to client. It also depends on a timestamp.

▶ The three-way scheme involves one message from client to server, a second message from server back to client, and a final message from client to server. It does not require a shared time reference.

For details on the three schemes, see Section 24.9 in Bruce Schneier's *Applied Cryptography*.

Samples

Figures 5.1 and 5.2 show the sending and the receiving of an encrytped message. Figures 5.3 and 5.4 show the sending and the receiving of a digitally signed message.

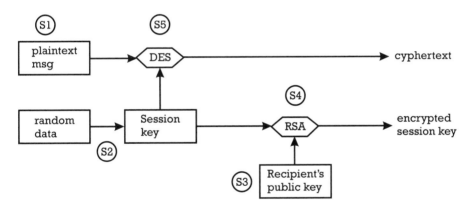

Figure 5.1 Sending an encrypted message.

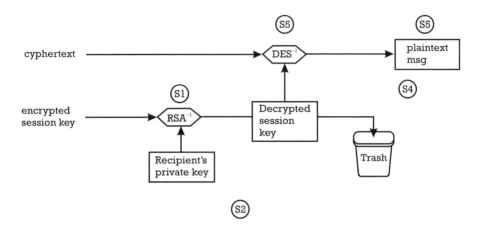

Figure 5.2   Receiving an encrypted message.

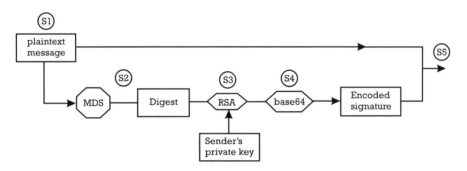

Figure 5.3   Sending a digitally signed message.

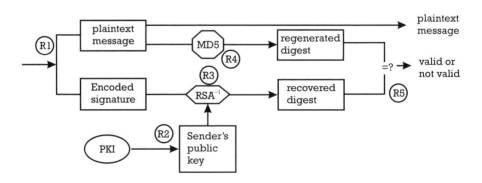

Figure 5.4   Receiving a digitally signed message.

Open e-mail versus proprietary systems

First, we should define the terms *open* and *proprietary*, especially because over the years many companies intentionally have misused *open* to give the appearance of openness without the reality of it.

It is easier to define *proprietary*, so we will start with that. Basically, if a computer software product or system is created, developed, and controlled by a single company, it is proprietary to that company. That can be achieved by treating various aspects of the design as "trade secrets" or through explicit legal protection in the form of patents and copyrights. For example, almost everything about the Apple Macintosh computer is highly proprietary. For many years, Apple took aggressive legal action against anyone who tried to make a system compatible with the Macintosh. IBM, on the other hand,

looked the other way when competitors made IBM-compatible computers (at least once Phoenix made the first noninfringing ROM BIOS). As a
result, the Macintosh remained a relatively small, closed market (with
high prices and limited choices) while the IBM PC-compatible market
exploded and prices dropped like a rock. Apple is now almost out of
business.

The details about an open system are available for anyone to read and
use, ideally without even paying royalties. A completely open system
(like TCP/IP-based networking) has all the technical details necessary to
build a fully functional, interoperable system readily available and free for
anyone to use. Some open systems (like TCP/IP) came about through a
clause that requires that any technology created as a part of certain U.S.
government-sponsored research to be released automatically into the
public domain. Other open technologies (like Extended Industry Standard Architecture (EISA) plug-in boards or Video Electronics Standards
Association (VESA) video standards) came about through joint efforts by
a group of companies trying to ensure that their products will interoperate. Other open standards came into use due to legal protection lapsing,
either through the owner being sloppy or a patent term expiring, as happened April 29, 1997, on certain key public key patents (i.e., the Diffie-
Hellman key exchange and the El Gamal algorithms, which are covered
by the same patent).

Standards bodies

Standards-creating bodies, funded by governments, universities, and
even private companies, have the primary purpose of carefully engineering and subjecting to widespread peer review elaborate specifications in
areas where interoperation is important, such as Electronic Document
Interchange (EDI) and interbank transfers. (This section describes seven
of those standards bodies.)

Many standards have been created by one organization, then adopted
essentially intact by one or more other organizations; hence, a standard
may be known by several different names.

When it comes to e-mail systems, both open and proprietary systems
have existed side by side for many years (Internet e-mail is not exactly
new, although recently the rate at which it is being enhanced and

deployed has accelerated greatly). However, until recently, most commercially developed e-mail systems have been proprietary (Internet e-mail was more likely to be user supported or freeware).

Because most commercial systems were proprietary, many people have made the wrong association that *open* connotes less sophisticated or less capable. Until recently, proprietary e-mail systems have been easier to use and had more features than standards-based e-mail. Today, many companies are creating supported, sophisticated systems based on open standards, and the standards themselves not only are catching up with the capabilities of proprietary systems but surpassing them greatly, especially in the area of scalability (the number of users that can be supported on a single server) and interoperability with the largest possible number of other e-mail users.

ISO

The ISO (International Standards Organization; see http://www.iso.ch) is based in Europe and is one of the most formal standards bodies. Among other things, it has defined the X.400 e-mail system and the closely related X.500 Distributed Directory Service. It also defines many elaborate computer graphics standards, such as the graphics kernel system (GKS) and the programmer's hierarchical interactive graphics system (PHIGS). Those standards are available for purchase in many languages at fairly high prices. ISO standards typically are worked out over one or more four-year cycles. X.500 originally was released in 1988, radically improved in 1992–1993, and finished in 1996–1997.

ITU

The ITU (International Telecommunications Union) is a treaty organization formed in 1865 to ensure interoperability of the rapidly expanding electronic communications systems. The ITU is now a specialized body within the United Nations (see gopher://info.itu.ch/1/).

CCITT

CCITT (an acronym for a French name that translated into English is International Telegraph and Telephone Consultative Committee) was formalized as a part of the ITU in 1956. The CCITT defines standards for telephones, faxes, and many other communications services, including things like the V series for modems (e.g., V.34) and the X series for network interfaces (e.g., X.25). The standards are available for purchase in many languages at fairly high prices.

IEEE

The standards of the IEEE (Institute of Electrical and Electronics Engineers; see http://www.ieee.org) are available at nominal charges (for online ordering, see http://stdsbbs.ieee.org).

ANSI

ANSI (American National Standards Institute; see http://www.ansi.org) was founded in 1918 by five engineering societies and three government agencies. It is a private, nonprofit organization. Among other things, it was responsible for defining ASCII. ANSI is the sole U.S. representative to the ISO and coordinates standards efforts with that organization. There is a nominal charge for a copy of any ANSI standards document.

NIST

The NIST (National Institute for Standards and Technology) is a government agency that is a part of the U.S. Department of Commerce (see http://www.nist.gov). It was established in 1901 as the National Bureau of Standards (NBS).

IETF

The IETF (Internet Engineering Task Force; see http://www.ietf.cnri.reston.va.us/home.html) is an industry group that oversees the standardization process for Internet technologies, including the RFCs, STDs (standards), and FYIs (for your information). All those documents are available via the Internet for free. The IETF standards related to e-mail are included on the CD-ROM accompanying this book.

A legacy proprietary e-mail system: MS Mail

For many years, the only e-mail system that Microsoft sold was MS Mail, which was originally a (pre-Windows) MSDOS-based system. It used a passive server design, which means there was no actual running software at the central site, just a shared file system (via whatever shared-file network system was already in place, e.g., Novell NetWare or Microsoft LAN Manager). The central shared-file system implemented not only the message store but also the e-mail account system (independent of any network account scheme) and the e-mail directory services.

Originally, the only client software was a primitive text-interface application for MSDOS. Later, a 16-bit Windows client (and a remote dial-in version) was created. Even a Macintosh version of the client was created. When Windows for Workgroups was released, the regular version (not the remote version) was included as a standard part of the package. In fact, it was possible to create a simple, single workgroup post office with only the free client (for peer-to-peer network workgroups). When Windows NT 3.1 was released, that same client was ported to 32 bits but otherwise left pretty much intact.

Using the peer-to-peer capabilities of Windows for Workgroups or Windows NT (without an NT Server to implement domain-style networking) and the free MS Mail clients included with those operating systems, many people have implemented simple workgroup e-mail systems with no additional cost (and little effort).

By purchasing the optional MS Mail Server Kit (about $800), you can add the following capabilities to such a workgroup e-mail system:

- A real administration program (albeit with MSDOS text mode user interface);

- The ability to add gateways to other e-mail systems (e.g., Internet Mail or X.400);

- Support for the remote mail client for asynchronous modem dial-in access;

- The ability to link several LAN-based systems into an enterprise system.

The MS Mail Server Kit includes an installation program that can create the initial shared-file system for a real (not just workgroup) post office. It also includes the administration program (admin.exe) to manage the shared-file system, the e-mail account database, workgroup address books, and the MTA. The original (v3.2) MTA itself (external.exe) is a kludge that requires a dedicated MSDOS computer on which to run. It simulates multitasking (poorly) with a program called dispatch.exe. There is also a program to implement directory synchronization. A dispatch program allows you to load/start/stop one or more copies of the MTA and the directory synchronization program as required (only one can be running at any given time). The MTA is for exchanging inter-post office messages (via WAN or dialup links) and to accept incoming dialup calls from the remote client.

To connect the system to other e-mail systems, you have to purchase optional gateway packages, such as the MS Mail SMTP Gateway or MS Mail X.400 Gateway. Those programs also run on dedicated MSDOS computers (the Microsoft SMTP gateway is notoriously difficult to install and not very reliable, not to mention expensive).

A typical installation with Internet access (in addition to a server to hold the shared-file system) includes two dedicated MSDOS computers at each site, one for the MTA/directory synchronization and one for the SMTP gateway. I installed and maintained such a system for the Asia/Pacific branch of Intergraph Corporation while working there, with sites in Hong Kong (the regional node), Seoul, Singapore, Sidney, Beijing, Taipei, and Tokyo. Some of the sites were connected via frame relay WAN and some by dialup asynchronous modems. There also was a WAN link back to the corporate headquarters in Huntsville, Alabama. It was not a pretty sight. We never did get the directory synchronization working correctly, except among subsets of the overall system. We wound up using the SMTP gateways to link most of the sites.

When Windows NT was released, an NT server was the ideal place to install and share the central shared-file system, but unfortunately neither the MTA from MS Mail v3.2 nor the SMTP gateway would run on NT (at least not reliably). A multitasking MTA that would run on OS/2 v1.3 was recommended for large or hub sites. When MS Mail v3.5 was released, the OS/2-based multitasking MTA was ported to run under the OS/2 subsystem of NT, as an NT service (like a UNIX daemon). At least one third-party MS Mail-to-Internet gateway is available for NT (from Consensys), but Microsoft never upgraded its gateway to run under Windows NT (at least as of MS Mail v3.5).

The MS Mail client program (the only client that can work with MS Mail) reads and writes files in the central shared-file system via either Microsoft or Novell file sharing. It supports binary attachments (with a proprietary scheme, but that is converted to uuencode when messages with attachments are sent through the MS Mail-to-SMTP gateway). There is no support for security (either digital envelopes or digital signatures). There is good support for a personal address book and a workgroup address book (maintained in the central shared-file system by the administration program). If you add the SMTP gateway, you also can support a shared SMTP address book (manually). If you implement the directory synchronization scheme, all of the workgroup address books will be available (as separate address books) to all users of any post office. Finally, at each site, you can merge all the shared address books into one giant global address book (using the administration program).

There is good support for return receipts (notification that a recipient has read a particular message). There is also good support for managing a local message store with hierarchical folders (with drag-and-drop from the currently displayed contents of a given folder into any other folder).

Overall, the system (at least by v3.5) is fairly complete (except for security) but difficult to install and manage. It also is highly proprietary, and some of the features do not work well (or at all) via many of the gateways (e.g., return receipts do not work via the SMTP gateway). In particular, the original Microsoft SMTP gateway is the weakest point of the system (although third-party solutions are significantly easier to install, are more reliable, and even support mapping the binary attachments to and from MIME rather than uuencode).

The central shared-file system is one of the biggest problems with MS Mail. For it to work, you have to make the entire central file subtree accessible to all e-mail users in read/write mode (called an exposed message store). Clever users might try to look inside the directories (to read each other's messages) or use the shares to obtain additional server storage. The design also leads to a lot of network traffic compared to a client/server design.

A given post office cannot support more than 500 users. For a large enterprise, the overall cost of the hardware and software can become quite high, and the administrative overhead is significant.

Lotus/IBM have a competitive system called cc:Mail, which is roughly comparable to MS Mail in most respects.

A modern proprietary e-mail system: Microsoft Exchange Server

When Exchange Server first saw the light of day as v4.0 (perhaps Microsoft was trying to convey the message that Exchange Server was the continuation of MS Mail), it corrected many of the deficiencies of MS Mail and added some significant new features, such as client/server architecture, security, and public folders. However, it is still clearly a proprietary architecture (loosely based on the MS Mail design).

One of the main advances is the change from a central shared-file system to an active server and a client/server design. The protocol used between client and server is MAPI 1.0 (a proprietary Microsoft messaging protocol and API that is layered on top of RPC, which in turn can layer on top of various transports available on NT, such as NetBEUI, IPX/SPX, or TCP/IP). The protocol supports not only the message transfer but also the directory services and other facilities. In theory, MAPI is open, in that Microsoft has published the specifications, but in practice there has not been a great deal of support for it, and it does not work very well over the Internet. MAPI is a large, slow, cumbersome protocol compared with standard protocols like SMTP and POP3.

The SMTP gateway was completely redone for Exchange Server and is now known as the Internet Connector (there is also a proprietary connector and an X.400 connector). All capabilities (except possibly security) will work via the connectors, if the proprietary clients are used at both ends. If communicating with an Internet UA, the new SMTP gateway can convert binary attachments to either uuencode or MIME on a recipient-by-recipient basis. The new gateway also is implemented as an NT service, which means you do not need a dedicated MSDOS computer to run it, and it is more reliable.

The administration in Exchange Server is radically improved and is implemented as native NT GUI utilities, well integrated with the native NT account system (it is possible to create Mail accounts for all existing NT accounts during installation and also to automatically create an Exchange

Server account when you create a new NT account), and all e-mail account management is done with minor extensions to the NT User Manager.

The directory synchronization also is radically improved (and implemented as native Win32 services, not a OS/2 subsystem).

Four completely new features have no analog in MS Mail: public folders, advanced security, electronic forms, and rules processing.

The public folders are similar in nature to (but incompatible with) Internet News Net Transfer Protocol (NNTP) Usenet newsgroups. The same client software (the Exchange Client) is used to work with both normal e-mail and the messages in public folders. It is possible to set up automated replication of public folders among sites. The public folder system is similar in many respects to the capabilities of Lotus Notes (one major difference is the granularity with which replication can be done; in Notes you can replicate individual fields in a message).

The advanced security features are complete (if highly proprietary and limited to use within an organization) implementations of digital envelopes and digital signatures using both asymmetric key and symmetric key technology (Nortel's Entrust technology, to be specific). There is also a KM Server, which implements all the facilities of a general key server (key generation, public key certificate management, CRL management, etc.). The KM Server is implemented as an optional NT service. It must be installed before the security feature of the Exchange Client will work.

The electronic forms are a scheme to allow nonprogrammers (or entry-level programmers) to create simple messaging applications in a visual programming environment. A utility called the Electronic Forms Designer assists in creating forms (which really are just simple Visual Basic programs, which means that a more advanced programmer can extend an Exchange form application with the full capability of Visual Basic if desired). That is roughly similar to the application development capability in Lotus Notes (but of course completely incompatible).

The rules processing allows you to preprogram various activities based on string matching in any field of incoming messages (e.g., if the message subject contains the string CEBU-L, file it in the folder named CEBU; if the message is from Dewey Cheatem & Howe Law Firm, reply with a blistering tirade about junk e-mail and discard it). Most vendors that have implemented such schemes do so completely in the client. In

Exchange Server, the user interface for setting up rules is in the client, but the actual rule processing is done in the server. While that does allow rule processing to take place whether your client is running or not, it also means that rule processing will not work if you use a client program other than Exchange Client. Furthermore, it adds significant execution overhead to the server (which affects scalability).

With the second release of Exchange Server (v5.0), Microsoft added considerably more Internet support. One new feature is an alternative POP3 protocol to read messages from the proprietary Exchange Server message store. Together with the Internet Connector (renamed the Internet Mail Service), that allows you to use any Internet e-mail client against Exchange Server (instead of using Exchange Client). However, if you do, many of the Exchange Server features are not available (security, rules, public folders, return receipts, electronic forms, etc.). Another powerful addition is the ability to retrieve directory information via LDAP.

Finally, Microsoft added support for NNTP (the protocol used to implement Usenet news groups). That allows public folders to be linked to Usenet newsgroups (so that new messages from Usenet will appear in the Exchange Server public folder, and messages posted to the public folder will be posted automatically to the corresponding Usenet news group).

Unfortunately, the overhead, complexity, and poor scalability of Exchange Server make it an ineffective way to implement an Internet Mail Server unless you specifically want the native capabilities of Exchange Server as well. A native Internet Mail Server easily can handle 10 to 100 times as many users on a given hardware platform as Exchange Server. On the other hand, if you are planning to install Exchange Server anyway for some of its powerful (but proprietary) capabilities, its support for Internet protocols does allow people who prefer to use SMTP/POP3 clients to use Exchange Server as if it were an Internet Mail server.

Relative merits of Internet Mail versus proprietary systems

So how do proprietary systems compare with Internet Mail, the dominant open e-mail system?

Because it is an open system, many products available from a variety of vendors (including freeware) work with Internet Mail. In comparison, aside from a few minor third-party add-ons that address deficiencies in the product (e.g., a better MS Mail-to-SMTP gateway or plug-ins that support S/MIME), both MS Mail and Exchange Server are the exclusive property of (and their evolution is tightly controlled by) a single company. While Microsoft has published the details of the client/server protocol (MAPI 1.0), few other companies have chosen to support it. Quite a few details of the architecture of both systems (e.g., the internal organization of the central shared files in MS Mail) remain undocumented (and companies wishing to create add-ons have had to reverse engineer such details to create interoperable products. Such companies have no guarantee that those details will not change completely in a new release.

Proprietary products (coming as they do from a background of supporting a fairly small number of users in an isolated LAN or at most a few LANs connected into a WAN) do not scale to very large numbers of users well (typically 500 or so users on a given server). In comparison, the lightweight protocols of Internet Mail scale well to tens or hundreds of thousands of users on a single server (AT&T Worldnet currently is running over one million users and eventually plans to scale up to tens of millions). As a truly distributed system, Internet Mail can scale to the entire globe.

To even the scales a bit, both MS Mail and Exchange Server today have significantly better support for address books. Exchange Server already has provided strong support for digital signatures and digital envelopes (at least within an enterprise and, with v5.0, even some support for cross-enterprise secure mail). Also, some nice groupware products (e.g., Schedule+) are based on Exchange Server. Finally, in Exchange Server, Microsoft has provided a very nice mechanism to allow companies to create their own messaging-based groupware applications using simple visual paradigm tools like Visual Basic (called the Electronic Forms Designer). They also have provided a good scheme for public folders with replication between sites.

There already are a few groupware products similar to Microsoft's (vCalendar and iCalendar)Schedule+ that layer on Internet Mail rather than Exchange Server. I am not currently aware of anything similar to Exchange Server's electronic forms, but it is only a matter of time before someone creates such a facility that layers on Internet Mail. As for the public folders, the Internet has a well-established global scheme (using NNTP) called Usenet. Already Internet Mail clients are including support for NNTP and Usenet newsgroups in a manner exactly analogous to Exchange Server's public folders (but already with a worldwide user base).

If you compare cost of ownership, Internet Mail can be deployed for dramatically less money (especially if you use some of the freeware products), but even a system based on supported commercial products is dramatically cheaper than either of the above systems).

One area in which Internet Mail is clearly superior to the above systems is support for multiple platforms. The server components of MS Mail can be deployed only on Microsoft or Novell servers and networks, and clients are available only for DOS, Windows 16, Windows NT, and Macintosh (no support for UNIX other than via SMTP gateways, through which the address books and delivery notification schemes do not work). Again, Exchange Server can be deployed only on Windows NT (Intel, Alpha, and PowerPC versions are supported; MIPS is being dropped). The Exchange Client currently is available for MSDOS, Windows 16, Windows 95, Windows NT, and Macintosh in quite a few variations and under several names (some of which come free with Windows 95 or Windows NT 4.0, or as part of Office 97). With Exchange server v5.0, more support is available for Internet Mail protocols (which means you also can use any SMTP/POP3 client, even UNIX ones). However, the advanced features (address books, public folders, delivery status notification, and security) work only when the real Exchange clients are used.

In comparison, essentially every computing platform in existence that supports TCP/IP has some kind of Internet Mail client, most of which support MIME attachments. Also, there is considerably more choice available in terms of server platforms. Internet Mail servers are available for all UNIX platforms and even some mainframes and supercomputers. Even though Windows NT is an adequate, cost-effective platform for most organizations that need Internet Mail, the largest sites, like AT&T Worldnet, still require "big iron" (e.g., eight-CPU Sun Sparc systems, with a gigabyte of RAM, running UNIX). The availability of such high-end platforms to run Internet Mail servers that can still be used by the simplest, free SMTP/POP3 client is one of the strongest arguments for the adoption of this technology for a universal, worldwide e-mail system.

From the beginning, the basic design philosophy of Internet Mail was to keep the core (the basic protocols and server architectures) simple and efficient and to keep the sophistication and innovation mostly at the periphery (e.g., MIME, S/MIME, and PGP). That (together with the protocols being open) makes it possible for the core technology to be deployed on a variety of platforms, including very high end ones, and ensures that such servers will scale very well. In comparison, most proprietary systems have concentrated most of the complexity and need for performance at the core and kept the periphery simple (the assumption historically being that client systems were underpowered, which is no longer the case). For example, in Exchange Server, all the rules processing is done by the server, which is a substantial part of the overhead of a server. In Internet Mail, rules processing always is done in the client, so that very high overhead is distributed.

There are several areas in which proprietary systems historically have been more sophisticated (and more attractive to users) compared with Internet Mail, such as address books, delivery status notification, security, and the polish of the clients. There are several reasons that Internet Mail has lagged in those areas. Internet Mail was developed as freeware (in universities), and there was no financial pressure driving its evolution as there was in proprietary systems. Only recently (with the explosion in the size of the Internet) have commercial companies and large financial incentives entered the picture in standards-based products. Also, Internet Mail from the beginning attempted to address the largest possible user community (conceivably every computer user on Earth), which severely complicates things like address books and security.

Each of those areas is being addressed by recent developments in the Internet Mail standards. In the area of address books, X.500 and LDAP are being developed and deployed to make possible a worldwide directory service (including distribution of public key certificates for security). In the area of security, PGP and S/MIME are fighting for market share, but either will be able to deliver on the promise of private e-mail resistant to tampering and fraud in ways not even possible prior to e-mail. In the area of commercial polish, the financial rewards for a successful Internet Mail vendor are attracting some of the best talent in the computer field and large infusions of venture capital. The pioneering days of user-supported freeware (at least as the most widely used products) are rapidly drawing to a close. Even the venerable Sendmail program is going commercial.

One early trend in that direction was for each vendor of a proprietary system to provide gateways to Internet Mail. That allowed their users not only to communicate with direct users of Internet Mail but also to use the Internet as a backbone to connect islands of their own (and even other vendor's) systems. However, the gateways used to connect systems tend to be notoriously unstable and somewhat opaque to advanced features like address books, delivery status notification, and security, in some cases even when both the sender and the receiver are using the same proprietary technology (but the mail between them must traverse two gateways and part of the Internet).

The next logical phase of e-mail evolution (which already is well underway in bellwether institutions like Harvard University) is for the Internet backbone to be extended all the way to everyone's desktop. With the addition of universal address books and security and continued innovation in Internet Mail client software (UAs), the writing is on the wall for proprietary systems.

Other open e-mail standards

Internet Mail is not the only contender for an open, universal e-mail standard. Its primary competitor is the X.400 Message Handling System devised by ISO. Certainly all the arguments for open e-mail systems apply just as well to X.400-based systems. X.400 already has very well thought out solutions for universal address books (X.500), security (X.509 and the digital signature and encryption aspects of X.400), and delivery status notification. So why doesn't everyone just adopt it? Why is Internet Mail winning this war?

Basically X.400 is a complex system that is difficult and time consuming to implement; hence, products based on it tend to be large, expensive, and difficult to deploy. Several commercial X.400-based systems are available (e.g., Hewlett Packard's Open Mail). The major problem is that any X.400-based system depends heavily on having the ISO Open System Interconnect (OSI)–style networking underneath. That networking standard has met with some acceptance in Europe, but the rest of the world (especially the United States) has gone almost exclusively with the underlying Internet networking standards (based on IP and DNS). The explosive growth of the World Wide Web (which is definitely based on Internet networking standards) pretty much has sealed the fate of OSI-style networking. In effect, Internet Mail is riding to global domination on the coattails of the World Wide Web.

Actually, Microsoft tried to cover its bets by making Exchange Server able to interoperate with both Internet Mail and X.400-based systems (both through gateways), but there has not been as much interest in its ability to interoperate with X.400 systems as Microsoft might have thought there would be.

TCP/IP: the Internet protocol suite

It would be easy to write an entire book on TCP/IP; in fact, several authors have. I do not intend to cover every aspect of this complex subject, just enough so you will be able to follow the material on the higher-level protocols we are really interested in, such as SMTP, POP3, IMAP, and DNS.

Most books on networking start with a discussion of the seven-layer model of the ISO OSI networking standard. While useful from a pedagogical viewpoint, the creators of TCP/IP were not familiar with that standard (it did not exist at the time TCP/IP was designed). They used a somewhat simpler four-layer model. Because we are far more interested in the Internet protocol suite than in the ISO OSI protocol suite, we will use the four-layer model.

The Internet four-layer model divides the functionality of the Internet protocols into the following layers (from the lowest, or closest to the physical connection, to the highest, or most abstract):

▶ Network access layer (Ethernet, PPP);

▶ Internet layer (IP, ICMP, ARP, RARP);

▶ Host-to-host layer (TCP, UDP);

▶ Process/application layer (FTP, TELNET, SMTP, POP3, etc.).

Figure 7.1 shows a TCP/IP stack.

Network access layer

The network access layer is the lowest layer, in the sense of being the least abstract, the closest to the real, physical world. It is concerned

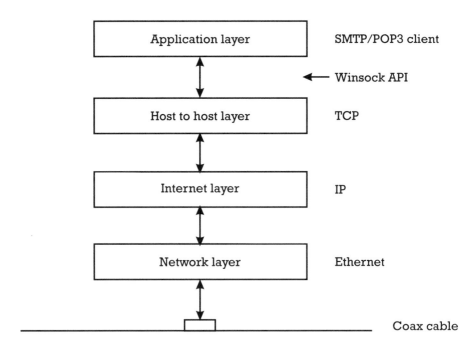

Figure 7.1   The DOD four layer network model.

with the physical connection between nodes. That connection can be accomplished with any number of technologies, such as Ethernet (which itself has many variants), PPP (a way to establish a network connection between two nodes via a pair of asynchronous dialup modems and a voice-grade telephone line), or something like frame relay (a WAN connection technology available from most phone companies that you typically connect to via an X.25 interface). More recently, frame relay may actually be implemented with an RF broadcast scheme called wireless networking. The key point is that any of the technologies provides a mechanism for packets of information to be exchanged between at least two nodes, in a standard way, using well-defined physical network addresses (media access control (MAC) addresses).

Ethernet itself may be physically implemented with so-called multidrop (many nodes connected to a single physical wire) using either of two sizes of coaxial cable (thinnet/10base2 or thicknet/10base5) or with a star architecture using four conductor twisted pair (similar to phone wiring), called 10baseT. In addition to the wires, of course, there must be an interface between the node and the wire that can put packets onto the wire and receive them from it. This device is called a network interface card (NIC). Every ethernet NIC ever made has a unique MAC address, which is a 48-bit value assigned by the manufacturer at the time the card was made. Typically, MAC addresses are written as six 8-bit bytes in hexadecimal, separated by dashes. (The MAC address for the NIC in my home computer, e.g., is 00-04-05-16-7C-14.)

Some writers use colons instead of dashes to separate the bytes of the MAC address (e.g., 00:04:05:16:7C:14). There is no particular advantage to either convention. Windows NT uses dashes. The first three bytes (00-04-05) identify the company that manufactured the NIC. A list of the codes for all manufacturers is available as an RFC should you happen to want to find out who made a given NIC card. The last three bytes (16-7C-14) were assigned by that manufacturer for the particular NIC I own.

Just like the 32-bit IP addresses (which we will cover later), that representation is merely for the convenience of humans. The hardware and low-level software think of it as a 48-bit binary value. No other ethernet NIC in the world has (or at least is not supposed to have) this MAC address. With 48 bits, there are some $2.8 \cdot 10^{14}$ possible addresses, which is enough for every human alive today to have over 40,000 distinct addresses (in other words, we are in no immediate danger of running out of MAC addresses). If you are interested in finding out the MAC address in your NIC, and you happen to be running Windows NT, just type the command "ipconfig /all" in a DOS Command window (there is also a GUI version of this program in the NT Resource Kit). A MAC address will be listed for each NIC in your computer (if more than one is listed, you have a multihomed computer).

The original ethernet scheme supports a maximum data rate of 10 million bps, although throughput between any two network devices never can reach that level, due to overhead in the packets (addresses, packet type codes, checksums, etc.). In practice, with 30 or 40 nodes all shouting on the same wire in a given LAN, collisions will occur (two or more parties starting to talk at the same time), requiring all but one of the parties to wait their turn, further lowering the practical achievable data rate between any two nodes. However, the theoretical maximum bit rate characterizes the technology, which accounts for the 10 in all three ether-net schemes described later in this section. More recent schemes support a maximum bit rate of 100 million bps and so are characterized with 100 as the first part of their names.

It would be possible to modulate a high-frequency carrier signal (in the same way that simple modems modulate between two audio tones in the roughly 1-KHz range with the digital signal). With most ethernet systems, however, the digital signal is put directly on the wire, which is called baseband signaling (the frequency at any given instant is either 0 Hz or 10 MHz, as opposed to switching between two higher frequencies, say, 10 MHz and 20 MHz). With standard ethernet then, the frequency range is from 0 Hz to 10 MHz. In a modulated system, it might go from 100 MHz to 110 MHz. That accounts for the "base" in the names of the ethernet variants described in this section.

Different cable diameters have different propagation characteristics (smaller cables lose more signal in a given length than larger ones), so the final part of the name characterizes the cable size in terms of the maximum useful length of a given segment. 10base2 is limited to approximately 200m (well, actually 185). 10base5 can have segments 500m long. The T in 10baseT refers to twisted-pair wiring, which has totally different characteristics than coax cables (but a given segment is limited to 100m, so to be consistent, perhaps it should have been called 10base1).

Ethernet 10base2

Ethernet 10base2 is a baseband signaling scheme with a maximum data rate of 10 million bps, using RG-58U (thinnet) coaxial cable (which allows a maximum segment length of about 185m). To connect nodes to a 10base2 cable, you need a BNC tap on the NIC itself and a T connector onto which you connect two lengths of thinnet coax cable to the two computers closest to yours. (For those of you who have always wondered, BNC stands for British Naval connector.) If your computer is the last one on either end, you connect one side of the T connector to a 50Ω terminator. New computers are added by removing the cable connected to one side of the T connector on an existing node, connecting that cable to one side of the T connector on the new node, then adding a new coax cable between the old tap from which you removed the first cable and the other side of the T connector on the new computer.

The sum of the lengths of all the cables is the segment length (the total length is what may not exceed 185m). The cables between nodes should not be less than 0.5m long. You cannot connect more than 30 nodes on a given cable segment. If more than 30 nodes must be connected, you must connect multiple segments via repeaters (of which you can have no more than four, which limits you to, at most, 150 nodes on a given 10base2 LAN and a total length of 925m). If your NICs have a BNC connector on them (a roughly 3/8-inch-diameter by 1/2-inch-long hollow metal cylinder with two small stubs on either side), and you have a small number of nodes, 10base2 is by far the cheapest and simplest way to connect your nodes. It also is the least reliable because each new node adds at least one connection (every connection is a potential point of failure for the entire network). This setup is called a bus architecture.

Ethernet 10base5

Ethernet 10base5 is similar to 10base2, except that a thicker coaxial cable is used (which allows maximum segment lengths of 500m). Connected nodes require a DIX connector (which looks like a 9 pin RS-232 connector but has holes for 15 pins). You run a multiwire attachment universal interface (AUI) cable from the connector to an external transceiver. The transceiver itself is clamped onto the thick coaxial cable (usually with a so-called vampire tap). The transceivers must be a minimum of 2.5m apart. A given network segment cannot exceed 500m, and the entire LAN may not exceed 2,500m. The AUI cables (from the thicknet cable to the NIC) can be as short as required but may not exceed 50m. There can be a maximum of 100 nodes on a given segment. Unlike 10base2, one end of each segment must be terminated to ground. Again, up to five segments can be combined with repeaters. This setup also is a bus architecture.

Ethernet 10baseT

Very different from the two coaxial cable schemes, Ethernet 10baseT uses a star architecture to connect each node with a distinct four-conductor unshielded twisted-pair cable to a central hub or concentrator. The NIC must have a 10baseT female connector (an RJ45 socket that looks like a wide modular telephone cord socket). Typically, network grade (e.g., CAT-5) twisted-pair wiring is brought though the walls to a socket plate near the node, and a network-grade cable with a male RJ45 connector at each end is used to connect the NIC to the wall plate. Central 10baseT hubs are available from 4 node to 32 node (or even larger). Typically in a large system, multiple 32-node units are connected at the central site. A given 10baseT LAN can support up to 512 nodes, each of which must be no more than 100m from the closest hub. It is possible to put numerous 10baseT hubs around a plant and connect the hubs with a 10base2 (or even 10base5) coaxial cable, for a hybrid system. Today, most NICs have 10baseT connectors and possibly also a 10base2 or 10base5 connector.

Ethernet 100baseT

Ethernet 100baseT is a recent technology that closely resembles 10baseT but supports a maximum data rate of 100 million bps. Both the NIC and the hub must support the higher data rate (dual-mode NICs are available that can work with either 10baseT or 100baseT hubs). You must use category 5 (or cat 5) wiring for 100baseT links. Currently, 100baseT hubs are dramatically more expensive than 10baseT hubs, while dual-mode cards are not significantly more expensive than single-mode 10baseT ones. For installation of new twisted-pair cabling in a facility, you should always install cat 5 cables, even if you are planning to use only 10-Mb networking. In practice, cat 3 cables will work fine for 10-Mb networks, but the cost for cat 5 is only a little more than for cat 3. That will allow you to upgrade to 100-Mb networking later without an expensive rewiring job.

PPP

PPP (Point-to-Point Protocol) is basically IP-style packet communication between two nodes over a serial link, such as two asynchronous modems and a telephone line. PPP is the scheme used by most Internet service providers (ISPs) to allow home users to dial into their sites, at which point packets coming from them are fed into some higher bandwidth (e.g., T1) traditional network interface, and packets coming from that channel for them are directed down the PPP link. Windows NT remote access service (RAS), as a subset of its capabilities, can act as either a PPP client or a PPP host (or even a PPP-to-Internet router). A later chapter discusses both how to use a RAS client to dial into any ISP and how to use a RAS server to allow any PPP (or RAS) client dial into you as a network service provider (e.g., to allow dial-in access to your e-mail server).

Software

Network hardware (NICs and the wiring and hubs that connect them) are of little use by themselves. There must be many layers of software built on that foundation before we get up to the level of e-mail protocols. The lowest level of this software is highly device dependent (a different version is needed for each style of NIC, e.g., one for Novell NE2000-style interfaces and a different one for the Western Digital WD8003). There are several standards for such network interface device drivers (depending on the operating system. Microsoft uses Network Device Interface Specification (NDIS), created by Microsoft and 3Com. Specifically, Windows NT uses version 3 of that standard, which supports flat-memory model 32-bit access. Novell uses ODI. Vendors of NICs provide a range of different device drivers so you can use their cards with various higher level network software programs. Windows NT includes NDIS v3 drivers for most commonly encountered NICs and can recognize them automatically during installation. However, very recent cards may not be covered by the drivers that come with NT, so be sure any very recently released NIC includes a 32-bit NDIS driver for NT (along with the appropriate installation files used by NT during setup). You can often download these drivers from the vendor's Website.

The functionality provided by the NDIS software layer is at quite a low level and would not be of much use to a typical user. However, it does provide the basic functions that the next layer up requires, including things like reading the MAC address, sending a packet, and receiving a packet. A packet is just a block of data bytes plus various header information (e.g., address of sender, address of receiver, packet sequence number) and trailer information (e.g., checksum). Each abstraction layer typically adds more overhead information as packets work their way down from the application to the hardware or removes overhead information as the packets work their way up from the hardware to the application. This hardware-dependent layer could be used by any kind of networking software (TCP/IP, Novell, or Microsoft NetBIOS). All higher level layers described in this chapter are specific to the Internet protocol suite.

Above the device driver, but still below the Internet layer, is one of the following protocols:

 ▶ Ethernet II;

 ▶ IEEE 802.3;

 ▶ Token ring (IEEE 802.5);

 ▶ Metropolitan area network (IEEE 802.6).

An ethernet-based system might use Ethernet II or IEEE 802.3 (the main difference is the exact manner in which the packets are framed). Neither system is clearly superior, but you must be consistent in choice throughout a given LAN. If you have token ring hardware, the 802.5 layer must be used. If you have a MAN-style fiber network, 802.6 must be used.

Internet layer

Sitting on top of the network access layer is the lowest layer of device-independent software, the lowest level Internet suite protocols: IP, ICMP, ARP, and RARP.

IP

IP (Internet protocol) is concerned with the addressing of nodes and the fragmentation of packets. IP provides a datagram service between communicating hosts. This level does not provide any end-to-end reliability or flow control (those are provided at higher levels).

ICMP

Internet Control Message Protocol (ICMP) is concerned with transporting error and diagnostic information for IP. The ping utility uses ICMP to determine whether a functioning path exists between two nodes and, if so, what the turnaround time is. ICMP is also used to implement Source Quench (a flow-control mechanism that tells the sender that the receiver's cup overfloweth). ICMP can be used to inform the sender that a better route to the receiver exists. It can be used to tell the sender that a given packet's time to live (TTL) has expired, which leads to that packet being discarded. It can be used to tell the sender that a packet has a bad entry in the IP header bytes. Finally, it can be used to obtain information on packet timing (e.g., turnaround time in ping).

ARP

Address Resolution Protocol (ARP) maps 32-bit TCP/IP addresses to and from the 48-bit MAC addresses. Basically, the ARP code maintains a table of physical addresses and their corresponding IP address. IP address resolution proceeds as follows:

1. The IP layer invokes the network access layer to send a datagram (that contains a 32-bit IP address). The network access layer looks up the IP address in an internal table. If it finds that address, it sends the packet to the corresponding 48-bit MAC address from the table, and we are done.

2. If the network access layer cannot find an entry for the IP address, an ARP broadcast is sent requesting the MAC address of the node with that IP address.

3. All computers on the network segment check to see if the IP address happens to be theirs. The guilty party (if any) responds to the requester with its MAC address.

4. The requester receives the response, updates its internal table, then sends the packet to the newly determined MAC address.

This same protocol is used to resolve addresses outside a local network segment through interaction with a network router (that acts as proxy for outside nodes). The gory details of how a router does that are interesting but not specifically necessary for an understanding of how Internet e-mail works. You can just pretend that all nodes in the world are really on a single LAN segment. Aren't routers wonderful?

RARP

Reverse Address Resolution Protocol (RARP) does the reverse of what ARP does. If a diskless workstation knows its MAC address but does not know its IP address, is can use RARP to determine it. A similar but better protocol, called BOOTP, can do that more efficiently and also can be used to bootstrap-load an operating system onto a diskless workstation via the network. The BOOTP protocol also is required by Dynamic Host Configuration Protocol (DHCP), which will be covered later.

Host-to-host layer

The host-to-host layer is where TCP and User Datagram Protocol (UDP) live. These two very different approaches to exchanging data have uses in higher level network software. Usually, end users do not get down to the host-to-host level, but it typically is the lowest level at which programmers interact with the network (various APIs exist to allow programmers to use TCP and UDP, the most common being the socket API). To implement the POP3 protocol, you must understand TCP. To implement a DNS resolver in an SMTP server, you must understand UDP.

TCP

TCP (Transmission Control Protocol) is a reliable, connection-oriented transport. Reliable means that no packets are lost, duplicated, or received out of order. Connection oriented means that you establish a path, pump some amount of data through it (typically requiring many packets), then close the path. TCP is the preferred protocol for services like FTP, Telnet, and SMTP. When you open a socket in Stream mode, TCP is used. That means you can view the network as an arbitrarily long stream of bytes (you divide records in a way meaningful to you, independent of the packet size, or where division between packets falls).

UDP

UDP (User Datagram Protocol) is unreliable, connectionless transport. In this case, unreliable means that no checking is done for lost, duplicated, or out-of-sequence packets. It is assumed that if such packets are important, a higher level protocol will see to such things. Because it does not worry about those details, UDP is faster than TCP. Connectionless means that each packet is considered to be a separate entity (like a small child with no notion of past or future—only the present exists). This is the preferred protocol for NFS and DNS. When you open a socket in Datagram mode, UDP is used. That means you must always read and write packet-sized chunks of data (and handle lost, duplicated, or out of sequence packets).

Process/application layer

Most protocols that end users are aware of live at the process/application layer. Each one of the protocols at this layer is specific to a particular application. The commonly used protocols from this layer are FTP, Telnet, NNTP, HTTP, NFS, DNS, DHCP, SMTP, POP3, IMAP4, and LDAP.

FTP

File Transfer Protocol (FTP) is used to transfer files between systems. A server program, or daemon (typically called "fptd"), must be running on one system (unattended), and the client (typically called "ftp") is brought up on another (attended) system. The transfer is controlled from the client end. The client can request a list of files in the current directory, change the current directory, get a file from the other system, or put a file to it. Some clients are very simple command line interface (text mode) programs. It is also possible to create a GUI FTP client program that looks more like using Windows File Manager (or Explorer). Most Web browsers (e.g., Netscape Navigator and Microsoft Internet Explorer) also can function as FTP clients by specifying a URL like this:

ftp://ftp.microsoft.com

Various degrees of access control are possible with FTP, including wide-open or anonymous FTP (which is the only kind supported by Web browsers). It is possible to restrict access to certain accounts, with passwords required, or in some cases allow access to different sets of files for different users. A good FTP server and a simple command line client are included in Windows NT 4.0.

Telnet

Telnet emulates various popular "dumb" terminals, such as DEC VT100 via network links as opposed to the more common asynchronous serial link. Again, a server (typically called "telnetd") must accept TCP network connections from Telnet clients and connect them to something that knows how to accept input from and send output to a dumb terminal. On a UNIX system, that is simple, because dumb terminals are supported as the primary means of interaction. On Windows NT systems, dumb terminals are not really supported or very useful. Some third-party vendors have created Telnet servers that essentially allow use of a remote DOS command prompt window from a Telnet client. A Telnet client presents a (typically) 80-by-25-character screen (possibly within a window) that can access a Telnet server via the network. Typically, communications speeds are much greater than via dialup modems (limited only by network throughput). Telnet rapidly is becoming obsolete as a way of interacting with network services. Windows NT includes a good Telnet client but no server.

NNTP

NNTP (News Net Transfer Protocol) is the protocol used to implement Usenet, which is a worldwide system of newsgroups similar to Exchange

Server's public folders. There is an incredibly large and diverse hierarchical set of newsgroups (organized by topic). For example, rec.travel has information about travel under the general category of recreational newsgroups. A newsnet server program may manage one or more newsgroups of its own or exchange postings to and from other newsnet servers further upstream. In theory, a message posted to any NNTP server connected to the Internet will be replicated to all other NNTP servers in the world (at least those that subscribe to the newsgroup in question). An NNTP client can retrieve posted messages from (or post messages to) any newsgroup the server manages or subscribes to. Newsgroups are similar to e-mail, but messages are publicly available for all to read and respond to, as opposed to e-mail, which typically is one-to-one or at most directed to a specifically enumerated group of recipients (possibly via a mailing list). Windows NT does not include an NNTP server, but Microsoft Exchange includes what amounts to a public folder-to-NNTP gateway. The freeware Microsoft Internet Mail and News Client includes a good NNTP client (in addition to a fair SMTP/POP3 Mail client). Good NNTP servers for NT are available from third parties, such as MetaInfo (www.metainfo.com). An NNTP server is now included in the "Option Pack" for NT 4.0.

HTTP

HTTP (Hyper Text Transfer Protocol) is the protocol of the World Wide Web. A Web server manages a set of local pages in the Hyper Text Markup Language (HTML), possibly including text in various fonts, sizes, and colors; images; and even sounds or movies. An HTTP client (also called a Web browser) can retrieve such pages from anywhere on the Net and is essentially a glorified, multimedia extension of the gopher information publishing protocol. The Hyper Text part involves the ability to provide links to information on other local pages or even pages elsewhere on the Net. Clicking on such a link reconnects the Web browser to the referenced page (assuming that server is up and running and the page in question is still available). Microsoft Windows NT Server includes a good HTTP server, in the form of the optional Internet Information Server (IIS). The freeware Microsoft Internet Explorer is a good HTTP client for Windows NT (as is Netscape Navigator).

NFS

Network File System (NFS) is the network file-sharing scheme originally designed for UNIX. Windows NT uses a different scheme based on the NetBIOS network interface and Server Message Block (SMB) protocol. Novell uses yet another file-sharing scheme and protocol. Microsoft does not provide either an NFS server or an NFS client with Windows NT, but several third parties provide both. Intergraph, together with Sun (the inventors of NFS), created both an NFS server and an NFS client for Windows NT, marketed by Intergraph as DiskShare and DiskAccess, respectively. The client allows NT computers to access file systems made public from UNIX systems. The server allows NT file systems to be made public using the NFS protocol so that any NFS client can mount them (typically UNIX computers or DOS PCs running older network software, such as PC-NFS or PC/TCP).

DNS

DNS (Domain Naming Service) is a hierarchical symbolic naming scheme that was created to help users of the Internet keep track of the millions of nodes connected to it, with the highest level dividing the Internet into a few top-level domains. The domain names include "com" (for commercial enterprises), "edu" (for educational institutions), "net" (for network related companies), "gov" (for government agencies), and "mil" (for military institutions). Below each of those are organizations of those types. For example, under "com" is "ingr," for Intergraph Corporation, and "dec," for Digital Equipment Corporation. Under those organizations can be further divisions and even additional levels as appropriate (e.g., "asi-ahq" for the Asia/Pacific headquarters of Intergraph Corporation). Full domain names consist of those names separated by periods, with the highest level at the end, for example:

sales.megacorp.com

A particular node for a computer connected to the Internet would have a name (that need be unique only within its domain), for example, "msmail." A fully qualified domain name consists of the nodename followed by the complete domain name, for example:

msmail.sales.megacorp.com

The DNS protocol is used to resolve such symbolic nodenames into 32-bit numeric IP addresses. For example, the above nodename might resolve to 148.53.150.2. Even though that node happens to be located in Hong Kong, the DNS system can find it and obtain the correct IP address in under a second.

DNS server programs (or daemons) are available for UNIX and Windows NT (among other platforms). Windows NT Server 4.0 includes a good DNS server that is integrated with the DHCP and WINS servers. DNS clients typically are built into operating systems or mail servers. The DNS protocol (typically implemented with UDP/IP) is used by DNS clients (also known as resolvers) to query the information in the distributed DNS database.

DHCP

DHCP (Dynamic Host Configuration Protocol) is used to automatically assign IP addresses (and essentially all other TCP/IP configuration information, such as default gateway and address of DNS server) to a node that supports TCP/IP at the time the computer first connects to a TCP/IP network. A DHCP server manages a pool of IP addresses and configuration data (a DHCP server is included with Windows NT Server). A DHCP client typically is a part of a client operating system.

SMTP

SMTP (Simple Message Transport Protocol), the primary protocol of Internet Mail, is used to send a message from a UA to an MTA (mail server) or to relay mail from one mail server to another. The server is a part of an Internet Mail server, typically called "smtpd" and are available for UNIX and Windows NT (among other platforms). Microsoft does not include an SMTP server with Windows NT, but Exchange Server includes a kind of SMTP server called Internet Mail Services (formerly Internet Mail Connector), which is really a gateway between Exchange Server and Internet Mail but has many of the characteristics of a full SMTP server (although it cannot be installed apart from Exchange Server). Microsoft has introduced a standalone SMTP server as part of the MCIS product, but it is available only to ISPs. SMTP clients are built into an Internet mail UA or an MTA. The freeware Microsoft Internet Mail and News Client is a fair SMTP/POP3 mail client. The Outlook Express mail client (part of Internet Explorer 4.0) is an excellent SMTP/POP3/IMAP/LDAP client that even supports S/MIME. It is also possible to install an add-in to Exchange Client to allow it to work directly with SMTP/POP3 mail servers. A simple SMTP server is included in the "Option Pack" for NT Server 4.0 (but no POP3 server).

POP3

POP3 (Post Office Protocol, version 3) is the most common protocol used by Internet Mail UAs to retrieve mail from the local MTA. The server is part of an Internet Mail server, typically called "pop3d." Microsoft does not include a POP3 server with Windows NT, but there is a POP3 alternative front end in Exchange Server 5.0 that allows a SMTP/POP3 mail client to retrieve messages from the Exchange message store. Microsoft has introduced a standalone POP3 server as part of the MCIS, but it is available only to ISPs. POP3 clients are built into an Internet Mail UA. See SMTP for comments about SMTP/POP3 clients from Microsoft.

IMAP4

IMAP4 (Internet Message Access Protocol, version 4) is a newer protocol (it is just coming into vogue) used by Internet Mail UAs to retrieve mail from the local MTA. The server is part of an Internet Mail server, typically called "imapd." The client is built in to an Internet Mail UA. Microsoft does not currently support IMAP on either the server side or the client side.

LDAP

LDAP (Lightweight Directory Access Protocol), a very recent Internet protocol based on the ISO X.500 DAP, is used to search and retrieve address book information (including various attributes, such as e-mail address and public key) from an LDAP server. The server may be a part of an Internet Mail server or a standalone generalized directory service. Microsoft does not include an LDAP server with Windows NT, but

Exchange Server 5.0 includes one. The client may be a standalone directory browser or built into an Internet Mail UA. Microsoft Exchange Client 5.0 includes LDAP client support, as does Outlook Express.

Example of address resolution

The following example illustrates how the different Internet layers work together to actually send a packet to the correct node:

1. An application, for example, ftp, wants to open a connection to a particular node given a fully qualified domain name, for example, ftp.microsoft.com.

2. The host-to-host layer queries DNS to resolve that symbolic nodename to an IP address, which happens to be 198.105.232.1.

3. The Internet layer uses the ARP protocol to find the MAC address for the IP address, which might be 00-00-01-12-34-56.

4. The network layer puts the MAC address into the appropriate field of the outgoing packet and sends it.

IP Addresses

An IP address is a 32-bit binary value. For convenience, we break it up into four 8-bit fields, write the value of each 8-bit field as a decimal number (from 0 to 255), then separate the fields by periods. For example, the IP address of the well-known node ftp.microsoft.com is commonly written as 198.105.232.1, but in more conventional notation, it is the 32-bit number 0xC669E801 (hexadecimal), 3,328,829,441 (decimal), or binary 1100 0110 0110 1001 1110 1000 0000 0001.

A given IP address is divided into two parts (typically into the leftmost n bits and the rightmost 32-n bits). A net mask (typically a pattern of bits with the leftmost n bits set to 1 and the rest set to 0) is used to split the IP address into its two components, which are a network number and a host number within the network. Common net masks have the first 8 bits (a class A address), the first 16 bits (a class B address), or the first 24 bits (a class C address) set to 1, and the rest set to 0. The overall address space is shown in Table 7.1.

The total number of possible 32-bit addresses (which happens to be 2^{32}, or 4,294,967,296) is divided into various address classes, based on the value of the first eight digits. Very large organizations (e.g., the U.S. government) were allocated an entire class A range (all IP addresses starting with a given 8-bit pattern in the range of 0 to 127).

There are 126 possible Class A address blocks (e.g., 123.x.x.x), each of which has some 16,777,216 distinct IP addresses in it (e.g., 123.0.0.0 to 123.255.255.255). All the Class A addresses together account for 50% of all possible IP addresses.

There are 65,534 possible Class B address blocks (e.g., 150.123.x.x), each of which has some 65,526 distinct IP addresses in it (e.g., 150.123.0.0 to 150.123.255.255). All the Class B addresses together account for 25% of all possible IP addresses.

There are 2,097,152 possible Class C address blocks (e.g., 200.123.111.x), each of which has 256 distinct IP addresses in it (e.g., 200.123.111.0 to 200.123.111.255). All the Class C addresses together account for 12.5% of all possible IP addresses.

Class D addresses are for multicast and account for 6.025% of all possible IP addresses.

Table 7.1
Address space divisions

Class	First 8 bits	Net mask	Address/ block	Address Blocks	Address
A	0–127	0xFF000000	16,777,216	128	2,147,483,648
B	128–191	0xFFFF0000	65,536	16,384	1,073,741,824
C	192–223	0xFFFFFF00	256	2,097,152	536,870,912
D	224–239	0xFFFFFF00	256	1,048,576	268,435,456
E	240–255	0xFFFFFF00	256	1,048,576	268,435,456
					4,294,967,296

Class E addresses are for Internet experimentation and account for the remaining 6.025% of all possible IP addresses.

In practice, network addresses with the node number field containing all 0s (which is used as the network number) or all 1s (which is used as the network broadcast address) cannot be assigned to actual nodes. For example, a Class C address really gives you usable IP addresses for only 254 nodes.

When an organization applies for a block of Internet addresses, it is assigned a complete block of addresses (really big organizations were granted Class A addresses, smaller ones were granted Class B addresses, and really small ones or individuals were granted Class C addresses). It is up to the organizations to whom the address blocks were granted to assign addresses (or even subnets) within their address blocks. Some time ago, all Class A address blocks were assigned, then all Class B addresses, and finally all Class C addresses have been assigned. However, most organizations with Class A and Class B addresses have not used all (or in some cases even a significant percentage) of their block of addresses. So today if you apply for a block of addresses you typically only get a handful of valid addresses, perhaps eight. See the section on IP v6, for the long-term solution to that problem.

IPv6 (also IPng)

Although 32-bit IP addresses (of which there are some 4 billion) would seem to be sufficient for all computers in the world, by using the above address class allocation scheme (which seemed like a good idea at the time) we have run out of addresses. The current IP protocol and address scheme is Internet Protocol, version 4 (IPv4). A new version of the IP protocol (called IPv6, or IPng, for "next generation") with larger addresses (128 bits, or 4 times the current number of bits, which should be sufficient for a very long time) is currently being worked out. Over the next few years, we will have to go through a painful transition to the new IP address scheme, but at that point we will have addresses available for the foreseeable future.

Given that a painful, messy transition is inevitable, a great deal of thought has gone into minimizing the problems and even providing clean mappings from existing IPv4 addresses and even Novell IPX addresses to the new IPv6 address space. Also, the enormous new address space will allow for far more efficient hierarchical groupings of addresses (not currently practical with IPv4).

Microsoft NetBios over TCP/IP

No discussion of TCP/IP in a book about Windows NT would be complete without a discussion of Microsoft's implementation of NetBIOS over TCP/IP. First off, NetBIOS in the scope of this discussion is a network API, not a protocol. There once was a network protocol called NetBIOS (in IBM's earliest network products for PCs), but it evolved into NetBEUI. However, the programming interface (a set of subroutine calls that a programmer employs to use some facility, in this case network I/O) and the namespace has survived even in Windows NT 4.0.

Just as it is possible to write programs that do network I/O by making calls to the windows socket library (WinSock), there is another API called NetBIOS. When you define a computer name and NT domain name, they are actually in the NetBIOS namespace. If you install TCP/IP on Windows NT, you also can assign a computer name (in this case, it is called a nodename) and a domain name in the IP/DNS namespace (the one used by the Internet). While it is possible for the NetBIOS computer name and the IP/DNS computer name to be the same, it is not required that they be so. It is less likely that the NetBIOS (or NT) domain name would be the same as the IP/DNS domain name. For one thing, NetBIOS domain names are flat (e.g., SCDEV), and IP/DNS domain names are hierarchical (e.g., devel.software.com). Names in the IP/DNS namespace are mapped to network addresses by DNS. Names in the NetBIOS namespace are mapped to network addresses either by broadcast or by WINS.

Also, in the IP/DNS world, all network I/O eventually goes over IP. In the NetBIOS world, it might go over IP, IPX, or even NetBEUI at the lowest level, which means that name-to-address resolution in the NetBIOS world must be able to resolve to several kinds of network addresses, not just IP addresses.

All native Microsoft file and print sharing (i.e., that done by File Manager or Explorer and Print Manager) is done via the NetBIOS API and uses names in the NetBIOS namespace (you mount a shared-file system using the NetBIOS computer name, not the IP computer name, if they are different, e.g., \\SCDEV\data). If you install an NFS client, that uses the IP computer name to mount shared files (in fact, UNIX computers have only IP computer names).

The whole NetBIOS networking and namespace scheme and how to layer it over TCP/IP, in particular, are covered in two IETF RFCs, namely 1001 and 1002. Microsoft just happens to be the only vendor that has really supported that scheme. You may also see the term Server Message Block (SMB) used to refer to Microsoft's native file and print services. Microsoft is currently trying to define an Internet standard based on SMB-style file and print sharing, so that anyone can implement interoperable systems.

DNS

DNS (Domain Name Service) is one of the most important parts of the Internet infrastructure. Without it, you would have to keep track of the 32-bit IP address of every computer you want to access on the Internet. Fortunately, DNS does exist, so today we can remember to connect our Web browsers to ftp.microsoft.com instead of to 198.105.232.1 (or even worse, to node number 0xC669E801 in hex, or 3,328,829,441 in decimal). (To review IP addresses, see Chapter 7.)

History of name resolution on IP networks

DNS has not always been around. In the early days of TCP/IP and the Internet, nodenames, and IP addresses were kept track of manually in a file called "hosts" (in UNIX, the file is found in the directory /etc; in Windows NT, it is found in the directory%SystemRoot%\

system32\drivers\etc). The file is just a list of IP addresses and the corresponding symbolic name(s) for each of them. Table 8.1 lists an example.

For a site with only a few dozen nodes, it is not too difficult to keep track of the IP addresses of your own nodes and perhaps another few dozen outside nodes that your users often use. In fact, for very small, isolated networks (intranets), some people still use hosts files today. When a network application calls the system routine "gethostbyname" to map a symbolic nodename to an actual IP address, that routine searches through the hosts file (on that computer) and returns the corresponding IP address, if the specified name is found.

But what if you have hundreds or thousands of nodes in your network? Or your users routinely access outside sites all over the world? What if you have a node that you want outside users to be able to locate by a symbolic name (perhaps an FTP, SMTP, or HTTP server)?

In the first case, once you pass a certain threshold (in terms of total nodes on your network), it becomes more difficult, if not impossible, to manage all the symbolic names and IP addresses manually in a central authoritative hosts file, let alone keep the hosts files on all computers in synchronization. Various schemes have been implemented to replicate automatically the authoritative hosts file to all nodes. (Accessing a central hosts file via a network shared-file connection involves the chicken and the egg paradox: How does a given node find the node containing the authoritative copy of the hosts file before it has access to the hosts file?)

In the second case, while a few years ago there might have been only a handful of sites outside your local net with which users wanted to access to exchange files via FTP or E-mail via SMTP, today users want to be able to access nodes all over the world. They may even require the IP addresses of dozens, even hundreds of nodes in a single Web browser session.

Table 8.1
Example of hosts file

127.0.0.1	localhost	
123.45.67.1	lehnts.bronwen.com	www.bronwen.com
123.45.67.2	msmail.bronwen.com	
198.105.232.1	ftp.microsoft.com	

Keeping track of all those symbolic names and their IP addresses in a hosts file is out of the question.

In the third case, you may want people all over the world to be able to find your nodes for e-mail and FTP and especially to access your Web server. You certainly do not want them to have to use IP addresses, and there is no way you could keep the hosts files on their computers updated with information about the names and addresses of your public nodes.

Fortunately, this kind of "bookkeeping" is one of the things that computers do very well. Imagine, if you will, a network application that maintains a database of symbolic nodenames and their corresponding IP addresses for all the computers in the world (or at least all those connected to the Internet). That grand vision was first conceived by Paul Mockapetris (now CTO of Software.com) and exists today in the form of the global DNS. It is physically implemented as a set of database files similar to the hosts files, together with servers and clients and an IP-based protocol that allows those servers to communicate with each other. No one computer could possibly manage the entire set of names and IP addresses, let alone the billions of network queries it would have to process every day. Even if it could, simply managing that much information (keeping it current) in a single place would be a hideously complex task. DNS is implemented as a distributed system, with at least two DNS servers for each domain. Each primary DNS server maintains its tiny piece of the overall database (the nodenames and IP addresses in the domain for which it is "authoritative"). The responsibility for managing that piece of the database is delegated to the person in charge of that DNS server.

If a DNS server is available, the "gethostbyname" call actually does a query of the configured DNS server to resolve the symbolic name to an IP address (no change is required in the application; it works equally well with hosts files or DNS).

A DNS server requires a "real" multitasking operating system that supports the concept of "daemons" (e.g., NT services) and TCP/IP. MSDOS (even with Windows 16 and Windows 95 extensions) just did not have the necessary facilities. On the other hand, UNIX and Windows NT are well suited to running a DNS server. Prior to Windows NT 4.0, Microsoft did not provide a commercial-grade, supported DNS (there was an unsupported, limited function DNS server included with the NT Resource Kit, but it was unsuitable for real-world users). Most people either got their ISP to provide DNS services for them or installed at least one UNIX computer on their net to run BIND (for Berkeley Internet Naming Daemon), a freeware implementation of a DNS server for UNIX. For those companies that wanted to migrate entirely to Windows NT, several companies ported the BIND source code to run under Windows NT. Metainfo (www.metainfo.com) created a supported commercial version for about $495. The Internet Software Consortium (ISC) released an unsupported freeware part of BIND 4.x for NT. This is available from Software.com's Website, www.software.com.

With the release of Windows NT 4.0, Microsoft has released a commercial-grade, supported DNS server that runs only on NT Server 4.0. It has some interesting features, including integration with DHCP and WINS, plus an NT-style GUI enterprise administration console. Even though there is no extra charge for the DNS server, it must be installed separately, after the basic NT Server is up and running. Of course, that server must have TCP/IP installed and correctly configured prior to installing DNS. Because the server is adequate for servicing even large NT sites and it comes free with NT Server 4.0, this is the server we will describe and go through the installation of in this book. If you choose to use another DNS server, many of the concepts and configuration details described in this chapter can be applied. See the documentation provided with your DNS server for details on how to perform the equivalent procedures.

Jargon

Node   A node is some object that supports the TCP/IP protocol and that has an IP address. Typically, a node is a computer, but it also could be a terminal adapter (e.g., an X.25 PAD), a router, a printer, or even a lab instrument. If a given computer is multihomed (i.e., has mutiple network interfaces, each with its own IP address), then that computer will appear to DNS as multiple nodes (one per IP address).

Domain   A domain is a group of nodes whose symbolic names all have the same final set of fields (after the nodename itself), for example, able.acme.com, baker.acme.com, and charlie.acme.com. The symbolic names and IP addresses for all members of the group are managed by two or more DNS servers (each of which manages the same set of information).

Domain name   The domain name is the part of a fully qualified domain name after the nodename, for example, acme.com in the examples listed above. Domain names must be registered with Internic, and the DNS servers must be connected into the overall distributed DNS scheme, if that domain is a part of the Internet. The name itself is part of a hierarchical scheme, with the highest level listed last. Subdomains extend to the left in decreasing order. For example, in stc.nato.int, the top-level domain is "int," "nato" is a subdomain of "int," and "stc" is a subdomain of "nato.int." The top-level domain names (defined in RFC 1591) currently must be from the set listed in Table 8.2 (although the list is being extended greatly to deal with recent abuses of this system).

Table 8.2
Top-level domains

Organizationally based domains	
com	commercial enterprise, e.g., microsoft.com
edu	educational institution, e.g., fsu.edu
gov	government organization, e.g., whitehouse.gov
mil	military, e.g., arl.mil (Army Research Laboratory)
net	network organization, e.g., internic.net
org	noncommercial, nongovernmental organization, e.g., ieee.org
int	international (transnational) organization, e.g., nato.int
Geographically based domains:	
au	Australia
be	Belgium
ca	Canada
ch	Switzerland
de	Germany (Deutschland)
dk	Denmark

Table 8.2 (continued)

Geographically based domains:	
es	Spain (Espana)
fi	Finland
fr	France
gr	Greece
it	Italy
jp	Japan
kw	Kuwait
nl	Holland (Netherlands)
no	Norway
nz	New Zealand
se	Sweden
uk	United Kingdom (England, Ireland, Scotland, Wales)
us	United States

The "us" top-level domain is further subdivided (as defined in RFC 1480) into state-level domains (e.g., ca.us, for California, and fl.us, for Florida).

DNS server   A DNS server is a server component in the distributed DNS system that manages the symbolic names and IP addresses in a particular IP network domain. It requires an operating system that supports TCP/IP, daemons, and preemptive multitasking. A DNS server can answer queries from DNS clients or other DNS servers concerning information it manages (or has cached from recent queries) and relay queries to other DNS servers in the Internet if the query is for information not in the set it manages. Relayed queries can be done either by telling the client where to look elsewhere or by looking elsewhere on the client's behalf (a recursive query). DNS servers can do forward resolutions (map any symbolic name onto the corresponding IP address) or reverse resolutions (map an IP address onto the corresponding primary symbolic name). They also can return various other kinds of information about a given domain, such as the preferred mail server(s) (from MX records in its database).

DNS client   A DNS client (also called a DNS resolver) is a client component that can query a DNS server to map a symbolic name onto an IP address, to map an IP address onto a symbolic name, or to retrieve information about a particular domain, such as the preferred mail server(s). A DNS client usually is built into an operating system or an SMTP mail server. The debugging utility nslookup is a standalone DNS client.

DNS protocol   The DNS protocol is a set of syntactically well defined queries and possible responses that can take place between a DNS client and a DNS server (or between two DNS servers, one of which temporarily is acting as a client for that particular connection). Available queries allow resolution of a symbolic name to an IP address, resolution of an IP address to a symbolic name, and retrieval of information from various other records (e.g., preferred mail server as specified in MX records). The DNS protocol can be implemented over TCP or UDP, but it most often is done over UDP (for performance reasons). The response may contain the desired information, a recommendation of a better place to look (the address of another DNS server elsewhere on the Internet), or an authoritative answer that there is no such symbolic name or IP address.

DNS namespace   A DNS namespace is the hierarchical tree of domains and nodes in an entire IP Internet, starting at the root (the "."), working down through the top-level domains (e.g., com, edu), all the way to nodenames (e.g., ftp.microsoft.com).

DNS zone   A DNS zone is the portion of the overall DNS namespace managed by a particular set of data files on a DNS server. A DNS zone can be an entire subtree starting at a given domain or an entire subtree minus one or more subtrees lower than the starting point (which have been delegated to other DNS servers). For example, a small organization might have only a single DNS zone, even if it includes one or more subdomains, as long as all of them are managed by a single set of database files. A larger organization might carve off one or more subtrees of their own domain space into different zones (e.g., development.acme.com and marketing.acme.com might be carved out of the overall acme.com zone to make two new zones; what is left of the original overall zone still would be a zone). A given DNS server can manage one or more zones, but each zone is managed independently (has its own set of database files) and has no knowledge of any other zone managed by that DNS server other than possibly having parent/child relationships.

Primary DNS server   A given DNS zone has one primary DNS server, which owns and uses the authoritative set of database files for that zone. The primary DNS server also caches information from other servers as it resolves requests.

Secondary DNS server   To be connected to the Internet, any zone must have at least one other DNS server configured to replicate its database files from the primary DNS server's database files via zone transfer operations that keep the multiple copies of the database files in synchronization. The secondary DNS server also caches information from other servers as it resolves requests. A secondary DNS server should be at least on a different IP subnet (if the site has more than one subnet) or even at a different physical site (the idea is to achieve stability and robustness via redundancy).

Caching DNS server   It is possible to have a DNS server that is neither primary nor secondary (in that it does not have its own copies of the database files) but functions only in a mode of caching information from other servers as it resolves requests. The idea here is to improve performance and do load balancing in a large or very active site.

Start of authority (SOA) record   A start of authority (SOA) record is a record type (in the zone database file) that indicates that the DNS server is authoritative for that DNS zone. It must be the first record in the zone file. It also includes various timing constants (e.g., time to live, or TTL) to be used in that zone.

A record   An A (address) record is the basic record type (kept in the zone database file) that associates a symbolic name with an IP address. One A record should exist for every node in the zone.

CNAME record   A CNAME record is a way to specify additional aliases for nodes in the zone, for example, "ftp," "mail," or "www." CNAME records are optional, but they are useful for providing widely known aliases for nodes that host publicly available services. They also make it easier to move the services around from one machine to another.

PTR record   A PTR ("pointer") record is the mirror image of the A record (kept in the reverse zone database file) and is used to map an IP address to a symbolic name. One PTR record should exist for every managed IP address.

NS record   An NS (name server) record is a record type in the database files that identifies the DNS servers for the domain. One NS record should exist for every DNS server in the zone, or in zones immediately subsidiary to the current zone.

MX record   An MX (mail exchange) record is a record type in the database files that specifies a host running an SMTP server, together with a priority (lower numbers are more preferred than higher numbers). There should be one MX record for each SMTP server in the DNS zone. Other SMTP servers with mail for someone in your domain will query your DNS server to determine the preferred SMTP server in your domain for incoming mail. The procedure is to try nodes in decreasing order of preference (from lower to higher priority numbers) until a working server is found. If multiple working servers all have the same priority, the sending SMTP server should randomly select among them. It is possible for a large site to have its DNS server present different priorities in response to each subsequent query (called a DNS rotor) to force load balancing for incoming mail among multiple SMTP servers.

Zone file   A zone file is a database file that contains resource records (e.g., A, CNAME, MX) for nodes in the zone for which the DNS server is authoritative. A typical zone file for domain bronwen.com might be called bronwen.com.dns.

Reverse zone file   A reverse zone file is a database file that contains PTR records to help map IP addresses to symbolic nodenames. A typical reverse zone file for the block of IP addresses from 123.45.67.00 to 123.45.67.255 would be 67.45.123.in-addr.arpa.dns.

Cache file   A cache file contains NS and A records for your Internet root DNS servers. An up-to-date copy of this file for the real Internet can be obtained at any time from ftp://ns.internic.net/domain/named.cache

In Microsoft DNS, this file is called CACHE.DNS.

Nslookup   Nslookup is a standalone DNS client that is used to debug or verify a DNS installation.

Preparing to install Microsoft DNS Server

You must do several things before you can jump into installing the Microsoft DNS Server. First, you must have a correctly installed Windows NT Server 4.0. The DNS server will not work on earlier versions of Windows NT, nor will it work on NT Workstation 4.0. In particular, you must have installed TCP/IP correctly, including setting the pointer to the DNS server to your own nodename (even though it does not yet exist).

If you are connecting your site to the real Internet, you must obtain a block of IP addresses (e.g., 123.45.67.00 to 123.45.67.255) from an ISP (there are thousands of ISPs around the world; see your telephone company's yellow pages directory). The ISP will explain the details and costs involved. You also must reserve a unique domain name (from Internic). To do that, connect a Web browser to http://www.internic.net, then go to Registration Services. There, you can see what domain names already are taken and by whom. With the Whois option, enter a domain name; if it is not found, you can reserve it for yourself. You will need to supply the nodenames and IP addresses of the two DNS servers (one primary, one secondary) that will be managing your part of the DNS distributed database. Fill in the template available from this site and return it via physical mail (with your payment). There is a one-time startup fee and an annual yearly fee to keep an Internet domain name reserved for your exclusive use (both are in the $50–$100 range). You may want to contract with your ISP to provide one or both of those services (for a charge). If you are connecting a large site to the Internet, you will be installing your own DNS servers and will need only to arrange for your ISP to provide a parent site for your DNS servers to use. Often, two organizations will work out a deal to provide the secondary DNS site(s) for each other, at no cost to either. Once they have implemented their own DNS server, the other company's zone(s) easily can be managed by those same servers (a single DNS server can manage many zones).

If you are installing your own intranet (not connected to the Internet), you will be creating your own DNS root servers and can work out any domain naming hierarchy and IP addresses you want. A sophisticated network engineer might need only a handful of real Internet IP addresses and hide a corporate Intranet behind a firewall, yet still allow controlled access to Internet facilities from inside the firewall, as desired. In that case, they would put some public servers (including DNS and, optionally, WWW, FTP, or SMTP) outside the firewall for outsiders to access and to act as the forwarder for DNS and SMTP. Those topics go beyond this simple discussion of DNS.

Finally, you need the Windows NT Server 4.0 distribution files from which to install Microsoft DNS Server, either on the NT Server 4.0 CD-ROM or from a network share.

Installing Microsoft DNS Server

To install the Microsoft DNS Server, perform the following steps:

1. Install the files for DNS Server. Bring up the Control Panel (Start → Settings → Control Panel). Start the Network applet. On the Network Properties page, select the Services tab. Under the list of current Network Services (which should not include Microsoft DNS Server) are four buttons. Click the Add button. From the list of available services, select Microsoft DNS Server, then click the OK button. If asked, enter the path of the NT Server 4.0 distribution files (e.g., E:\I386). The list of current network services now should include Microsoft DNS Server. Click the Close button. The Binding Analysis will then proceed. When asked, restart the computer.

2. Start the DNS Manager. To do that, follow:

   Start → Programs → Administrative Tools → DNS Manager

3. In the initial DNS Manager, only the newly installed DNS server will be listed. Click that server to select it. In a few seconds, the statistics for that server should appear in black. Under the new server, there may be several lines:

lehnts.bronwen.com
   Cache
   0.in-addr.arpa
   127.in-addr.arpa
   255.in-addr.arpa

You may want to explore those lines by expanding them (click each entry, including any new entries that open up below them). The entries are there for the following reasons:

▶ Cache contains the resource records required to connect to the root DNS name servers.

▶ 0.in-addr.arpa prevents the passing of reverse lookup requests for IP address 0.0.0.0 to the root server(s).

▶ 127.in-addr.arpa prevents the passing of reverse lookup requests for the loopback IP address (127.0.0.1) to the root server(s).

▶ 255.in-addr.arpa prevents the passing of reverse lookup requests for broadcast name queries (x.x.x.255) to the root server(s).

4. Configure the server (set the IP address of the DNS server and optionally specify the address of the forwarder). To do that, right click the new server name (e.g., lehnts.bronwen.com) and select Properties from the popup menu. The IP address already may have been set correctly if you previously entered the address of the node for the DNS server in the TCP/IP configuration.

5. Create your "forward" zone. To do that, select the DNS server by right clicking its name (e.g., lehnts.bronwen.com), then selecting New Zone from the popup menu. In the Creating New Zone for x.y.z dialog box, select Primary Zone. Enter the zone info when requested. For zone name, enter your domain name (e.g.,

bronwen.com). The zone filename should fill in automatically (e.g., bronwen.com.dns) by tabbing to it or clicking it. It should have an NS record and an SOA record automatically created in it, for example:

bronwen.com	NS	lehnts.bronwen.com
bronwen.com	SOA	lehnts.bronwen.com., LHughes

6. Create your reverse zone. To do this, select the DNS server by right clicking its name (e.g., lehnts.bronwen.com), then selecting New Zone from the popup menu. In the Creating New Zone for x.y.z dialog box, select Primary Zone. Enter the zone info when requested. For zone name, enter the reverse lookup format for your IP address (e.g., for 123.45.67.xx, enter 67.45.123.in-addr.arpa). The zone filename should fill in automatically (e.g., 67.45.123.in-addr.arpa.dns) by tabbing to it or clicking it. It should have an NS record and an SOA record automatically created in it, for example:

67.45.123.in-addr.arpa	NS	lehnts.bronwen.com
67.45.123.in-addr.arpa	SOA	lehnts.bronwen.com, LHughes

7. Manually add in the PTR record for the node running DNS (nodes added in the future should add corresponding PTR records automatically). To do that, right click the reverse zone file, then select New Record from the popup menu. In the New Resource Record dialog box, select the record type to be PTR record. Enter the IP address of the node (e.g., 123.45.67.1) and the nodename (e.g., lehnts.bronwen.com). The new record should show up in the reverse zone records.

123.45.67.1	PTR	lehnts.bronwen.com

8. Now enter a second node. Right click the forward zone name (e.g., bronwen.com). Select New Host in the popup menu. In the New Host dialog box, enter the nodename without the domain name (e.g., msmail) and IP address (e.g., 123.45.67.2). Check the

Create Associated Pointer Record option, then click the Done button. A new A record should appear for the specified node. Verify that the corresponding PTR record was added into the reverse zone file by clicking it (e.g., 67.45.123.in-addr.arpa). You may have to refresh the view to see the new record (right click the name and select Refresh).

123.45.67.2   PTR   msmail.bronwen.com

9. Add an MX record to specify the preferred SMTP server(s) for domain bronwen.com (whether or not you already have an SMTP server running). Right click the forward zone name (e.g., bronwen.com). From the popup menu, select New Record. In the New Resource Record dialog box, select the record type to be MX Record. Enter the name of the node that will be running the SMTP server as the Mail Exchange Server DNS Name (e.g., lehnts.bronwen.com). Enter a Preference Number (e.g., 100). Click OK. The new MX record should appear in the list of records:

bronwen.com   MX   [100]   lehnts.bronwen.com

10. Add an alias for the main node, so people can connect to a Web server running on it by using the name www.bronwen.com. To do that, right click the forward zone name (e.g., bronwen.com). From the popup menu, select New Record. In the New Resource Record dialog box, select the record type to be the CNAME record. Enter the Alias Name to be www (no domain name). Enter the For Host DNS Name to be the nodename you will be running your Web server on (e.g., lehnts.bronwen.com). The alias should appear in the list of records:

www   CNAME   lehnts.bronwen.com

The basic installation of your Microsoft DNS Server is now complete. You may want to enter additional nodes or other records.

The preceding installation steps result in the creation of the following database files, found under %SystemRoot%\system32\DNS:

file "bronwen.com.dns":

```
;
;Database file bronwen.com.dns for bronwen.com zone.
;      Zone version:  21
;

@         IN   SOA  lehnts.bronwen.com.LHughes.bronwen.com.(
               2              ; serial number
               3600           ; refresh ;
               600            ; retry
               86400          ; expire
               3600           ); minimum TTL


;
; Zone NS records
;

@             IN   NS      lehnts


;
; Zone records
;

@             IN   MX      100   lehnts
lehnts        IN   A       123.45.67.1
msmail        IN   A       123.45.67.2
www           IN   CNAME   lehnts
```

file "67.45.123.in-addr.arpa.dns":

```
;
; Database file 67.45.123.in-addr.arpa.dns
; for 67.45.123.in-addr.arpa zone.
;      Zone version:  21
;

@             IN SOA lehnts.bronwen.com. LHughes.bronwen.com.(
               2              ; serial number
               3600           ; refresh
               600            ; retry
```

```
            86400        ; expire
            3600      )  ; minimum TTL


;
; Zone NS records
;


@            IN   NS        lehnts.bronwen.com.


;
; Zone records
;


1            IN   PTR       lehnts.bronwen.com.
2            IN   PTR       msmail.bronwen.com.
```

Verifying your Microsoft DNS server installation

The simplest tests are done using the "ping." Make sure you no longer have a "hosts" file (you may wish to rename it to something else, like "hosts.old"). In a DOS command prompt window, try issuing the following commands (substitute your nodenames and domain name, as appropriate):

- ping 123.45.67.1

  This should ping the IP address without using DNS:

  Pinging 123.45.67.1 with 32 bytes of data

- ping lehnts

  This should expand the name to the full domain name:

  Pinging lehnts.bronwen.com [123.45.67.1] with 32 bytes of data

- ping lehnts.bronwen.com

  This should produce the same results as lehnts

- ping msmail

  This should expand the name to the full domain name:

  Pinging msmail.bronwen.com [123.45.67.2] with 32 bytes of data

▶ ping www.bronwen.com

This should map www.bronwen.com onto lehnts.bronwen.com:

Pinging lehnts.bronwen.com [123.45.67.1] with 32 bytes of data

▶ ping -a 123.45.67.1

This should do a reverse lookup on the IP address:

Pinging lehnts.bronwen.com [123.45.67.1] with 32 bytes of data

If all the preceding ping tests work, try using nslookup to test the MX record:

```
C:\nslookup
```

you should see something like the following:

```
Default Server: lehnts.bronwen.com
Address: 123.45.67.1
```

Tell nslookup to return all records:

```
set type=all
```

Query your domain name:

```
bronwen.com
```

You should get back something like the following:

```
Server:   lehnts.bronwen.com
Address:  123.45.67.1
          primary name server = lehnts.bronwen.com
          responsible mail addr = LHughes.bronwen.com
          serial   = 3
          refresh  = 3600 (1 hour)
          retry    = 600 (10 mins)
          expire   = 86400 (1 day)
          default TTL = 3600 (1 hour)
bronwen.com MX preference = 100, mail exchanger = lehnts.bronwen.com
lehnts.bronwen.com      internet address = 123.45.67.1
lehnts.bronwen.com      internet address = 123.45.67.1
```

The final test is to do the nslookup test from another viewpoint in the DNS hierarchy (e.g., from the viewpoint of one of the AT&T Worldnet DNS servers). To do that, tell nslookup to connect to a DNS server other than the default (local) one. With the nslookup that comes with NT 4.0

DNS, you specify two parameters, the first being a minus sign and the second being the IP address of the remote DNS server (e.g., 204.127.129.1), as follows (be sure there is a space before and after the minus sign):

```
nslookup - 204.127.129.1
```

Do the same commands described previously. If you get the same results, then not only have you configured your DNS database files correctly, you also have linked your DNS into the worldwide DNS system correctly. There are two pointers that must be set correctly, one uplink (from your DNS server to its parent) and one downlink (from the parent DNS to your DNS server). To set the uplink pointer, right click the DNS server name (in DNS Manager) and select Properties. On the Forwarders tab, select "use forwarders" and enter the IP address of the parent DNS server (your ISP will tell you which address to use). The downlink pointer must be entered as an NS record in the parent's DNS database files (pointing to your DNS server). When you sign up for service from your ISP, you likely will get a block of valid IP addresses (perhaps eight), one of which is for you to assign to your DNS server (i.e., the ISP already has created an NS record for that address). If the ISP does not tell you which IP address to assign to your DNS server, tell the provider what address you assigned to your DNS server so they can add the necessary NS record. The technical folks at any reasonable ISP should be able to help you with those details.

Other DNS servers

Two other popular DNS servers for Windows NT are ports of the UNIX BIND program. One, from Metainfo (see www.metainfo.com), is a supported, commercial program with considerable original work to simplify the configuration and management via GUI administration programs. The other is a fairly vanilla but free implementation from the Internet Mail Consortium, available from Software.com (see www.software.com, "freeware BIND"). Other servers doubtless also are underway or even available for NT.

For further study

For further information on DNS and BIND, see the following:

▶ Microsoft Technet CD-ROM (e.g., November 1996), *NT Server 4.0 Networking Guide*, Chapter 9, "Managing MS DNS Servers."

▶ Albitz & Liu, O'Reilly & Associates, *DNS and BIND*, 2nd edition, ISBN 1-056592-236-0.

▶ Washburn & Evans, *TCP/IP—Running a Successful Network*, Reading, MA: Addison-Wesley, ISBN 0-201-62765-5. In particular, see Chapter 14, "Working with names."

▶ Website www.internic.net.

There are quite a few IETF RFCs concerning DNS, but the most important ones are 1876, 1591, 1536, 1183, 1101, 1035, 1034, and 1033. Following is the complete list. (All the RFCs are on the accompanying CD-ROM.)

2065	PS	D. Eastlake, C. Kaufman, "Domain Name System Security Extensions," 01/03/1997. (Pages=41) (Format=.txt) (Updates RFC1034) (Obsoletes RFC1035)
2052	E,	A. Gulbrandsen, P. Vixie, "A DNS RR for Specifying the Location of Services (DNS SRV)," 10/31/1996. (Pages=10) (Format=.txt)
2010	II	B. Manning, P. Vixie, "Operational Criteria for Root Name Servers," 10/14/1996. (Pages=7) (Format=.txt)
1996	PS	P. Vixie, "A Mechanism for Prompt Notification of Zone Changes (DNS NOTIFY)," 08/28/1996. (Pages=7) (Format=.txt) (Updates RFC1035)
1995	PS	M. Ohta, "Incremental Zone Transfer in DNS," 08/28/1996. (Pages=8) (Format=.txt) (Updates RFC1035)
1912	I	D. Barr, "Common DNS Operational and Configuration Errors," 02/28/1996. (Pages=16) (Format=.txt)
1886	PS	S. Thomson, C. Huitema, "DNS Extensions to Support IP, version 6," 01/04/1996. (Pages=5) (Format=.txt)
1877	I	S. Cobb, "PPP Internet Protocol Control Protocol Extensions for Name Server Addresses," 12/26/1995. (Pages=6) (Format=.txt)
1876	E	C. Davis, P. Vixie, T. Goodwin, I. Dickinson, "A Means for Expressing Location Information in the Domain Name System," 01/15/1996. (Pages=18) (Format=.txt) (Updates RFC1034)
1816	I	F. Networking Council (FNC), "U.S. Government Internet Domain Names," 08/03/1995. (Pages=8) (Format=.txt) (Obsoletes RFC1811)
1811	I	F. Networking Council, "U.S. Government Internet Domain Names," 06/21/1995. (Pages=3) (Format=.txt) (Obsoleted by RFC1816)

1794	I	T. Brisco, "DNS Support for Load Balancing," 04/20/1995. (Pages=7) (Format=.txt)
1713	I	A. Romao, "Tools for DNS Debugging," 11/03/1994. (Pages=13) (Format=.txt) (FYI 27)
1712	E	C. Farrell, M. Schulze, S. Pleitner, D. Baldoni, "DNS Encoding of Geographical Location," 11/01/1994. (Pages=7) (Format=.txt)
1706	I	B. Manning, R. Colella, "DNS NSAP Resource Records," 10/26/1994. (Pages=10) (Format=.txt) (Obsoletes RFC1637)
1664	E	C. Allocchio, A. Bonito, B. Cole, S. Giordano, R. Hagens, "Using the Internet DNS to Distribute RFC1327 Mail Address Mapping Tables," 08/11/1994. (Pages=23) (Format=.txt)
1637	E	B. Manning, R. Colella, "DNS NSAP Resource Records," 06/09/1994. (Pages=11) (Format=.txt) (Obsoletes RFC1348) (Obsoleted by RFC1706)
1611	PS	R. Austein, J. Saperia, "DNS Server MIB Extensions," 05/17/1994. (Pages=32) (Format=.txt)
1591	I	J. Postel, "Domain Name System Structure and Delegation," 03/03/1994. (Pages=7) (Format=.txt)
1537	I	P. Beertema, "Common DNS Data File Configuration Error," 10/06/1993. (Pages=9) (Format=.txt)
1536	I	A. Kumar, J. Postel, C. Neuman, P. Danzig, S. Miller, "Common DNS Implementation Errors and Suggested Fixes," 10/06/1993. (Pages=12) (Format=.txt)
1535	I	E. Gavron, "A Security Problem and Proposed Correction With Widely Deployed DNS Software," 10/06/1993. (Pages=5) (Format=.txt)
1480	I	A. Cooper, J. Postel, "The US Domain," 06/28/1993. (Pages=47) (Format=.txt) (Obsoletes RFC1386)
1464	E	R. Rosenbaum, "Using the Domain Name System To Store Arbitrary String Attributes," 05/27/1993. (Pages=4) (Format=.txt)
1401	I	Internet Architecture Board, L. Chapin, "Correspondence Between the IAB and DISA on the Use of DNS Throughout the Internet," 01/13/1993. (Pages=8) (Format=.txt)
1386	I	A. Cooper, J. Postel, "The US Domain," 12/28/1992. (Pages=31) (Format=.txt) (Obsoleted by RFC1480)
1383	I	C. Huitema, "An Experiment in DNS Based IP Routing," 12/28/1992. (Pages=14) (Format=.txt)
1348	E	B. Manning, "DNS NSAP RRs," 07/01/1992. (Pages=4) (Format=.txt) (Updates RFC1035) (Obsoleted by RFC1637)
1183	E	R. Ullman, P. Mockapetris, L. Mamakos, C. Everhart, "New DNS RR Definitions," 10/08/1990. (Pages=11) (Format=.txt)

1123	S	R. Braden, "Requirements for Internet Hosts—Application and Support," 10/01/1989. (Pages=98) (Format=.txt) (STD 3)
1101		P. Mockapetris, "DNS Encoding of Network Names and Other Types," 04/01/1989. (Pages=14) (Format=.txt) (Updates RFC1034)
1035	S	P. Mockapetris, "Domain Names—Implementation and Specification," 11/01/1987. (Pages=55) (Format=.txt) (Obsoletes RFC0973) (STD 13) (Updated by RFC1348)
1034	S	P. Mockapetris, "Domain Names—Concepts and Facilities," 11/01/1987. (Pages=55) (Format=.txt) (Obsoletes RFC0973) (STD 13) (Updated by RFC1876, RFC1101)
1033		M. Lottor, "Domain Administrators Operations Guide," 11/01/1987. (Pages=22) (Format=.txt)
1032		M. Stahl, "Domain Administrators Guide," 11/01/1987. (Pages=14) (Format=.txt)
0974	S	C. Partridge, "Mail Routing and the Domain System," 01/01/1986. (Pages=7) (Format=.txt) (STD 14)
0973		P. Mockapetris, "Domain System Changes and Observations," 01/01/1986. (Pages=10) (Format=.txt) (Updates RFC0882) (Obsoleted by RFC1034, RFC1035)
0921		J. Postel, "Domain Name System Implementation Schedule, Revised," 10/01/1984. (Pages=13) (Format=.txt) (Updates RFC0897)
0920		J. Postel, J. Reynolds, "Domain Requirements," 10/01/1984. (Pages=14) (Format=.txt)
0897		J. Postel, "Domain Name System Implementation Schedule," 02/01/1984. (Pages=8) (Format=.txt) (Updates RFC0881) (Updated by RFC0921)
0883		P. Mockapetris, "Domain Names: Implementation Specification," 11/01/1983. (Pages=73) (Format=.txt)
0882		P. Mockapetris, "Domain Names: Concepts and Facilities," 11/01/1983. (Pages=31) (Format=.txt) (Updated by RFC0973)
0881		J. Postel, "Domain Names Plan and Schedule," 11/01/1983. (Pages=10) (Format=.txt) (Updated by RFC0897)
0819		Z. Su, J. Postel, "Domain Daming Convention for Internet User Applications," 08/01/1982. (Pages=18) (Format=.txt)
0799		D. Mills, "Internet Name Domains," 09/01/1981. (Pages=6) (Format=.txt)

Internet directory services

The introduction to address books in Chapter 3 should have convinced you of the need to manage and make available to users of mail clients a list of names and addresses at a higher level than just a personal address book. This can be done at the mail server or LAN level. For example, when a new mail account is added to a given mail server, it also is added (automatically) into an address book that all users of that mail server can access. It also can be done at the enterprise level, where some directory synchronization scheme propagates (and maintains up-to-date) copies of each server-level address book to all other servers. In that case, mail clients should be able to browse or search a hierarchical list of everyone in the entire enterprise. The ultimate goal (for e-mail at least) would be a globally distributed database that would contain all e-mail users anywhere (at least those reachable via the Internet), that could be browsed or searched by any user in the world.

At least one technology, X.500, is actually capable of reaching that lofty goal.

Directory services, as applicable to the Internet, is a rapidly changing area. LDAP is emerging as the protocol of choice and is being supported by many server and client vendors. The issue of how to do lookups for people outside your local organization is still very much in dispute. It currently appears that LDAP will be extended to allow any number of servers to work together as a distributed directory service, much as DNS does for the much simpler nodename-to-IP address resolution problem. There is still considerable resistance in the Internet world to implementing native X.500 servers to perform that function, due in part to its overwhelming complexity and in part to its dependence on an underlying OSI protocol stack.

Actually, once you have a mechanism that is general and powerful enough to serve as a global address book, it seems you also could use it to keep track of a number of other useful things. There is already a perfectly good scheme for keeping track of Internet domain and node names and their associated IP addresses, called DNS. In addition to a user's full name and e-mail address, the following items would be excellent items to keep track of in such a directory.

E-mail–related items would be:

▶ Preferred e-mail attachment type (uuencode, MIME, etc.);

▶ Extended e-mail capabilities (rich text, audio, images, etc.);

▶ Public key certificates;

▶ CRLs;

▶ Mailing lists.

Other computer-related items would be:

▶ Fax telephone number;

▶ Shared resource location;

▶ Shared files (via FTP, NFS (Network File System), Microsoft SMB or Novell);

▶ Shared printers (via UNIX lpd (line printer daemon), Microsoft SMB (Server Message Block), Novell);

▶ World Wide Web pages (URLs);

▶ Gopher/archive documents;

▶ Shared data bases;

▶ Other shared computing services;

▶ Distributed object components;

▶ Java applets.

You could also keep track of noncomputing-related items (although over time, even some of these could be used by computer-based applications).

Personal information could include the following items:

▶ Home mailing address;

▶ Home telephone number;

▶ Inventory of personal property;

▶ List of interests, hobbies.

Job related information:

▶ Company name;

▶ Company department;

▶ Mail code;

▶ Work voice telephone number and extension;

▶ Job title;

▶ Areas of expertise;

▶ Inventory of equipment.

History

Directory services is one area in which proprietary e-mail systems have long held an advantage over Internet e-mail, due to the more restricted user community that they are designed to support. Most of these systems restrict their scope to a single LAN or at most one organization (enterprise) with a number of LANs connected in a private WAN. A typical system supports the following three levels of address books.

- ▶ Personal address book. Each user of the e-mail system has a local address book managed and used only by their UA (mail client program). Typically, each user enters the entries into his or her personal address book manually. Some UAs have the ability to extract the e-mail addresses from incoming and outgoing messages and automatically add them to a personal address book. No one but the user can view, manage, or use a personal address book.

- ▶ Workgroup address book. This address book is managed by the workgroup e-mail administrator. Typically, users are added to it automatically when e-mail accounts are created for them. Many e-mail systems support one or more other shared workgroup address books that are created manually and managed by the workgroup e-mail administrator (e.g., list of common addresses available via an Internet gateway).

- ▶ Enterprise address book. This is really the set of all workgroup address books that exist on the LANs in a WAN, with an automated replication scheme called directory synchronization, which ensures that copies of (and changes to) all address books are distributed automatically to all workgroups in the enterprise. When a workgroup e-mail administrator adds a new e-mail account (or a manually created entry in one of the other shared address books), the new address book entry is soon propagated to all e-mail workgroups in the enterprise.

When it comes to the Internet, the replication scheme used in proprietary e-mail systems does not scale well to the global level. Imagine if participating in the global directory scheme meant you had to have a computer with enough disk storage to contain (and a CPU fast enough to process) the addresses of all e-mail users in the world (either directly on the Internet or accessible via gateways connected to the Internet). Even today that number is in the tens of millions. Soon it will be hundreds of millions. That is a good size database even for mainframes. What is required is a real distributed database, in which pieces of the database itself are distributed over any number of servers. In such a scheme, the UA issues a query to the nearest server agent. If that server agent happens to have the answer (which it would if the query were for someone in the same workgroup), it immediately replies with it. Otherwise, the server finds the answer by asking other directory server agents around the network. DNS is an example of such a distributed database, but that system is difficult or impossible to extend beyond its current task of mapping computer names to the corresponding IP address. DNS can resolve queries like "what is the nodename on which the preferred SMTP mail server for domain x.y.z exists?" That is done with MX records. In theory, a new record type could be added into DNS, a directory exchange (DX) record, that could the preferred directory server agent (DSA). That could help speed the process of obtaining directory information if you happen to know what Internet domain the user is in (in general, though, that is not a valid assumption).

Several directory service schemes have been proposed for the Internet and even implemented on a limited scale. Examples of such services for the Internet include Finger, Ph, Whois, and more recently Whois++. Most of those schemes are little more than workgroup address book servers, with no real way to allow users from other workgroups to use them to find you. Whois++ is the scheme most likely to be able to support a global address book. Unfortunately, none of these schemes has been widely adopted. The most likely scheme to be adopted is X.500 with clients accessing it via the LDAP protocol.

Finger

The Finger protocol originally was defined in RFC 742. It evolved through RFCs 1194 and 1196 and is currently defined by RFC 1288. The author of the RFC is David Paul Zimmerman (at Rutgers). It is a client/server design, using TCP over port 79. Most UNIX systems include both Finger servers and Finger clients. The protocol is so simple that the implementation of a standalone client or even a server is an appropriate student exercise for anyone trying to learn socket-style network programming. The tricky part is trying to provide the service without opening a major breach in your security. The service achieved some notoriety as one of the principal points of entry by the Morris "worm," that infected a significant percentage of the UNIX computers connected to the Internet.

A typical Finger query proceeds as follows:

1. Client requests a TCP connection to a Finger server at port 79.

2. Server accepts connection from the client (acceptance could be based on the domain or IP address of the client).

3. Client sends a one-line query (ASCII characters followed by CR,LF).

4. Server returns an answer, then closes its end of the connection (kind or level of information returned also could be based on who is making the query).

5. Client receives answer, senses the server has closed its end of the connection, then closes its end.

On UNIX, the information returned typically is account information from /etc/passwd, which contains the UNIX login name and full name, among other things. Typically, on UNIX, the e-mail name is the same as the login name. It is possible for much more information to be returned, such as physical address, telephone number, and so on. However, there is no standard for the format in which that information is returned or even what is returned, because the response is intended to be displayed to be read by humans, not by programs. When supported in contemporary software, for example, Eudora Pro, you typically would cut and paste an e-mail address from the response to set a To: field in a compose-message form.

A request consisting of the string /W (followed by CR,LF) returns a complete list of users of that system. It is acceptable for a server to refuse such a request for security reasons (by responding with something like "Finger online user list denied").

A request consisting of /W followed by a valid user's name (which can be obtained from the above request if supported; otherwise, it must be known) returns detailed information about that user. Some servers may accept some ambiguity in the specified name. For example, if the specified string matches any part of a name, include it in the reply. Hence, /W Hugh might return information about Hugh Smith and Howard Hughes.

A simple chaining mechanism allows you to obtain information indirectly from Finger servers on other machines, by including an @hostname in the request (where "hostname" is the name of some other computer). So even though you may be connecting to node flintstone, you can obtain information about user Barney on node Rubble with the request /W Barney@Rubble. Again, for security reasons, the mechanism may be disabled by the system administrator, which would result in such a request getting the response "Finger forwarding service denied."

As you can see, the Finger service is primitive and not particularly suitable as a real directory service for e-mail. Ideally, the information returned and the format of it should be specified precisely enough so that a UA could use it directly to fill in a To: or Cc: field from a query or browse. A good service also should allow access to that information without creating serious security breaches.

Windows NT includes a standalone client (%SYSTEMROOT%\system32\Finger.exe) but no Finger server. Several public domain Finger servers are available for Windows NT, and some e-mail servers (e.g., Software.com's Post.Office) include a Finger server that responds with e-mail account information. The Eudora Pro Internet Mail UA includes a Finger client, in addition to its other address book facilities.

Ph

Ph (short for "phone book") has not yet made it all the way to being an official RFC. As of the writing of this book, it is still a draft proposed standard (draft-ietf-ids-ph-01) and, therefore, a work in progress, subject to change without notice. The authors are Roland Hedberg, Steve Dorner, and Paul Pomea (the last two being from Qualcomm, the creators of the Eudora Internet Mail UA).

Ph originally was developed at the University of Illinois at Urbana-Champaign. It is intended to be a nameserver for people and objects. It was designed to maintain (and make available to clients) a relatively small amount of arbitrary information on a relatively large number of people or objects.

Even though it is designed for use on the Internet, the draft specification specifically defines Ph as local, which means that "no server is supposed to be able to refer a client to another server which might hold the

wanted information" (automatically). It also has no provision for retrieving information from any other server on behalf of a client. However, a
given Ph server may contain a list of other Ph servers that a client might
try manually to see if they contain the desired information.

Like Finger, Ph is a client/server design, using TCP (via port 105).
Unlike Finger, multiple queries can be made during a given connection.
One interesting aspect of the protocol is the support for a very weak
encryption scheme (based on a "three rotor Enigma engine") to achieve a
casual level of security.

The command syntax is significantly more complex than Finger, and
the responses are well structured enough that it can be used by a UA to
automatically fill in To: and Cc: fields from a given search. Again, the
Eudora Pro UA includes support for a Ph server.

Unfortunately, Ph still falls far short of being a generalized directory
service and is specifically not a distributed scheme. It also is not widely
used. It is adequate primarily as a simple workgroup-level address book
server, with no provision for scaling beyond that level.

NICNAME/Whois

Another early client/server directory service for the Internet was originally defined in RFC 812 and updated in RFC 954. NICNAME/Whois
originally was designed to provide online information (full name, mailing
address, telephone number, and e-mail address) for "each individual with
a directory on an ARPANET or MILNET host, who is capable of passing
traffic across the DoD Internet." When that was written, the number of
such individual was manageable for a single server. All users of the Internet were asked to register their information, which was maintained on a
single-server Network Information Center (NIC, not to be confused with
a network interface card) at SRI International. Today there are just too
many Internet users to keep track of with Whois, let alone on a single
server. The date of the most recent specification is October 1985.

Whois is yet another TCP client/server design (port 43 this time). Like
Finger, the client connects to a server and sends a one-line query. The
server sends a response to that one query and then drops the connection.
Again, the output is designed for humans, not programs, to read.

Whois++

Whois++ is an evolution of the older Whois service. Originally defined by Deutsch et al. in RFC 1835 (Aug. 1995), Whois++ recently was updated in RFC 1913 (Feb. 1996), by Chris Weider (Bunyip Information Systems), Jim Fullton (MCNC Center for Communications), and Simon Spero (EIT). A companion RFC, 1914, is "How to Interact with a Whois++ Mesh." RFC 1914 is from Patrik Faltstrom and Chris Weider (both of Bunyip Information Systems) and Rickard Schoultz (KTHNOC, Sweden).

Whois++ is an elaborate client/server distributed system that is intended to be a generalized directory service for the Internet. It addresses the issue of maintaining multiple views into the same directory information (e.g., by geography and by industry), on which X.500 is still a bit weak. For further details on Whois++ and Bunyip's implementation of it, called "Digger" (only for UNIX at this time), see www.bunyip.com.

Unfortunately, Whois++ has one major problem: It is not X.500. That is what most people are working on integrating with the Internet, and setting up a worldwide infrastructure for. Already, many companies and organizations in both the United States and Europe have entered hundreds of thousands (by now perhaps millions) of names into sample directory information bases. There also are a larger number of (more mature, compared to Whois++) implementations of X.500. X.500 also already addresses the security issue in a comprehensive way.

Present-day directory services

X.500

The primary competition to the Internet standards is the networking and messaging systems devised by the ISO. Corresponding to the lower level Internet protocols (e.g., IP and TCP), the ISO has its own protocols, called the OSI protocol suite. The ISO e-mail component (which the ISO calls a message handling system) is defined by the standard X.400 (actually, several standards in the X.400 to X.499 range). X.500 is an elaborate, sophisticated system. Consequently, it is difficult to implement and requires enormous resources to run. It also requires the underlying OSI network protocols. Fortunately, Internet RFC 1006 defines a way to tunnel OSI protocols over IP, so you can think of it as a "short stack" that, to anything running on top of it, looks like the OSI protocols but that does its work by calling routines in an existing Internet protocol stack. An RFC 1006 implementation is also known as a TP0 protocol. RFC 1006 has been implemented for Windows NT.

Unlike the Internet e-mail system, the ISO recognized the need for a directory system from the start and created the X.500 standard (again, really a set of standards in the range X.500 to X.599), which is elaborate and complex. Many things about X.400 were designed to make use of an X.500 directory service, and many things about X.500 were designed to be used by an X.400 message handling system. However, X.500 is really a generalized distributed database directory service that can function well (and be quite useful) apart from X.400. In particular, it definitely can function as a global address book for Internet e-mail. Fortunately, RFC 1006 allows such a system to run on a computer that otherwise has only Internet network protocols (TCP/IP) installed.

ISO creates working groups that hammer out standards over one or more 4-year cycles. The X.500 standard initially was created from 1984 to 1988. Unfortunately, much of the real work (especially in the areas of interserver references and security) was not complete at the end of the first 4-year cycle, so a second cycle was required, from 1988 to 1992. It actually ran a bit late, so that second version is referred to as the 1993 version. A number of companies implemented what had been defined during the first cycle, X.500-1988 compliant products. Many of those have been or are in the process of being upgraded to the X.500-1993 standard. Even further refinements are soon to be released from the third 4-year cycle as X.500-1997.

A number of existing directory service products are based on at least the ideas, if not the actual protocols, of X.500-1988. For example, Novell NDS (the big addition in v4.0 of NetWare) is very similar to X.500 (such systems are called "X.500-like") but will not interoperate with it (in this case, it runs over the Novell network protocols SPX/IPX). At the time this book was being written, Microsoft was about to release early versions of some of the parts of Windows NT 5.0, including Active Directory, which is another X.500-like directory service. It remains to be seen if that directory service will interoperate with X.500 either directly or via gateways.

Architecturally, X.500 is a client/server system similar to Internet e-mail. There is an active server component similar to an Internet MTA, a DSA, and an active client component similar to an Internet UA, called a directory UA (DUA). There is also a passive database component that is called either the directory information base (DIB) or the directory information tree (DIT), depending on whether you are talking about the physical or logical organization of the data.

Directory server agent

The DSA, the server component of X.500, runs on a powerful central computer, which must be available at all times, and performs the following functions:

▶ Accepts and processes directory queries from DUAs;

▶ Manages a piece of the distributed DIB;

▶ Queries other DSAs for information not in its local piece of the DIB;

▶ Answers queries from other DSAs concerning information in its piece of the DIB.

Directory user agent

The other active component of X.500 is the DUA, the client part. While it definitely is possible (and would be useful) to create a standalone utility program that has only client functionality, such functionality typically is implemented as a part of some other more complex application, such as an e-mail UA. Of course, an end-user application or even a server component that needs to submit queries to a DSA could implement that functionality. For example, an Internet MTA might need to make directory queries, and so it could contain DUA functionality. Whether we are talking about a standalone program or one component of a more complex application or service, the DUA performs the following functions:

▶ Assembles and submits directory queries to a DSA;

▶ Does something with the returned directory information (e.g., displays it or uses it to address an outgoing e-mail message).

Directory Information Base

The DIB is the sum total of all the pieces of directory information managed by all the DSAs in the network. Part of the magic of a distributed database is that clients can treat the information as if it all were physically present on the computer running the DSA to which it makes queries.

Directory Information Tree

The DIT is actually the same as the DIB, but it is logically organized, that is, in a hierarchical manner. There is an overall root of the entire DIT, below which are subtrees for each country, for example, C = US, C = Japan, C = United Nations.

Below the top level is the organization name, for example, O = Acme Corp., O = Government.

Even further down is the organizational unit name, for example, OU = Research & Development, OU = Defense.

There can be many layers of organizational units, but finally at the bottom is the common name, for example, CN = John Q. Public.

X.400 mail addresses comprise the country name, the organization name, the organizational unit name, and the common name, for example,

C = US, O = Government, OU = White House, CN = William J. Clinton

There are ways to hide that complexity from the Internet e-mail users of an X.500 directory service via hierarchical browsers or mappings between X.400 and Internet style e-mail addresses (as covered in several RFCs).

The communication between DSAs is done using one or more of the following protocols:

▶ Directory Server Protocol (DSP), specified in X.500-1988 and X.500-1993;

▶ Directory Information Shadowing Protocol (DISP), newly specified in X.500-1993;

▶ Directory Operational Binding Management Protocol (DOP), newly specified in X.500-1993.

The communication between a DUA and its nearest DSA is done using DAP (Directory Access Protocol), specified in both X.500-1988 and X.500-1993.

A number of projects are well underway (both in Europe and the United States) to create X.500-based directories, in both government and large corporations. Some of these directories already have hundreds of thousands of names in them.

LDAP

There are two primary disadvantages to using the real X.500 DAP from a typical client application (e.g., to provide global address book capability to an e-mail UA):

▶ The real DAP requires an OSI protocol stack (or at least an RFC 1006 "short stack" on top of IP); it cannot be used on a platform that supports only TCP/IP.

▶ The full DAP as defined in X.500 is a complex protocol, and much of the flexibility that complexity provides is not really necessary for typical uses.

To address both of those issues, a new protocol, LDAP (Lightweight Directory Access Protocol) was created (in RFC 1777). LDAP is still evolving (version 3, which incorporates support for features coming in X.500-1997, is in draft form as of this writing). It has many of the same capabilities of DAP, with the following two exceptions:

▶ No OSI protocol stack is required; it will work fine on TCP over IP or on UDP over IP.

▶ Some of the more complex aspects of DAP have been eliminated, which makes it much simpler and faster to implement, while still addressing the bulk of situations for which it might be needed.

Much of the work on LDAP was done at the University of Michigan. Netscape recently hired some of the principal people involved in that work. As this book is being written, they are implementing an address book server based on LDAP. Note that any product based on the University of Michigan code base is not a full X.500 server that supports LDAP and will require some kind of LDAP-to-X.500 gateway product to interoperate with the coming worldwide X.500 infrastructure. Also, the University of Michigan code was implemented for only one UNIX platform, so companies using it as a starting point face a long and difficult porting process, especially for Windows NT. Finally, that code base is what developers refer to as "college code" (i.e., designed and written by clever, but not necessarily experienced or professional, software developers while still in college).

ISODE X.500 DSA with LDAP

ISODE started life as a non-profit consortium (a joint venture sponsored by a number of companies or institutions) whose task was to create reference implementations of various ISO standards. They quickly decided to concentrate on the messaging (X.400) and directory services (X.500) standards. They recently (December 1996) became a for-profit company and changed their name from the ISODE Consortium to ISODE Limited. However, they still do not plan to market their implementations directly. Rather, they license their source code to other companies, which customize it in any way they see fit (e.g., tie the X.500 server into an existing Internet Mail server) and "private label" it (i.e., sell it as their own products under their own trademarked product names).

This server is a completely general X.500 DSA that also supports LDAP. ISODE already has ported it to several UNIX platforms and to Intel Windows NT. Compared to the University of Michigan code, the ISODE code base includes full X.500 functionality and was created by experienced professional developers. ISODE is one of the primary drivers of the LDAP specification (Steve Kille is both CEO of ISODE and one of the three authors of the LDAP RFCs). There are some powerful advantages of the ISODE approach compared with the University of Michigan approach:

▶ You get the full functionality of X.500 (including X.509 for management of public key certificates) to use as a fully generalized directory service.

▶ You get interoperation with the coming worldwide X.500 infrastructure for free, because of the support for the DSP, DOP, and DISP protocols.

▶ The X.500 design is known to be able to scale to global levels (hundreds of millions of names), which still is unproved for an LDAP-only design

Microsoft Exchange Server v5.x

Having recently gotten religion, as it were, about the Internet, Microsoft has been adding Internet support for most of its products. An "Internet Protocol only" mail server is part of Microsoft's Normandy project (released as MCIS), but that is being kept to a tightly targeted distribution channel (ISPs that need significantly more performance than Exchange Server can provide but do not need all its additional features). They are trying to make a silk purse out of the Exchange Server v4.0 sow's ear. Among other things, the new version has an alternative POP3 front end and an LDAP server. Version 5.5 added IMAP support.

In Exchange Server 4.0, the SMTP gateway (the conduit between Exchange and Internet e-mail) was called the Internet Mail Connector and was an extra-cost option. In Exchange Server 5.0, that component was renamed Internet Mail Services and is included in all versions of the product. One of the major changes was the addition of a "wizard" that makes installation of the gateway about an order of magnitude easier than it was in Exchange Server 4.0. Note that even in 4.0, once the SMTP gateway was installed, like any Internet Mail MTA, it could accept connections from an Internet Mail UA to support transfer of messages into the message store (via SMTP). What was missing to allow an Internet-only UA (e.g., SMTP/POP3 but no MAPI) as a client to Exchange Server was support for the POP3 protocol (to support retrieval of messages from the message store).

The POP3 protocol is implemented in Exchange Server 5.0 as an alternative access mechanism to MAPI. Once installed, mail can be retrieved from the message store by either a MAPI client or a POP3 client. Of course, a MAPI client will have access to a lot more functionality (e.g., MAPI address book) than will a pure SMTP/POP3 client. In Exchange 5.5, the IMAP4 access protocol is supported, yet another alternative to retrieve messages from the message store.

More to the point of this chapter, Exchange Server 5.0 also provides server-side support for the LDAP protocol. It is enabled in much the same way that the POP3 access protocol is enabled. Once enabled, it provides access to Exchange Server's address book information via LDAP in addition to via MAPI. Essentially, Exchange Server 5.0 with LDAP appears to an LDAP client as a general-purpose LDAP server with the names, e-mail addresses, and related information of all accounts on that

Exchange Server. An Internet Mail UA that supports SMTP, POP3, and LDAP will have much the same functionality as a MAPI client. With the support of IMAP in Exchange 5.5, an Internet Mail UA that supports SMTP, IMAP, and LDAP will have almost all the functionality of the MAPI client.

In a recent release of its new Internet Mail and News Reader, Microsoft included a third component (in addition to the SMTP/POP3 UA and NNTP reader), called the Windows Address Book. Versions of this product since 4.70.1160 have had the standalone version of the Windows Address Book. Essentially, it is a standalone LDAP client that comes with support to access the enormous address books of several national directory services (previously available only via manual retrieval with an HTTP browser), such as Bigfoot (see http://www.bigfoot.com) and Four-1-1 (see http://www.four11.com). Those services have hundreds of thousands even millions of white pages listings (a reference to telephone directory white pages as opposed to commercial listings in the yellow pages). The conventional way to search those listings is via a Web page-based search engine. That method is slow and cumbersome, however, and cannot be tied directly to your e-mail UA (found addresses must be cut and pasted into the To: field of your UA's send-mail form). The Windows Address Book can use LDAP to search the listings and allow a retrieved address to be added to your personal address book. It also is possible for the Windows Address Book application to invoke your default e-mail UA (which could be Eudora Pro instead of the associated Microsoft Mail and News Reader) and feed the retrieved name directly into the To: field of the UA's send-mail form.

These online directory services (Four-1-1 and Bigfoot) even now are trying to figure out how to make money off LDAP access to their directories. Conventional Web page access is paid for by advertisements on the Web pages, but there is no way for such advertising to be done via Windows Address Book. Possibly the directory services will come up with some way for individuals to subscribe to the service (e.g., a few dollars per month) or be billed a tiny amount for each retrieval (e.g., using e-cash).

If you configure your Exchange Server 5.0 to support LDAP, the addresses of all the accounts on that server can be made available to any LDAP client, such as the Windows Address Book, as another source of directory information.

Client support for LDAP

One of the main things driving LDAP and Internet directory services is the fact that most Internet Mail clients soon will include LDAP support, at least to provide a global address book capability. In some cases, they also plan to use that support to obtain public key certificates for S/MIME.

The Windows Address Book from Microsoft is really a standalone personal address book manager and LDAP client that can be used to add LDAP directory search capability to almost any e-mail UA. Most current e-mail UA vendors, however, are working hard to incorporate LDAP access capability directly into their UAs. As of this writing, most vendors had plans to do that, but none was available for beta testing other than the Microsoft Mail and News Reader client.

Most e-mail server vendors are trying to add LDAP server capability into their products. Many are using the combined X.500/LDAP approach, which will allow them to participate in the coming global directory services. Some prefer to avoid X.500 and likely will find themselves either isolated in the years to come or scrambling to catch up with those who based their systems on X.500 early on.

With the release of Internet Explorer 4.0, Microsoft has included an upgrade of the Mail and News Reader, called Outlook Express. Outlook Express integrates the functionality of the Windows Address Book (LDAP client), as well as supports the IMAP protocol (as an alternative to POP3). It also supports S/MIME for secure messaging. All those features, together with a fairly nice user interface (not to mention the fact that it is free), make Outlook Express a strong contender for the Internet e-mail client of choice. The full-blown Outlook client (which has nothing in common with Outlook Express other than part of its name) is many, many times larger and slower and still does not include support for IMAP or S/MIME (although both are scheduled for the near future). To be fair, Outlook does support MAPI, Schedule Plus calendaring, and "forms." None of those is compatible with or required for native Internet messaging.

Contents

Relevant Windows NT 5.0 innovations

The 5.0 release of Windows NT will contain several new key technologies important to Internet e-mail, including Active Directory and Distributed Security Services (and the related Internet Security Framework). Both technologies are important to Internet e-mail in the Windows NT environment.

Because of those new facilities, NT 5.0 will be an amazingly powerful platform for Internet messaging, and forward-looking companies will be quick to capitalize on it.

(*Caveat:* Windows NT 5.0 beta 1 has just been released as of the writing of this chapter. What is finally released, currently expected in mid- to late-1998, may or may not bear any resemblance to what is described here. Microsoft has been known to change its programs between first beta and final release.)

Active Directory

Active Directory is the most important of the new technologies slated to be released as part of Windows NT 5.0. It also is one of the areas in which Novell can claim to have had a real competitive advantage over NT Server with their NDS, starting with NetWare v4.0. Other vendors, such as Banyan Systems with their StreetTalk, also have been ahead of Microsoft in this area. To be frank, the directory services that Microsoft has included in NT 4.0 and before have been fairly weak in comparison.

Novell's NDS is based (*very* loosely) on the concepts of X.500 but is in no way interoperable with it. Banyan's StreetTalk is not even in the X.500 ballpark. StreetTalk is a highly proprietary scheme that until recently was not even available for TCP/IP; it is based on Banyan's proprietary Vines IP transport protocol. Both schemes are very interesting and somewhat useful technology, but their proprietary nature doomed them from the start.

Microsoft (prior to Windows NT 5.0) has contented itself with providing a hodgepodge of directory services distributed among a number of parts of NT, including:

- User account information in the Security Account Manager (SAM) database;

- NetBIOS domain and nodenames (flat, in a different namespace from IP domain, and nodenames, managed by broadcast or WINS);

- Universal Naming Convention (UNC) filenames or device names, a kind of early Uniform Resource Locator (URL) that allowed specification of files, printers, and so on, in terms of a network path. The first item is the host computer's NetBIOS name (e.g., \\SERVER01). The second item is the NetBIOS sharename on that computer (e.g., ACCTFILES). Any further items are a valid path (zero or more subdirectory names optionally followed by a filename) downward from the shared directory. A full UNC might look like this:

\\SERVER01\ACCTFILES\ar\jan97.dat

- DNS for Internet/intranet things like FTP, Internet Mail, and so on;

▶ Hierarchical filenames and directory names, with multiple roots (logical drive names, like C:);

▶ Two different Remote Procedure Call (RPC) locator services (Microsoft proprietary and Distributed Computing Environment (DCE) compatible);

▶ Configuration information in the registry;

▶ Internet URLs for Web objects, including Java applets.

Some of those mechanisms have schemes for replication in a distributed environment, system-call APIs for use from applications, and so on; some do not. Those that do tend to each have their own schemes for such things.

The ideal for Microsoft's Active Directory is to incorporate all objects that you might want to keep track of and make use of from applications under a single umbrella with schemes that are consistent for management, access, and replication.

For some time, Windows NT has supported multiple transport protocols (e.g., NetBEUI, TCP/IP, IPX/SPX) and multiple file/print redirection schemes (Microsoft client, NetWare client, etc.). These allow NT to act as the glue that integrates multiple environments into a seamless whole. Active Directory will provide similar ways to integrate existing directory schemes into the more inclusive Microsoft scheme so that existing NDS directory trees can exist as subtrees within an Active Directory scheme. Client software can be written to a generalized directory API that can map calls onto any embedded directory scheme. For example, a developer can write an application using Active Directory Services Interface (ADSI) rather than NDS system calls directly and make use of directory information anywhere in the Active Directory scheme, including within an NDS subtree. ADSI is to directory services what ODBC is to database access. It essentially provides a vendor- and product-independent way for a program to do key functionality, which means that the implementor is not locked into a single vendor's products.

Active Directory also will provide the ability to manage significantly more objects than any of the existing NT directory systems can handle (potentially millions of objects). What is very important is that Active Directory will eventually be able to interoperate with X.500 in most ways, via subsets of the real X.500 DAP, DSP, DISP, and/or LDAP.

A recent RFC (1823) defines a recommended C API for use with LDAP. That API has become widely used and is supported in Active Directory. That means application programs written under Windows NT 5.0 will be able to incorporate LDAP functionality without having to implement the LDAP protocol itself (which is fortunate, because LDAP is radically more difficult to work with than most Internet protocols, such as SMTP, POP3, etc.). An excellent book on the subject is by Timothy Howes and Mark Smith entitled *LDAP—Programming Directory-Enabled Applications with Lightweight Directory Access Protocol* (MacMillan, 1997, ISBN 1-57870-000-0). Directory-enabled applications developed using this API should work with any LDAP-compliant directory service, including Active Directory.

The database technology used in Active Directory originally was used in Exchange Server and is based on the Microsoft JET engine. It is capable of handling far more objects far more efficiently than the technology used in current NT directory services (primarily the Registry). When NT 5.0 is first deployed, there will be some duplication of directory services for current Exchange users, but a later release of Exchange Server will be able to use the new NT directory services instead of its own, which will eliminate the redundancy.

One of the objects you will be able to manage and make available with Active Directory is each user's X.509 version 3 public key certificate. That will make it possible for applications (such as secure e-mail clients) to obtain other users' public keys to send them encrypted messages or to verify their digitally signed messages. It will be up to the application to verify the authenticity of such a certificate by verifying the signature in the public key certificate. If the certificate was not signed by a directly trusted authority, the public key of that authority also can be obtained and *its* digital signature verified. This chaining can be repeated until some directly trusted authority is reached. The mechanism of course will be most useful once intersite Active Directory references are supported, via some future evolution of LDAP or perhaps a Lightweight Directory Server Protocol (LDSP).

Distributed Security Services

Distributed Security Services is a new feature that radically improves the security model used in Windows NT 4.0 and earlier. Among other things, it will add Kerberos v5.0 authentication to the existing NT/LAN Manager (NTLM)–style authentication (possibly replacing it in time). (For details on Kerberos, see Chapter 5.) The DSS also includes management of X.509 public key certificates as another object accessible via the Active Directory.

The Distributed Security Service actually keeps much of its information in the new Directory Services database and can make certain parts of it (e.g., public key certificates) available via the Directory Services protocols (e.g., LDAP).

One important aspect of the Distributed Security Services is the CryptoAPI v2.0, which will provide a simple-to-use (and possibly royalty-free) way to add encryption technology into products, including public and private key encryption, digital signatures, authentication, and so on. You can think of that as a standardized system subroutine library that can perform the various security functions described in Chapter 5 in a vendor-independent manner. Various vendors of cryptographic technology (e.g., RSA) either already have provided or soon will be providing the underlying encryption technology (a Cryptographic Service Provider, or CSP, module) that can provide any required and appropriate strength of technology. For example, the same program could be provided with appropriate-strength cryptography for domestic or export markets, depending on which underlying CSP is installed on the platform on which the product is running. Also, as future, stronger cryptographic technologies become available, in theory the underlying CSP could be upgraded and all applications based on the CryptoAPI would be able to make use of it.

Another important part of the Distributed Security Services is support for SSL, version 3.0. You can think of SSL as a thin layer that sits on top of the socket abstraction layer (the system calls used by applications to do network operations in most e-mail systems), which adds authentication and privacy. This technology is used to provide a secure connection between Web (HTTP) servers and browsers. However, it is really more general than that, and it is possible to make SSL versions of any socket-based software (of course, both client and server components must support SSL). For example, FTP, Telnet, and SMTP all can be implemented in SSL versions to provide secure communications.

Microsoft Internet Security Framework

Closely related to (and built on) the Distributed Security Services is the new Microsoft Internet Security Framework (MISF). Many pieces of MISF already have been released in various Microsoft products, such as Internet Explorer and IIS and are just the beginning of this important new infrastructure. With the release of NT 5.0, many other network-aware applications will be able to take advantage of these facilities.

A number of pieces go together to make up MISF:

▶ Authenticode is a digital signature technology useful in verifying the supplier of software or components downloaded over a network (important for Java and similar technologies). Basically, every downloadable component is sealed with a public key certificate that allows the client, before installing software, to know for certain where it originated and that it has not been tampered with. Only software from trustworthy sources can be accepted.

▶ Certificate services are a system server component that can generate signed public key certificates (using the X.509 v3 standard, similar to those generated by Verisign or shortly to be available from the USPS). A certificate service allows an organization to establish and manage its own "trust" domains, if desired. This component also manages the CRL necessary for clients to determine if a given certificate is known to have been compromised. Clients can access the certificate server via various IPC schemes, such as RPC, HTTP, or e-mail. This component is critical to SSL or Private Communications Technology (PCT), because both require public key certificates.

▶ The secure channel is the actual SSL v3.0, PCT v1.0 and secure electronic transaction (SET) technologies. SSL can enhance any network-aware application based on sockets with strong authentication and/or privacy (currently used mostly on HTTP and soon to be widely used with other protocols such as FTP, Telnet, SMTP, etc.). SET is the scheme Visa and MasterCard devised and agreed on for private, untamperable exchange of sensitive information, like credit card numbers, over the Internet. Again, SET currently is used mostly in Web products, but it could be used in a variety of areas, such as e-mail.

▶ CryptoAPI v2.0 is the underlying technology used to implement the preceding facilities (although it could be used directly by sophisticated applications). It is a system call library with support for X.509 v3 public key certificates, X.509 v2 CRLs, PKCS 10 certificate requests, PKCS 7 digital signatures and envelopes, and lower level digital signatures and encryption. It also includes the ability to validate Authenticode certificates.

▶ Microsoft Wallet is a scheme for secure storage and cross-platform transfer of the various certificates and credentials used in the preceding schemes. It is based on the Personal Information Exchange (PFX) protocol.

Relevance of new NT 5.0 technology to e-mail

Many of the new capabilities of NT 5.0, such as authentication and directory services, are integral to robust, secure e-mail systems. Prior to NT 5.0, those capabilities would have to have been implemented by the e-mail vendor or acquired from a third-party vendor as a prerequisite to installation (e.g., a Kerberos server). With the introduction of NT 5.0, those facilities will be present as a part of the base OS, resulting in three primary benefits:

▶ The e-mail vendors will not have to create the technology themselves or license it from another vendor; they can just use what is already in place.

▶ Due to that first benefit, it is more likely that e-mail vendors will incorporate this kind of facility in their products

▶ Because these facilities are available in the underlying OS, there will be less duplication of effort and fewer redundant systems (hence, redundant accounts, account management tools, passwords, etc.). The single logon will be a reality

In theory, an Internet e-mail UA running on Windows NT 5.0 should (if it is written correctly) be able to use Active Directory to obtain workgroup and enterprise address book information, retrieve public key certificates, and so on. It should be able to use Distributed Security Services to do strong authentication with "Kerberized" e-mail servers, including secure exchange of a session key for a protection mechanism (encryption of all messages in both directions, starting immediately after authentication). It also should be able to obtain any necessary public keys for sending private messages or validating incoming signed mail.

An e-mail server designed to run under Windows NT 5.0 should be able to make use of the Active Directory and Distributed Security Services to allow use of the NT login password as the e-mail password and/or for strong authentication with a Kerberized UA. It also should be able to use Active Directory to provide a workgroup (and possibly enterprise or even global) address book.

Both POP3 and IMAP4 have optional cryptographic authentication (and protection mechanisms) defined (see Chapters 16 and 17 for details). The most common technology used to implement those schemes is Kerberos v5, which is a part of Distributed Security Services. Kerberos is difficult and expensive for e-mail vendors to implement themselves (or to license from a third party). Also, today, developers of e-mail products tend to specialize in clients or servers. Supporting Kerberos, of course, involves a high degree of cooperation between vendors if it is not a part of the underlying OS. Since Kerberos will be available "free" in NT 5.0 (to all vendors), complete with built-in administration tools, you can expect to see e-mail vendors take advantage of it.

Prior to NT 5.0, SSL technology has been difficult or expensive to add to an existing application. For applications written to run under Windows NT 5.0, the SSL capabilities will be just another part of the standard APIs available for use by any developer, just as the basic WinSock API has been in earlier operating systems (like NT 4.0). If you go back far enough (e.g., 16-bit Windows 3.1), even the TCP/IP stack and WinSock API (not to mention the development tools) were expensive, extra-cost add-ons. Much of the explosion of network-aware applications began with the inclusion of those components as a standard part of the operating system, starting with Windows for Workgroups v3.11 and accelerating in Windows NT 3.1 and Windows 95. You can expect a similar explosion of secure network applications with the release of Windows NT 5.0.

An important technology, EDI (Electronic Document Interchange), requires many of the technologies described in this chapter to be effective. The basic idea is to replace existing paper transactions within and between companies with electronic versions. A whole range of standardized electronic forms (e.g., invoices, purchase requisitions, payments) have been designed in ways that allow programs to parse the information reliably. However, if such messages are sent over existing insecure e-mail systems, they are subject to possible interception, tampering, and other kinds of fraud. In fact, it actually is far easier for such fraud to be committed (and on a far larger scale) than with paper systems. With the addition of appropriate parts of the technologies discussed here, EDI actually can be far more secure than any paper system. Also, you likely will see an explosion of EDI systems with the introduction of Windows NT 5.0.

Contents

Firewalls and proxy servers

In the olden days (thousands of "Internet years" ago, say, the late 1980s and early 1990s), when most networks were isolated LANs, often with no dial-in access (i.e., they consisted only of computers directly connected to an Ethernet cable), security was not typically a serious problem. To intercept data or messages, let alone commit fraud, perpetrators had to obtain physical access to the inside of a building or plant and often inside information about the network layout (e.g., a login and a password on one of the computers). A sophisticated hacker might be able to connect a network "sniffer" (a line monitor typically used for debugging) in a wiring closet. Break-ins were rare.

The sophistication and the accessibility of networks have escalated dramatically in the last few (calendar) years, and most of them have been physically connected to the Internet or provided with dialup network access (e.g., PPP or RAS). Figuratively speaking, we have left the doors wide open for perpetrators virtually anywhere on earth. A class of information-age criminals, espionage agents, and joy-riding hackers has emerged, complete with electronic versions of the traditional burglar's toolkit for breaking into computers and networks (a toolkit readily available for free on the Internet itself, if you know where to look). As bad as it is, for the most part, the bad guys are still in their own learning curves. In the future, really mature threats similar to (and perhaps allied with) traditional organized crime (or "organized government"?) likely will be an unpleasant fact of online life. Highly intelligent people in economic disaster zones and information-age terrorists will find the Internet and the immense wealth and easy pickings of advanced economies irresistible targets. Those easy pickings are as near and as convenient as a home computer with a modem. Already, some countries like France, Israel, and Japan have retrained their cold war spies as economic espionage agents and equipped them with sophisticated tools and major funding.

For a rather frightening glimpse into the electronic reefs and shoals, I highly recommend a fascinating book by an old friend of mine (we once tried to peddle a secure communications system to the IRS), Winn Schwartau. The book is *Information Warfare* (Thunder's Mouth Press, ISBN 1-56025-132-8). For a taste of this book and what it contains, see Schwartau's Website at http://www.info-sec.com.

If you think Schwartau's book can be dismissed as a voice in the wilderness, check out another book called *Cryptography's Role In Securing the Information Society* (National Academy Press, 1996, ISBN 0-309-05475-3). The book (notice the acronym formed by the title) was compiled by the U.S. government's own Computer Science and Telecommunications Board and the National Research Council. Contributors include some of the most respected names in government (including the military), research, academia, and the computer industry.

By this point, if you are not scared into wanting to rip out your connection to the Internet and any dialup lines, either you are one of the fortunate few that already have secured their networks from such attacks or you simply do not understand the problem.

But, you say, you already depend on such connectivity. Perhaps you already are sending some 30% of your mail in electronic form (the current national average). Perhaps you are attracting new business and high-quality employees or maybe even selling goods or services on your

Website. Perhaps your executives (and even staff) regularly access your network via dialup lines from anywhere in the world and maybe even telecommute from home. How do you keep all those benefits without leaving your underbelly vulnerable?

It clearly would be more secure and would cut way down on burglaries and even employee theft to brick up all the doors and windows in your physical buildings, but few people would consider that an option. Likewise, we cannot retreat into the physically isolated LANs of the 1980s. Over the years, we have come up with ways to prevent the most egregious threats to our physical facilities while still allowing employees and customers to come and go at will. We have created card access systems, locks and keys, security guards, and other ways to protect our physical facilities while still allowing legitimate access. There are analogous ways to secure our electronic facilities while still allowing legitimate access. An entire subindustry in the computer field is emerging to create and implement solutions to the problem.

The tools of the new computer network subindustry have some obvious and familiar-sounding names, like *firewall*, and some odd new names, like *proxy server* and *screening router*. This chapter attempts to convey some of the concepts behind those tools. Entire books have been written on just this subject, and I strongly recommend that you (or someone in your organization) read them immediately. One of the classics (albeit with a strong UNIX bias) is *Internet Firewalls and Network Security*, by Karanjit Siyan (New Riders Publishing, 1995, ISBN 1-56205-437-6). The knowledge you will gain from this and other such books is basic survival gear in today's world. In the 1990s, we must protect ourselves from unsafe computing.

Jargon

First, we will define a few terms.

Packet   A packet is the fundamental (indivisible from the network point of view) parcel of information transmitted over a network. A given packet might consist of from 50 to 1,500 or so bytes of data, complete with some header information (e.g., address of originating node, address of destination node, number of data bytes). It typically also includes some trailer information for error detection and correction (e.g., CRC). With UDP, each datagram is a separate packet. With TCP, what appears to the application as a continuous stream of data bytes is broken up by the TCP layer into discrete packets for transmission and reassembled into a continuous stream on receipt. With either TCP or UDP, when the data is actually on the wire, it always is split up into packets. For the purposes of filtering and security, a packet is the smallest unit of information that can be accepted or rejected, based on the address information in the packet header.

Network layers   Network software is divided into logical layers that handle different aspects of the total problem of how to get data from one point to another (see Chapter 7 for a detailed explanation of the four layers of TCP/IP). Security schemes typically are implemented at one layer of the overall model and can accomplish different things based on the layer in which they function.

Router   A router is an electronic device used to couple a LAN to an internet. It is beyond the scope of this book to delve too deeply into the details of exactly how a router works. Briefly, it does on-the-fly mapping of network addresses and keeps packets that originate and terminate within the LAN (typically, by far the bulk of messages on the LAN) from getting out into the overall Internet, while allowing packets that should cross the boundary (originate inside and terminate outside or vice versa) to pass through the router. A standalone router basically is a dedicated, special-purpose computer that does nothing but that. It also is possible to install multiple NICs onto a general-purpose computer (e.g., a PC-compatible running Windows NT) and install a software router program that does the same thing. A simple software router is included in Windows NT, and a more sophisticated one is available as the new "Routing and Remote Access Server" (RRAS). Search the Microsoft Website on RRAS for more recent information.

Screening router   A screening router is a fancy router that can have rules programmed into it to help it determine which packets to pass through and which to not pass through (i.e., to screen packets). For example, you may want to allow packets from only specific other LANs to cross into your LAN. Alternatively, you may want to allow any packets intended for nodes on your network except for ones from specific (blacklisted) networks (or nodes) to pass. It also is possible to grant or deny access based on which port the packet originated or is intended. That kind of screening typically has a large impact on router performance (in terms of total packets it can process per second) and can be fooled by IP address "spoofing" (hackers putting bogus acceptable origination addresses into packets). In spite of its limitations, a screening router can be a good component of an overall security solution.

Proxy   A proxy is an application that does something on behalf of (or stands in for) another application and typically is implemented as a pass-through design, as opposed to a store-and-forward one. A proxy is another way to intercept network traffic as it crosses over the boundary into (or out of) your "protected" network. For example, a proxy SMTP server could look just like a real SMTP server to outside MTAs, but its real function would be to allow only acceptable SMTP traffic to be passed through to (or from) an internal (conventional store-and-forward) SMTP server. Think of it as an electronic Janus, with one face always pointing toward the outside world (via one NIC going directly to the router) and the other face pointing inward (via a different NIC connected to your protected LAN). Because this kind of filter runs at the application layer, it has access to higher level information (things relevant to application-level protocols like SMTP) than would an IP layer filter (e.g., a screening router).

Virtual private network   A virtual private network (VPN) is a scheme that uses encryption between sophisticated routers to achieve private links that "tunnel" through the (insecure) Internet. All routers involved must support the scheme, and cryptographic keying relationships must be set up in each participating router. A "real" private network is one in which you rent your own WAN links (e.g., X.25 or frame relay) from a common carrier like British Telecom, and you are the only one using those lines or with any access to them. That could be used to securely connect several branch offices to headquarters. It tends to be expensive (and a lot of trouble to set up and manage) compared to a link into the Internet via an ISP. The downside of linking your offices via the Internet is the fact that anyone can intercept your packets. By encrypting all packets that travel between your own offices, you have the equivalent of an actual private network, at much lower cost than a real one. Because fairly strong encryption typically is used (e.g., 56-bit DES or better), this technology can introduce some interesting (and possibly unsolvable) problems for links crossing national boundaries. That is because the governments of the countries involved cannot snoop on your communications any better than your competitors can, which tends to annoy them.

Firewall   A firewall is any scheme designed to shield an internal LAN from outside security threats while still allowing access to external services from the internal LAN. You can think of a firewall as an electronic semipermeable membrane (like those that make osmosis possible) or an information-age Maxwell's demon that selectively allows data packets, instead of molecules, to cross in each direction. A firewall may include other functions as well, such as caching, logging, and restriction to external services by user, by IP address, by domain, by protocol, and so on. Often a firewall consists of several components operating at different levels of the network stack. The term *firewall* comes from the sturdy concrete walls in a buildings or the strong steel barrier between the engine and the passenger compartment in a car that are intended to keep fire from spreading between compartments of the overall structure. The packets that a network firewall are intended to keep out can be just as destructive as a wildfire. If you have ever faced the wrath of a fire marshall because someone drilled "just one tiny hole" through a real building firewall (say, to run a shorter network cable), you have a feeling for just how important firewalls (and tiny chinks in that armor) are. Figure 11.1 shows the typical firewall architecture.

We will now describe two examples of this type of product to give you a better feeling for the issues and technology involved and just what they can do for you.

Microsoft Proxy Server 2.0

Proxy Server 2.0 is a fairly new product from Microsoft. It has two primary components, a Web Proxy and a Winsock Proxy. Proxy Server can provide the necessary isolation from the Internet for a small LAN (say up to 50 users), Internet access to clients that have only IPX/SPX network drivers, and in-LAN caching of popular Web pages.

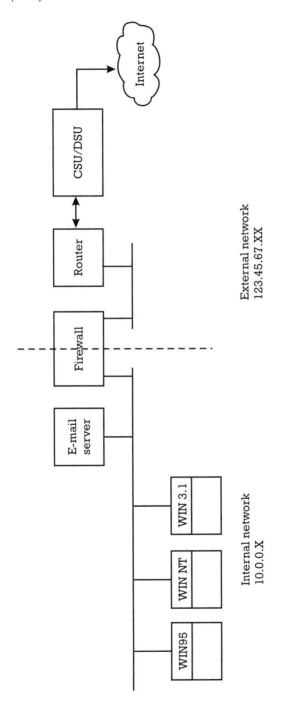

Figure 11.1 Example of a typical firewall.

Proxy Server should be installed on a reasonably fast PC-compatible computer running Windows NT Server (ideally installed as a standalone server, not as a PDC or BDC). For its usual function as a simple firewall, there must be two NICs, one connected only to the internal LAN and one connected only to the Internet. By default, Windows NT will route packets between two such NICs, which pretty much defeats the purpose of a firewall (Proxy Server itself should be the only route by which packets can cross between the two NICs), so the first step is to disable the IP routing on the computer that will run Proxy Server. You can configure TCP/IP or IPX/SPX protocols on the NIC connected to the internal LAN, but only TCP/IP on the NIC connected to the Internet. It is possible for the Internet side to connect to a local router (which in turn is connected to the Internet), or it can use RAS for on-demand dialup access to an ISP (one that supports PPP) via normal telephone lines (e.g., 28.8 modem) or ISDN. The qualifier *on-demand* means that Proxy Server normally can be in the disconnected state but can automatically dial into the ISP when an internal system tries to access the Internet or on a programmed schedule. After a specified period of inactivity, it automatically will disconnect.

The Web Proxy supports only the HTTP protocol (as used by the WWW). However, it is a sophisticated proxy and supports caching of popular Web pages on the proxy server computer, for enhanced performance and reduced traffic between the LAN and the Internet. It also can log all Web accesses and limit access on a variety of criteria. Unlike the Winsock Proxy, no special client software is required if you use Netscape Navigator or Internet Explorer for your browser, because both of those include the necessary extensions (called CERN proxy protocol).

The Web Proxy is a useful tool, but this chapter really is about firewalls and security. The Web Proxy is limited in terms of providing general security between a LAN and the Internet, unless HTTP is the only way your users ever access the Internet (which is unlikely). The Winsock Proxy is really more of a general firewall product and even includes support for HTTP (although not in a sophisticated manner as the Web Proxy).

The Winsock Proxy consists of a set of application proxies, including ones for the following protocols:

- HTTP (Web);

- FTP;

◗ Real audio (streaming audio);

◗ VDO live (streaming video);

◗ Internet Relay Chat (IRC);

◗ SMTP (Note: You normally do not want POP or IMAP to cross a firewall);

◗ NNTP (Network News Transfer Protocol);

◗ Gopher (older, text-only protocol similar to the Web).

Unlike the Web Proxy, special software must be loaded on each client that will use the Internet via the Winsock Proxy. When that software is installed, it renames the real Winsock DLL (dynamic link library) (e.g., WSOCK32.DLL in Windows 95 and Windows NT) and inserts its own extended Winsock DLL in its place. The extended version is basically like a T-pipe connector with an intelligent switch. Socket accesses to resources on the internal network (as determined by the IP address being in the local address table as managed by Proxy Server) are relayed directly to the real (renamed) Winsock DLL and thereafter proceed as usual. Socket accesses to resources on the Internet (i.e., to IP addresses not on the local address table) are redirected to the Winsock Proxy (this relaying can take place over TCP/IP or IPX/SPX). Assuming the originator of the socket access has permission and the destination and protocol are acceptable, the Winsock Proxy will relay the redirected socket access to the Internet. Incoming packets can be screened similarly, to prevent unauthorized access to an internal network by outside hackers.

Because the Winsock Proxy operates at the application level, it is possible to control access by protocol (e.g., allow SMTP in one set of situations and FTP in another). An authenticated logon between the extended client Winsock DLL and the Winsock Proxy (using the logon credentials of the person using the client computer), can be used to allow screening by users.

The Proxy Server is administered via the same Internet Service Manager used to manage the Internet Information Server and other Internet components (from any system on the internal network). The original release (1.0) can support only Winsock v1.1 clients (very few Winsock v2.0 clients are available at this time); Winsock v2.0 clients are supported in a Proxy Server 2.0.

One interesting aspect of Proxy Server is that it "hides" internal IP addresses from the Internet. As packets are relayed outward, the IP address of the Proxy Server is inserted as the originating address. There are several implications to that behavior:

▶ You need only a single "real" IP address from InterNic (or your ISP) for your entire internal network. Internally, you can use the special private class A addresses (i.e., 10.x.y.z), so you never will run out of internal addresses. You can even set up multiple internal segments, as required (e.g., 10.1.y.z, 10.2.y.z).

▶ No external hacker can physically see individual computers "behind" the Proxy Server.

▶ Any servers that need to be accessible to outsiders (e.g., SMTP MTA, DNS) must be installed physically on the computer running Proxy Server or (unprotected) on the external network segment. Additional computers on the external segment, of course, require individual "real" IP addresses. If the resource (e.g., DNS server) also must be accessed from the internal net, it must be installed either on the computer running Proxy Server or on a dual-homed computer (with two NICs, one connected to the internal network and one to the external segment). Note that this applies to an Exchange Server Internal Mail Connector as well (it effectively is a special-purpose SMTP MTA).

The bottom line is that Proxy Server is a relatively inexpensive, fairly simple way to protect a small internal network from the big, bad outside world. The downside is that it is not particularly transparent, like conventional firewall products (you must install the extended Winsock DLL on all clients and maintain an accurate local address table on Proxy Server). On the other hand, it has the interesting plus of allowing IPX/SPX-only clients to access the Internet without having to install and configure TCP/IP. It also provides high quality Web coding.

Check Point Firewall-1

Originally developed for UNIX (and still popular in that market), Check Point Firewall-1 is available for Windows NT (The current version is 3.0b). This is a typical "real firewall" product, and other similar products are on the market (either already or soon to be available to run on Windows NT). Another of the early firewall vendors to support Windows NT is Raptor Systems (http://www.raptor.com). Firewall-1, however, definitely is the leader in Internet firewalls today (40% overall market share; the next largest share is 9%). Most of the features described next can be found in varying degrees of sophistication in other such products. (Check out Check Point's Website at http://www.checkpoint.com for more information on their firewall products.)

This level of product does not come cheap. A license for a copy that can protect a 50-node LAN runs about $5,000; a license for an unlimited number of internal nodes (more than 250) runs about $19,000 (just for the software). Starting with version 3.0, a 25-user license is available for about $3,000. If that seems a bit pricey, but you still want a "real" firewall (and do not need the advanced features or the high performance of Firewall-1), you may want to look into the KarlRouter, which runs on MSDOS-based computers. Karlnet, Inc., manufactures KarlRouter, which costs from $2,000 to $3,200, depending on options, and includes all hardware, ready to install. Karlnet also has a shareware version, "kits" for customers who want to construct their own firewall, even "wireless" versions. See Karlnet's Website at http://www.karlnet.com for more information.

Of course, there is no reason you could not run a firewall on a UNIX box even if there are no other UNIX boxes on your entire network. However, the cost for a typical UNIX-based system (including the OS) is considerably higher than that for an equivalent NT-based system, and you would need high-level (expensive) UNIX system administration expertise to install and maintain that one box (unless you already have such expertise on your staff for other reasons). Many companies are moving to exclusively Windows NT-based systems for that kind of infrastructure because of the lower cost and wider availability of both the hardware platforms and the necessary system administration expertise. Traditionally, UNIX-only vendors of this kind of infrastructure software are being deluged by customer requests for NT versions, and most vendors (completely coincidentally, the ones most likely to be around next decade) have scrambled to meet the rising demand. A few appear to be stuck in the UNIX stage of their development.

The well-known research firm IDC has the following interesting comment:

> Vendors will need to sell firewalls to a significant number of users—perhaps soon they will be the majority—who do not know or want to learn Unix. Yet, these users will want to integrate their firewalls into their enterprise so that they can easily maintain and manage them—ideally from the same system with which they manage the rest of their network.

For this reason, IDC believes that firewall firms that choose to "take the high road" and forgo Windows NT will be relegated to a niche within the broader firewall market. Such vendors will likely (and probably should) also focus on high-end security consulting and development. That excerpt is from IDC report #11138, *Internet Commerce—The Worldwide Firewall Market: 1995-2000*, which is included in the marketing collateral from Check Point. See IDC's Website at http://www.idc.com for other excellent market intelligence reports.

Unlike Microsoft's Proxy Server, Firewall-1 is transparent. When installed on a dual-homed gateway computer, it intercepts all packets between the data link (IP) and network (UDP/TCP) layers and can look "inside" the packets for information concerning higher layers of the network (e.g., SMTP protocol). No special network drivers are required on the clients, and no forwarding of packets is done on the gateway at the application layer (which is inefficient). Also, it is possible to do fully generalized mapping of IP addresses in both directions. That allows you to hide internal addresses from the network (e.g., for legacy IP addresses that are not acceptable on the Internet or for the "internal" 10.x.y.z address block).

The approach of Firewall-1 allows it to control even ICMP (used by the ping utility) and other protocols. It can extract and maintain "state information" even on UDP- or RPC-based services, which is difficult for a Proxy-based service to do. That allows Firewall-1 to spread the overhead associated with authentication and filtering on such packets over a large number of packets, rather than having to do it for each packet (that happens naturally with connection-oriented protocols, such as those using TCP).

Firewall-1 does include some proxy servers, if you want to use them, but they are not required for the basic firewall protection. It also supports high-speed (10 Mbps) encryption with either DES or FWZ1 and "public key" certificate management sufficient to create VPNs (assuming all participants use Firewall-1), without requiring encrypting routers. One snazzy feature allows a remote or home user to establish a secure dialup access channel by interoperating with the security framework.

The evaluation copy I have is of Firewall-1 v3.0b and includes versions for the following UNIX variants:

▶ Sun SPARC Solaris 2.3 / 2.4;

▶ Sun SPARC SunOS 4.1.3;

▶ Sun x86 Solaris 2.4;

▶ HP PA-RISC HP-UX 9.0x, HP-UX 10.x.

The current release of Firewall-1 also includes support for Windows NT v3.51 and v4.0 (NT versions have been available since v2.1). It also has added a number of new features, including content security, which allows the filtering of viruses from files transferred via FTP or via attachments in SMTP e-mail. The new SMTP proxy also has the "feature" of supporting only the basic SMTP commands, "to prevent unauthorized outside users from exploiting security holes associated with advanced SMTP commands." Since that feature could wreak havoc with modern Internet Mail servers, it should be used only in situations that definitely warrant it and only after considerable thought. Ideally, your Internet Mail server should provide the option of disabling any commands (e.g., VRFY or EXPN) that might be exploitable in such a way. That should be done by returning not illegal commands but useless information (e.g., "No such user").

Network troubleshooting

Entire volumes have been written on the topic of network troubleshooting, but we have room to examine only the following four particularly useful tools and techniques:

▶ Ping, the all-purpose utility for low level TCP/IP debugging;

▶ Telnet, which is useful for checking out many network protocols "by hand";

▶ Performance Monitor, which also can be used to track a number of network statistics;

▶ Network Monitor, a general-purpose software network "sniffer" for looking at packets and protocols on a LAN. It is available in two versions: basic, which is included free with Windows NT server, and advanced, which is included in System Management Server (SMS).

Although it is beyond the scope of this chapter, one of the rapidly evolving areas of

Windows NT is in SNMP-based network management tools. Currently, there is good support for SNMP "agents" (the pieces that collect data for use by an SNMP management console), but little if anything from Microsoft in the way of management consoles. Fortunately, excellent examples of such tools are available from other vendors (hosted on both UNIX and Windows NT). Also, more and more physical network equipment (e.g., hubs and routers) is available in "smart" versions (i.e., they have the hooks necessary for monitoring and managing by SNMP).

When Microsoft's SMS was about to be released, most network administrators were hoping it would be a serious SNMP management console system. Unfortunately, it turned out to be something quite different (and easily the weakest and least useful component of Back Office). Expect to see this product radically revamped in future releases. Until then, look to third parties for serious SNMP tools. Probably the most useful part of SMS is the handy Network Monitor component you get with it (the advanced version of the free tool that comes with every Windows NT server, that is). We cover that tool in this chapter but do not examine the rest of SMS. If you are interested in SMS, many good books on the subject are available. Most of SMS is of little or no use to someone installing or managing a Windows NT messaging system.

Ping

The ping utility tool is deceptively simple to use. In reality, it is one of the most powerful tools of a network implementor or manager and can be used to do the following things:

- ▶ Determine if TCP/IP is installed correctly and functioning (at least the lower layers) on your own node, including determining your own IP address and/or nodename;

- ▶ Determine if another node on your LAN is alive, has TCP/IP installed and is functioning correctly (again, at least the lower layers of it), and is physically connected to the network (with no IP address conflicts or other such problems);

- ▶ Determine the same thing for a given node anywhere on the Internet, if your LAN is properly connected to the Internet, including the time required for the round trip;

◗ Indirectly invoke DNS to determine the IP address of a fully given qualified domain name (FQDN) (e.g., mynode.megacorp.com) or the fully qualified domain name associated with a given IP address (reverse lookup), which helps verify that the DNS systems involved are configured and working correctly.

The term *ping* comes from an analogous use of a submarine's sonar unit. A submarine commander who wants to determine if another solid object (say, a Russian sub) is nearby in the water will "ping" the object (send one high-intensity burst of sound). If an echo is heard, the commander knows an object is there. The time it takes for the echo to return lets the commander determine the distance to the object. In the case of a submarine, the sound literally bounces off the remote object (unless it is a so-called stealth object). In a TCP/IP ping, physical reflections do not happen (or at least they are not supposed to). An electronic "mirror" generates an automatic reply to any incoming ICMP ping transmission. That is performed internally by the TCP/IP stack itself (no actual application need be running). If the ping comes back, you know that at least the lower layers of the TCP/IP stack are in place and working on the target system.

The time it takes for the ping to go to the target system and for the echo to return is not really a measure of distance (although with really long-distance pings, that does set a lower limit on the round-trip time; even at the speed of light, it takes a packet at least 65 msec to go halfway around the earth). By far, the majority of the round-trip time is in network propagation delays (i.e., the time it takes routers along the way to receive and retransmit the packets, the time it takes the target system to generate and send the "echo," etc.). It is, however, a good indicator of how long it will take "real" (data-bearing) packets to traverse the same path.

You can ping any node on any network to which you are connected by specifying its IP address directly, for example:

```
ping 123.45.67.89
```

To ping a node on your LAN using a symbolic name, you can specify just the nodename (with or without the domain name), for example:

```
ping mynode
```

```
ping mynode.megacorp.com
```

Using the FQDN form (mynode.megacorp.com) also tests your DNS.

To ping a node on a remote LAN to which your LAN is connected, you need to specify the FQDN form of the node name:

```
ping ftp.microsoft.com
```

The result from any of those ping commands will look something like this:

```
Pinging mynode.megacorp.com [123.45.67.89] with 32 bytes
 of data:

Reply from 123.45.67.89: bytes=32 time<10ms TTL=128
Reply from 123.45.67.89: bytes=32 time<10ms TTL=128
Reply from 123.45.67.89: bytes=32 time<10ms TTL=128
Reply from 123.45.67.89: bytes=32 time<10ms TTL=128
```

If DNS is working correctly, even if you specify only the nodename (e.g., mynode), the response will contain the FQDN form of the nodename (e.g., mynode.megacorp.com).

To do a reverse lookup (given the IP address, find the FQDN of the node), use the -a parameter:

```
ping -a 123.45.67.89
```

Telnet

Most people think that Telnet is only for connecting to a Telnet server (typically on a UNIX system), although third-party Telnet servers (daemons) are available for Windows NT. By specifying a nonstandard port number (a Telnet server uses port 23), however, you can use it to test many Internet servers "by hand." In particular, you can use it to test (or even perform complete sessions) with SMTP, POP3, IMAP4, even NNTP. At the very least, Telnet is handy for determining if servers for those protocols are up and running (and possibly see information like a sign-on banner, not typically visible when using a real client for that protocol). With knowledge of the available commands and their syntax (the chapters in this book on those protocols are more than sufficient), you actually can do entire sessions using Telnet and see exactly how the protocols work.

Different Telnet clients have various ways to specify nonstandard ports (ones other than the default 23). The standard Telnet client that comes with Windows NT can be started from a DOS command prompt window (the command line interface) or through the Windows NT GUI (e.g., double-clicking an icon or selecting from a startup menu). If you start this client from a command line, you can specify the nodename of the server as the first argument and optionally specify a nonstandard port number as the second argument (using the real SMTP server at AT&T's Worldnet, from anywhere):

```
telnet mailhost.worldnet.att.com 25
```

If you start the connection through the NT GUI, you should use the Connect and Remote System menu options to specify the nodename and port number. Regardless of how you start the Telnet client, you should enable the Local Echo option (Terminal, Preferences, Local Echo, or you will be typing blind (most protocol servers do not echo characters sent by a client).

The results of a connection might look like the following. Note that many such protocols support a Help command just for such testing. The S: and C: fields were added to indicate which lines came from the server and which from the client (the person typing into the Telnet client). By using the EHLO command, you can tell exactly what SMTP extensions are supported.

```
S: 220 mtigwc03.worldnet.att.net ESMTP server (post.office v2.0 0613)
      ready Mon, 21 Apr 1997 19:29:25 +0000

C: help
S: 214-This SMTP server is a part of the post.office E-mail system. For
S: 214-information about post.office, please see http://www.software.com
S: 214-
S: 214-   Supported commands:
S: 214-
S: 214-       EHLO   HELO   MAIL   RCPT   DATA
S: 214-       VRFY   RSET   NOOP   QUIT
S: 214-
S: 214-   SMTP Extensions supported through EHLO:
S: 214-
S: 214-       EXPN   HELP   SIZE
S: 214-
```

```
S: 214-For more information about a listed topic, use "HELP"
S: 214 Please report mail-related problems to Postmaster at this site.
C: ehlo lehnts.bronwen.com
S: 250-mtigwc03.worldnet.att.net
S: 250-HELP
S: 250-EXPN
S: 250-XREMOTEQUEUE
S: 250-PIPELINING
S: 250 SIZE 10485760
C: quit
S: 221 mtigwc03.worldnet.att.net ESMTP server closing connection
```

To use the real POP3 server at Worldnet (available only from "inside" Worldnet):

```
telnet postoffice.worldnet.att.net 110
```

The following session was captured in the same way (the account and the password have been changed, and the "S:" and "C:" were added; otherwise, it is verbatim). There is no Help command in (this) POP3.

```
S: +OK Intermail POP3 server ready.
C: USER asmith
S: +OK please send PASS command
C: PASS swordfish
S: +OK asmith is welcome here
C: stat
S: +OK 114 376614
C: list 1
S: +OK 1 2798
C: quit
S: +OK asmith Intermail POP3 server signing off.
```

Performance Monitor

Performance Monitor is a wonderful, general-purpose tool for capturing and displaying real-time statistics about a wide variety of objects in NT itself (including many network objects). It also can be set to do alerts, which means various things can be done if certain counters exceed specified thresholds (e.g., send e-mail to the system administrator if disk space on a given drive goes below 5%). A number of third-party vendors are starting to include performance monitor hooks in their applications to allow monitoring of various statistics related to their programs as well. Such a feature is a major advantage to a good Windows NT administrator.

Exchange Server (and specifically the Internet Mail service) has numerous Performance Monitor counters (things you can monitor). For details, see the Exchange Server manuals or use Performance Monitor against a server running Exchange Server. The new MCIS product also has numerous Performance Monitor counters defined. Isocor's N-Plex Internet Mail server for Windows NT also supports such counters.

Even if your messaging applications do not have Performance Monitor counters of their own, there are a number of counters defined for the underlying network layers (TCP, IP, network interface cards, etc.).

Before you can monitor network statistics with Performance Monitor, you must have loaded the optional Network Monitor Agent (or Network Monitor Tools & Agent, which also loads the Network Monitor tool). You can verify if that has been done; if it has not, load it now. Log in with an administrative account and run the Network applet in the control panel. Click the Services tab of the Network applet. If you see either Network Monitor Agent or Network Monitor Tools & Agent listed under Network Services, that has already been done. If not, click the Add button at the bottom of the Network applet and load either of them now (I suggest you load both, since the Network Monitor will be needed in the next section). You then need to reboot your system before the network counters are available. If you are monitoring network counters on other NT systems (e.g., mail or news servers), each of those systems needs to have the Network Monitor Agent loaded on it. There is no way for any UNIX-based systems to participate on either side (agent or performance monitor) of the scheme.

Once the Network Monitor Agent is loaded and running, you will find several new performance monitor counter categories available, such as NBT Connection, Network Segment, and NWLink IPX. Most of those categories have a Bytes Total/Sec counter, and many allow individual access to transmitted and received bytes per second, as well. The Network Segment category has a "% network utilization," which is also useful. These counters can be quite handy for determining throughput of various kinds of networking, via different network connections (e.g., one or more NIC, RAS links, and various protocols).

Review the appropriate sections of the Windows NT resource kit on performance monitoring for further details on using this powerful tool to measure network activity or performance.

Network Monitor

There is a device called a communications line snooper for asynchronous or synchronous RS-232–type connections. You can hook the snooper into a communications link (e.g., between a terminal and its modem). Basically, the device lets you open the lid and watch the bits running back and forth. Even a basic snooper would allow viewing the data in binary, hex, or ASCII. Control characters such as CR and LF might be displayed with special characters that are easy to recognize. An advanced snooper might be able to understand various protocols, such as SDLC, and even be able to interpret various character sets, such as American Standard Code for Information Interchange (ASCII) and Extended Binary Coded Decimal Interchange Code (EBCDIC) .

I once used such a snooper to figure out the serial line protocol used by a Control Data Corporation (CDC) remote job entry (RJE) machine. The device included a card reader, a line printer, and an alphanumeric console, connected by a 4800-baud synchronous line and modem to a CDC mainframe. By doing various operations (like reading in a card deck or sending a listing to the printer) and watching the bits going between the RJE station and the mainframe with the snooper, I was able to emulate the RJE station with a much cheaper and far more reliable microcomputer. It could even "read" card decks (ASCII text files) from a floppy disk that had been created on another, general-purpose microcomputer.

In the network world, there is an equivalent device, called a network sniffer. The leading vendor of such a device is Network General (telephone, 1-800-SNIFFER; Website, www.networkgeneral.com). Basically, their sniffers are notebook computers with a special high-performance NIC (that supports "promiscuous" mode) and some proprietary software that is similar to the Microsoft Network Monitor.

It is possible to provide exactly the same functionality with a software-only product, given the appropriate NIC. Novell has had such a product for some time (which is rather pricey—over $1,000 for just the software). Microsoft released a free one that has essentially all the features of the Novell product but runs only on Windows NT. A more advanced version is available but only bundled in as part of SMS (which is fairly expensive). The trick with any software sniffer is finding an NIC that supports promiscuous mode. A typical NIC (including one that has a promiscuous mode when it is in normal operation) filters out packets not addressed to it (in hardware), to minimize the load on the CPU. A NIC that supports promiscuous mode can be told to (temporarily) allow *all* packets to be passed through to the network driver software. That allows sniffer software to look at all packets on the network, even ones that do not originate or terminate at the node on which you are running the sniffer software. Many popular NICs do in fact support this mode, but it is sometimes difficult to find out if a given one does without actually trying it. NIC vendors will be glad to tell you if a given card supports this mode. (Most salespeople in retail computer stores will have no idea what you are talking about.)

To install the basic Network Monitor, bring up the Network applet in the control panel. If Network Monitor Tools & Agent is not already present (not just Network Monitor Agent), you need to load it. Click the Add button and select Network Monitor Tools & Agent from the available network applications.

To install the advanced Network Monitor, obtain SMS, which is a part of Microsoft Back Office. The monitor can be installed as a part of the SMS installation or by itself. To install just the advanced Network Monitor, go to the NMEXT directory under SMS and run SETUP.EXE in that directory.

If you plan to do any serious network troubleshooting and you have a notebook computer, be sure you get an NIC (ethernet adapter) that supports that mode. Install NT and the extended Network Monitor software (from SMS) on it, and you will have a very powerful tool indeed. Be very careful to secure such a system with passwords or keep it in your personal possession at all times—it is also a very powerful tool for an unscrupulous person to obtain network passwords, read e-mail, and so on (if your network is not already using cryptographic authentication and protection mechanisms).

It is beyond the scope of this book to cover how to use such a sophisticated tool as Network Monitor, but, briefly, it can perform the following functions:

▶ Capture all packets on the network for some period and interpret them by protocol;

▶ Selectively capture only packets that meet various criteria;

▶ Be triggered by some event (e.g., a particular protocol header) and capture the next n packets;

▶ Save captured data to a file for future analysis;

▶ Analyze captured data in various ways (e.g., resolve MAC addresses to IP addresses or nodenames, search for particular patterns).

I strongly recommend that you try installing at least the basic Network Monitor and read through the online Help. Try capturing some packets and view them. If you are a bit more adventurous, try setting up a filter or even a trigger. In a short time, you will get a fairly good idea of what you can do with such a tool. If you find that you cannot capture any data other than packets sent to or from your node, then your NIC does not support promiscuous mode.

SMTP: Simple Mail Transfer Protocol

SMTP is the fundamental protocol of Internet Mail. It is used to upload a message from a UA (mail client) to an MTA (mail server) and to exchange messages between MTAs. The first solid definition was in RFCs 821 and 822, and the basic message syntax still is referred to as RFC 822. MIME (see Chapter 14) could be considered to be a part of the SMTP standard, but it is complex enough to rate a chapter to itself. Both SMTP and MIME likely will continue to evolve for the foreseeable future, as Internet Mail is used by a larger and larger community, and the Internet itself continues to evolve in scope and sophistication.

For the pedants among you, we should clarify that RFC 821 actually describes a link protocol, which means it exists only between any two nodes in the overall path of a given message. In comparison, RFC 822 describes an end-to-end syntax (and with minor exceptions, such as the propounding of new

Received: headers), the RFC 822–level information remains intact over the entire path of a message from sender to receiver. I am going to gloss over that distinction in the rest of this chapter, but you should keep the distinction in mind.

Internet Mail message syntax (RFC 822)

RFC 822 defines the syntactical rules for the message text referred to in the description of the SMTP protocol (which is actually covered by RFC 821). An Internet Mail UA must be able to create a message text according to the rules described in RFC 822, then send it to an SMTP receiver (typically, an Internet Mail server, or MTA) according to the rules in RFC 821.

An RFC 822-compliant message text includes two sections, the message headers and the message body, separated by a so-called null line (i.e., two CR,LF pairs in a row). Message headers need not come in any particular order. Only a few are required, but there are many optional ones. In a typical modern e-mail client (UA), the headers are created for you from configuration information and special fields you enter when you create a new message (e.g., "addressee"). In most cases, the headers are not displayed directly but are interpreted for you and show up only in lists of messages and special areas at the top of a displayed message. Often, only information from a few select headers is displayed (although some clients allow you to configure what information from the headers is displayed, up to and including "all"). The message headers consist of lines of ASCII characters with a keyword followed by a colon, then parameters. For example:

```
From: Albert Smith  <asmith@guys.com>
To: Barney Jones  <bjones@guys.com>
Subject: Howdy from Hawaii
```

The message body is completely free format and may include any number of null lines, for example:

```
This is a fascinating place!
It's hot as hell and someone stole my wallet.
Wish YOU were here.
```

A message signature (which is very different from a digital signature) can be appended automatically to messages as they are sent out by some e-mail clients, but they are not specifically defined in RFC 822. As far as RFC 822 is concerned, a message signature (whether automatically or manually appended) is just the last part of the message body.

Defined headers according to RFC 822

An e-mail address can be specified in several ways. The syntax for specifying the e-mail address is fairly flexible. Some of the common variants used by e-mail clients (all are legal by RFC 822) are as follows:

```
Asmith@guys.com
Asmith@guys.com(Albert Smith)
<Asmith@guys.com>
Albert Smith<Asmith@guys.com>
"Albert Smith"<Asmith@guys.com>
```

If no substrings are delimited by angle brackets < and >, the entire string is interpreted as the address. If any substring is delimited by angle brackets, just that substring is taken as the address, and anything else is ignored (as far as the automated mail delivery is concerned; it still may be useful to both the sender and the recipient, who may be able to see the full string displayed). Also, anything between the left and the right parentheses is considered to be a comment and is ignored.

There also are a couple of places that a time stamp (date-time) can occur. This field is in a specific format, for example:

Wed, 15 Jan 1997 20:10:15 –0800

The final field is the local difference (+/-HHMM) from Greenwich mean time (GMT), for example -0800 would be used for mail received on the West Coast of the United States (Pacific Standard Time), which is 8 hours earlier than GMT. The seconds field is optional. The actual RFC 822 specification shows only two digits for the year, but most current software uses four digits, to avoid the year 2000 syndrome (but it should accept either two or four digits).

```
Received: [from domain]
[by domain] [via path]
[with protocol] [id msg_id]
[for addr_spec] ; date_time
```

This optional header is generated automatically as mail flows through the system. A new Received: header is added to the start of the message each time it is received by an MTA, to document what sites it went through (typically the sender's MTA and your MTA, if those are different, but possibly many MTAs in between) and the date and time it arrived at each MTA along the way. Not all the optional fields (the ones in square brackets) have to be included, but typically the "from" and "by" fields are. The *date_time* field is mandatory. The "with" field can specify protocol variants, such as "with ESMTP" and is the only field that can occur multiple times. The "with" field would be a good place to document that a message has gone through a secure link (e.g., "with SMTP/SSL"). In most clients, to see the full details of a header like Received:, you must enable the display of all headers.

An example of a Received: header is:

```
Received: from gals.com by mail.guys.com; Wed, 15 Jan 97 20:10:15 -0800
```

From: e-mail_address

This required header typically is inserted by your e-mail client at message creation time and defines who sent the mail. Ideally, the user should not have control over this header; the client should build it automatically based on the logon name and local domain. If a user can change the From: header, it is easier to impersonate another user. If there is no Reply-to: or Return-Path: header, this is the e-mail address that replies should be sent to.

An example of a From: header is:

```
From: Albert Smith <Asmith@guys.com>
```

Sender: e-mail_address

This optional header is the e-mail address of the person sending the mail (e.g., a secretary) if the actual sender is not the same person the message is "from." Few if any clients provide any way to make use of such a field (e.g., to determine to whom to reply) other than to possibly display it (you may have to select "display all headers" to see this header). Most MTAs automatically generate diagnostic messages (e.g., "mail undeliverable"). If a Sender: header exists, this is the address to which such messages should be sent.

An example of a Sender: header is:

```
Sender: "Sue Wiggins"<swiggens@dorfco.com>
```

Reply-To: *e-mail_address*

This optional header is the address that clients (or MTAs generating diagnostic messages like "mail undeliverable") should use to send replies. If this header is not present, the address specified in the From: header should be used for replies.

An example of a Reply-To: header is:

```
Reply-To: Albert Smith<Asmith@guys.com>
```

Return-Path: *route e-mail address*

This optional header basically is the same as Reply-To: but with an optional route. If this header is present, it is the address that clients (or MTAs generating diagnostic messages like "mail undeliverable") should use to send replies to. If neither Reply-To: nor Return-Path is present, the From: address should be used for replies. You should not have both Reply-To: and Return-Path: headers in a single message.

An example of a Return-Path: header is:

```
Return-Path: <Asmith@guys.com>
```

Date: *date_time*

This required header is the date and time that the message was composed (and probably sent to the sender's MTA). Some e-mail clients display this time in the list of message headers, while some display the time the message was retrieved from the recipient's MTA via POP3. It is even possible that some e-mail clients could show the time the message was received by the recipient's MTA (from the Received: header).

An example of a Date: header is:

```
Date: Wed, 15 Jan 1997 20:10:01 -0800
```

To: *e-mail_address*
[, e-mail_address] ...
[, e-mail_address]

This header is the e-mail address of the primary recipient (or recipients, separated by commas). The actual delivery is determined by the RCPT TO: command(s) in the SMTP protocol (see Section 13.3), but those commands typically are generated from the To: field at the sending end. The To: header itself basically serves as documentation of to whom the message was delivered. It could be the name of a mailing list to which you subscribe. It can be used by some e-mail clients to implement the "reply to all" function (in which a copy of the reply is sent not only to the sender but also to any recipients other than yourself; one of the places that additional recipients might be found is the To: header).

An example of a To: header is:

```
To: Albert Smith<asmith@guys.com>, Barney Jones<bjones@guys.com
```

Cc: *e-mail_address*
[, e-mail_address]...
[, e-mail_address]

This header basically is identical to the To: header, except that some e-mail clients may not include recipients in a "reply to all" function from the Cc: header.

(For younger readers who may be wondering where "Cc:" comes from: In the years B.X.—before Xerox—if we wanted to send the same letter to two people, we would insert a sheet of thin plastic that had one side covered with fine carbon particles between two normal sheets of white paper, then insert the resulting sandwich into an odd, mechanical word processing machine called a typewriter. Anything typed on the top sheet made smudgy but readable images on the second sheet of paper, which was referred to as a carbon copy. At the bottom of the letter, we would indicate that the carbon copy was going to be sent to someone else with a small "cc:" notation at the bottom of the page followed by the name of the second recipient.)

An example of a Cc: header is:

```
Cc: Alice B. Toklas<atoklas@gals.com>
```

Bcc: *e-mail_address*
[, *e-mail_address***]** ...
[, *e-mail_address***]**

This header basically is identical to the Cc: header, except that it is for so-called blind carbon copies, which means that no other recipient is aware that the names listed here are also receiving the mail. The Bcc: header is not sent along with the message to anyone but a Bcc recipient. A clever SMTP sender would allow each Bcc: recipient to see only his or her own name, not the names of all Bcc: recipients.

An example of a Bcc: header is:

```
Bcc: Big Brother<bbrother@nsa.gov>
```

Message-ID: *message_id*

This header is optional but typically it is automatically generated. It is a unique identifier for a specific message (replies and forwards should get different message IDs). The worldwide uniqueness of this ID is guaranteed by the host that generates it. This ID can be referred to in an In-reply-to: header. Typically, it is intended for a machine to read, not a human.

An example of a Message-ID: header is:

```
Message-Id:<3.0.32.19961231153128.009b9ec0@bangkok.software.com>
```

Subject: *string*

This required header contains the topic or the gist of the message. It typically is displayed along with the sender, the date/time, and whether the message has been read or not in lists of messages (e.g., an inbox).

Although it is not defined in RFC 822, it is common practice for mail clients to copy the subject from the message to which you are replying into the new subject, with the string Re: added before the old subject (unless there is already one there, to prevent annoying Re: buildup). ("Re:" is an abbreviation for "with respect to.") For example, if the original subject was:

```
Howdy from Hawaii
```

The automatically generated subject in a reply would be:

```
Re: Howdy from Hawaii
```

Likewise, most clients will prepend the string Fw: to the subject of a message you are forwarding to someone else (again, unless the old subject starts with Fw:).

```
Fw: Howdy from Hawaii
```

An example of a Subject: header is:

```
Subject: Strange goings-on in the copier room
```

Other headers

A few other headers are defined, but most are rarely (if ever) used, including In-Reply-To:, References:, Keywords:, Comments:, and Encrypted:.

It also is possible for many of those keywords to be prefixed with the string Resent- in a forwarded copy of a message. A Resent-From: header would indicate who had forwarded the message, while the From: header would indicate the originator of the message, for example:

```
Resent-From: <bboop@maxfleischer.com>
```

Any header keyword not officially defined by RFCs should be prefixed with an X- (e.g., X-Sender:). There is no guarantee that other products will interpret keywords the same way your product does, and they likely will ignore them. However, your product can use keywords to do things above and beyond the current RFCs.

Example of an RFC 822-compliant message

A fairly elaborate RFC 822-compliant message follows:

```
Return-Path:   <Lawrence.Hughes@bangkok.software.com>
Received: from bangkok ([10.3.101.9]) by bangkok.software.com
         (Post.Office MTA v2.2 ID# 0-0U10) with SMTP id AAA303
         for <lawrence.hughes@bangkok.software.com>;
         Thu, 19 Dec 1996 09:50:26 -0800
Message-Id: <3.0.32.19961219095025.009bd2b0@bangkok.software.com>
X-Sender: LHugh@bangkok.software.com (Unverified)
X-Mailer: Windows Eudora Pro Version 3.0 (32)
Date: Thu, 19 Dec 1996 09:50:26 -0800
To: lawrence.hughes@bangkok.software.com
From: Lawrence.Hughes@bangkok.software.com (Lawrence Hughes)
Subject: sample RFC 822 format message

This is the message body
Now is the time for all good men to come to the aid of their neighbor
The quick brown fox jumped over the lazy dog's bones, 0123456789
```

Basic SMTP Protocol (RFC 821)

The basic SMTP protocol (i.e., without any extensions) is fairly simple. It is connection oriented and built upon TCP, using the "well-known service" port 25. Once a connection is established and the server sends a greeting, the entire session is controlled by the client. The client sends a command (which is an ASCII string followed by CR,LF), then the server replies with a result (which is also an ASCII string followed by CR,LF). The one departure from this model is when the client says, "Here comes the message text" (DATA command), and the server says, "Fire away" (354 response), after which the client blasts away with the entire message text without waiting for replies from the server. At the end of the message text, the client sends a line consisting of just one period (and CR,LF), to which the server replies, "Got it" (250 response).

Each command from the client to the server consists of a four-letter code (e.g., HELO, DATA, etc.) usually followed by a short ASCII string. Each response from the server consists of a three-digit numeric response code, followed by optional explanatory text (e.g., 250 OK).

It is possible for a simple sender to interpret just the first digit of the response, as follows:

```
2xx Request accepted and processed
3xx Ready to receive message text
4xx Some service unavailable, possibly temporarily
5xx Error, request rejected
```

The complete list of responses is as follows:

```
211 <system status or system help reply>
214 <help message for human using server interactively>
220 <domain>Service ready
221 <domain>Service closing transmission channel
250 Requested operation completed successfully
251 User not local, will forward to <forward path>

354 Start mail input, end with <CRLF>.CRLF>

421 <domain>Service not available, closing transmission channel
450 Request not completed, mailbox unavailable (busy)
451 Request not completed, error in processing
452 Request not completed, insufficient system storage (mailbox full)

500 Syntax error, command unrecognized
501 Syntax error in command parameter(s)
502 Command not implemented
503 Command sequence error
504 Command parameter not implemented
550 Request not completed, mailbox unavailable or access denied
551 User not local, please try <forward path>
552 Request not completed, exceeded storage allocation
553 Request not completed, mailbox name not allowed
554 Transaction failed
```

The following subsections list the commands in basic SMTP, as sent by the sender, and the possible responses from the receiver to those commands.

(new network connection)

A network connection is not really a command, but when the receiver accepts a new connection from the sender, it should reply with code 220, followed by one space and then the receiver's domain name, one more space and then a literal string, such as "Service ready." An example is:

```
220 guys.com Service ready
```

Possible responses are:

    Success: 220
    Failure: 421

HELO [*host.*]*domain*

This command identifies the sender's domain to the receiver. It must be the first command following the establishment of the connection. It is possible for the receiver to reject the connection based on the specified host name. An example is:

    HELO lehnts.bronwen.com

Possible responses are:

    Success: 250
    Failure: 500, 501, 504, 421

MAIL FROM: *e-mail_address*

This command identifies the sender and optionally the path by which the mail arrived at the current receiver (it can be a list of nodenames, e.g., @node1:user@node2). On receipt of this command, the receiver should clear its reverse-path buffer, its forward-path buffer, and the mail data buffer. The reverse-path buffer is set to the specified argument.
    Possible responses are:

    Success: 250
    Failure: 552, 451, 452
    Error: 500, 501, 421

RCPT TO: *e-mail_address*

This command identifies an individual recipient of the following message. Any number of recipients can be specified (one per RCPT TO: command). On receipt of this command, a server should set its forward-path buffer to the parameter. If a list of hosts is specified, it is a source route and indicates that the mail should be relayed to the next host on the list. The receiver has the option of relaying the message onward or rejecting it with code

550 (depending on configuration). If it does relay the mail, it must remove its own name from the front of the forward path and add its own name to the front of the reverse path, for example, if node HOSTA.GALS.COM receives a message with the following commands:

```
MAIL FROM: <ALBERT@HOSTX.GUYS.COM>
RCPT TO: <@HOSTA.GALS.COM,@HOSTB.GALS.COM:ALICE@HOSTC.GALS.COM>
```

The server should relay the message with headers changed as follows:

```
MAIL FROM: <@HOSTA.GALS.COM:ALBERT@HOSTX.GUYS.COM>
RCPT TO: <@HOSTB.GALS.COM:ALICE@HOSTC.GALS.COM>
```

This style of relaying is a holdover from pre-DNS days, when routes had to be specified explicitly by the sender. Today, the sender's MTA typically uses DNS to connect directly to the recipient's MTA. However, a typical MTA can easily determine if a recipient is local; if it is not, the MTA relays the message directly to the correct MTA (some servers allow independent enabling of relaying within their own domain and relaying to nodes in other domains). For example, if an MTA on node HOSTA.GALS.COM receives a message with the following commands:

```
MAIL FROM: <ALBERT@HOSTX.GUYS.COM>
RCPT TO: <ALICE@HOSTC.GALS.COM>
```

The mail server should realize that the message is not for this node and relay it directly to node HOSTC.GALS.COM (assuming that relaying within the local domain is enabled). This behavior allows an MTA to be used as a proxy or a gateway.

Possible responses are:

```
Success: 250, 251
Failure: 550, 551, 552, 553, 450, 451, 452
Error: 500, 501, 503, 421
```

DATA

This command requests the receiver to prepare to receive the entire text of the message (headers, body, and optional signature) in a continuous burst, to be ended with a period on a line by itself. If the receiver replies with 354, it is OK to blast away as fast as you can (the TCP protocol will provide you the flow control transparently).

If the receiver is ready and 354 was sent, the following responses to end-of-message text are possible:

```
Success: 250
Failure: 552, 554, 451, 452
```

If the receiver is not ready to receive data, the following responses are possible:

```
Failure: 451, 554
Error: 500, 501, 503, 421
```

RSET

This command aborts the current mail transaction. Any processed sender, recipients, and mail data must be discarded, and all buffers and state tables cleared (return to state immediately after HELO).

Possible responses are:

```
Success: 250
Error: 500, 501, 504, 421
```

NOOP

This command does nothing other than verify that the server is still alive or keep it from timing out.

Possible responses are:

```
Success: 250
Error: 500, 421
```

QUIT

This command terminates the session. The receiver should send a 250 response and then close the network connection. This command is *not* optional (although the mail client in Netscape Navigator seems to leave it off most of the time). A sender should not close the connection until it sends a QUIT and receives the 250 response. If a receiver detects that the connection has been dropped without a QUIT, it should act as if an RSET command had been received. If a sender detects that the connection has been dropped before sending a QUIT, it should act as if the command in progress had received a 4xx temporary error response.

Possible responses are:

```
Success: 221
Error: 500
```

VRFY *string*

This optional (maybe) command confirms that the string identifies a valid local mail account. If it does, then the full name of the user and the fully specified mail address are returned; if not, it may inform you that it can forward or possibly just inform you of the correct address. There is some confusion about whether this command is optional (it is defined both ways in different RFCs). Some administrators want to disable the command for security reasons. Since some popular e-mail UAs require the VRFY command to exist, the server should implement an "enable/disable VRFY" switch to actually toggle between releasing valid information and always replying "550 unable to verify" rather than replying to a VRFY command with a "500 Unknown Command." Some examples of its use follow.

  ▶ For a known, local user:

```
VRFY Hughes
250 Lawrence Hughes <Lawrence.Hughes@software.com>
```

  ▶ For a nonlocal user where the server knows the address and is able to forward:

```
VRFY Clinton
251 User not local, will forward to <BillClinton@whitehouse.gov>
```

  ▶ For an unknown user:

```
VRFY Satan
550 Unknown user
```

▶ for a nonlocal user where the server knows the address but is not able to forward:

```
VRFY Clinton
551 User not local; please try <BillClinton@whitehouse.gov>
```

▶ Possible responses are:

```
Success: 250, 251
Failure: 550, 551, 553
Error: 500, 501, 502, 504, 421
```

EXPN *string*

For security purposes, it should be possible to disable or restrict this optional command, which is used to expand a mailing list name to a list of valid mail addresses. The response is typically multiline, each of which is one mail address, for example:

```
EXPN Accounting

250-Albert Smith<Asmith@guys.com>
250-Barney Jones<BJones@guys.com>
250 Alice B. Toklas<Atoklas@gals.com>
```

or for a restricted list:

```
EXPN Board-of-directors
550 Access denied to you
```

Possible responses are:

```
Success: 250
Failure: 550
Error: 500, 501, 502, 504, 421
```

HELP *string*

This optional command is intended for use by someone accessing the server interactively via Telnet on port 25. It typically lists the available commands with syntax. If a string is specified, it might display detailed help on that command.

Possible responses are:

```
Success: 211, 214
Error: 500, 501, 502, 504, 421
```

SEND FROM: *e-mail_address*

This optional (but rarely implemented) command is an alternative to the MAIL FROM: command. It delivers mail via some immediate delivery mechanism (e.g., via a popup window on the user's workstation or terminal) rather than into a mailbox (for later retrieval via POP).

    Possible responses are:

```
Success: 250
Failure: 552, 451, 452
Error: 500, 501, 502, 421
```

SOML FROM: *e-mail_address*

This optional (but rarely implemented) command is an alternative to the MAIL FROM: command. It delivers mail using an immediate delivery mechanism (e.g., via a popup window), if available *or* into a mailbox (for retrieval via POP), if the immediate mechanism is not available.

    Possible responses are:

```
Success: 250
Failure: 552, 451, 452
Error: 500, 501, 502, 421
```

SAML FROM: *e-mail_address*

This optional (but rarely implemented) command is an alternative to the MAIL FROM: command. It delivers mail using an immediate delivery mechanism (e.g., via a popup window) if available *and* into a mailbox (for retrieval via POP).

    Possible responses are:

```
Success: 250
Failure: 552, 451, 452
Error: 500, 501, 502, 421
```

TURN

This command is optional (see following note). If it is implemented and the receiver has mail for the original sender, on receiving the TURN command, the receiver sends an OK (250) response, then changes roles to be the sender (of course, the original sender on sending the TURN command and getting a 250 response must change roles to be a receiver). If the receiver does not have any mail for the original sender or if the TURN command is not implemented, the receiver sends a refusal (502) and remains in the role of the receiver. This command is particularly useful in dialup connections ("OK, that's all I had for you. Do you have anything for me?").

NOTE *The TURN command has been "officially deprecated" by recent RFCs. In English, that means, "Oops! We really blew it with this one—it's a real security hole. If your server currently supports it, you really should take it out." The TURN command should be replaced with the rather more secure ETRN command.*

Possible responses are:

```
Success: 250
Failure: 502
Error:   500, 503
```

Command sequencing and syntax rules

There are a few rules as to the order in which commands can be used.

The first command must be HELO, but HELO can be used again later if necessary.

After the first HELO and before the final QUIT, the NOOP, HELP, EXPN, and VRFY commands can occur anywhere.

The MAIL, SEND, SOML, and SAML commands begin a mail transaction. Once begun, a transaction consists of one or more RCPT commands and a DATA command. A mail transaction can be aborted with an RSET command. There can be zero or more transactions in a session.

If the MAIL, SEND, SOML, or SAML commands are not acceptable, a 501 response is generated, and the receiver should remain in the original state. If the commands in a transaction are out of order, a 503 response is generated, and the receiver remains in the original state.

Each session must end with a QUIT command, which cannot be used anywhere else in a session.

Neither commands nor keywords are case sensitive.

Here is an example of a session using basic SMTP. In it, a message is sent from Atoklas at gals.com to Bjones and Cjohnson (both at guys.com), who are known users, and also to Dschwartz at guys.com, who is not a known user (i.e., does not have an account on this mail server). (R: and S: indicate recipient and sender, respectively.)

```
S: (opens network connection)
R: 220 guys.com SMTP Service Ready
S: HELO hosta.gals.com
R: 250 guys.com
S: MAIL FROM: <Atoklas@gals.com>
R: 250 sender OK
S: RCPT TO: <Bjones@guys.com>
R: 250 recipient OK
S: RCPT TO: <Cjohnson@guys.com>
R: 250 recipient OK
S: RCPT TO: <Dschwartz@guys.com>
R: 550 No such user
S: DATA
R: 354 Start mail input, end with <CRLF>.<CRLF>
S: Date: 14 Jan 1997 2345 PST
S: From: Albert Smith <asmith@guys.com>
S: To: <Bjones@guys.com>, <cjohnson@guys.com>,
<dschwartz@guys.com>
S: Subject: Test Message
S:
S: Now is the time for all good men to come to the aid
S: of their neighbors
S: .
R: 250 OK
S: QUIT
R: 221 signing off
S: (drops network connection)
```

Note that inside the message text (following the DATA command) no line can be a period all by itself on a line. When sending, each line of the message body is checked to see if it starts with a period. If one does, an additional period is propounded to that line. On receipt, each line is checked to see if it starts with a period. If it does and is followed immediately by an end of line (CR,LF), then that is the end of the message text. If not, the first period is discarded.

Extended SMTP (ESMTP) (RFC 1869)

Although the basic message syntax defined in RFC 822 and the basic protocol defined in RFC 821 were fairly well designed and worked well for several years, people soon wanted to add new features. To keep this orderly and to prevent the chaos of multiple incompatible variants of the protocols from emerging, a mechanism was created to extend basic SMTP in a controlled way that would ensure backward compatibility. This is known as Extended SMTP (ESMTP) and was originally defined in RFC 1651 (July 1994), then refined in RFC 1869 (Nov. 1995).

The basic idea is to introduce a new extended HELO (EHLO) that identifies a sender as supporting ESMTP, to which the receiver responds with a list of the extensions it supports. Some of the extensions introduce new commands. If a receiver advertises that it supports a particular extension, then it is fair game for the sender to use the corresponding new command(s), but those are the only extensions the sender can use.

If an ESMTP sender connects to a basic SMTP receiver, the sender starts off with the new EHLO, which the receiver will not recognize, so it will reply with a 550 (illegal command). When the sender gets the 550 response, it must drop back into basic SMTP (starting with a "real" HELO command, which the receiver *will* understand and respond to with a 250 OK). An example follows:

```
S: (opens network connection)

R: 220 guys.com SMTP Service Ready
S: EHLO hosta.gals.com
```

```
R: 550 unknown command
S: HELO hosts.gals.com
R: 250 OK
S: MAIL FROM:<Atoklas@gals.com>
R: 250 OK
S: RCPT TO: <Bjones@guys.com>
R: 250 OK
```

On the other hand, if a basic SMTP sender connects to an ESMTP receiver, that sender starts off with the old-fashioned HELO (instead of EHLO), and the receiver will realize it is talking with a basic sender and restrict itself to basic SMTP (specifically, the HELO, MAIL, RCPT, DATA, RSET, VRFY, NOOP, and QUIT commands, as defined in RFC 821).

```
S: (opens network connection)

R: 220 guys.com ESMTP Service Ready
S: HELO hosta.gals.com
R: 250 guys.com
S: MAIL FROM:<Atoklas@gals.com>
R: 250 OK
S: RCPT TO:<Bjones@guys.com>
R: 250 OK
```

Finally, if an ESMTP sender connects to an ESMTP receiver, the sender starts off with the new EHLO command, which the receiver recognizes and responds with a multiline answer. The first line contains the traditional "250 <receiver's domain>"; succeeding lines list the ESMTP extensions the receiver supports, one per line, as "250 <extension name>." Note that all the response lines but the final one must follow the 250 with a hyphen instead of a space to indicate that at least one additional response line will follow. Assuming that an ESMTP receiver understands only the SIZE and EXPN extensions, and an ESMTP sender connects to it, the negotiation might look like the following:

```
S: (opens network connection)

R: 220 guys.com ESMTP Service Ready
S: EHLO hosta.gals.com
```

```
R: 250-guys.com
R: 250-SIZE
R: 250 EXPN
S: MAIL FROM:<Atoklas@gals.com>
R: 250 OK
S: RCPT TO:<Bjones@guys.com>
R: 250 OK
```

Although in the examples we have shown the opening greeting to indicate whether a server supports ESMTP (SMTP server ready versus ESMTP server ready), that is not required and is not mentioned in the standard. In practice, server implementors can put anything they want after the domain name following the 220.

The extensions listed in Table 13.1 (actually optional commands listed in RFC 821) are defined in RFC 1869. The service extension is the name that the extension is referred to in the RFC. The keyword is what would be advertised by an ESMTP receiver in reply to an EHLO. The verb is the new command (that could be used by an ESMTP sender if the receiver advertised support for the corresponding extension). The initial set happens to use the same string for the advertised keyword and the command verb, but that is not required and is not the case in later extensions. The initial set defined here basically provides a way for servers to know which of the option RFC 821 commands the other server has implemented.

RFC 1869 also defines a way for new extensions to be registered, as well as a way to create proprietary or local extensions (just use a keyword

Table 13.1
ESTMP extensions

Service extension	EHLO keyword	Verb
Send	SEND	SEND
Send or Mail	SOML	SOML
Send and Mail	SAML	SAML
Expand Group	EXPN	EXPN
Help	HELP	HELP
Turn	TURN	TURN

that starts with an X, which means that no registered keyword can start with that letter). It also provides for an optional parameter to follow the keyword (see Section 13.6 for an example), but none of the initial extensions requires one.

SIZE Extension (RFC 1870)

The first extension beyond the basic set of optional commands from RFC 821 was the Message Size Declaration extension, first defined in RFC 1427 (Feb. 1993), refined in RFC 1653 (July 1994), and currently defined in RFC 1870 (Nov. 1995). It is widely implemented. There are two primary aspects to this extension. The first is an indication whether an ESMTP receiver supports the SIZE extension and what the maximum size message is that it will accept. The second is an indication of the size of a particular message (which is implemented by adding a new parameter to the MAIL command rather than defining a new command verb).

The first aspect involves adding the keyword SIZE to the list of extensions an ESMTP receiver advertises in response to an EHLO command from an ESMTP sender. With or without a parameter, this indicates that it is OK for the ESMTP sender to include the SIZE parameter in the MAIL command (and optionally in the SEND, SOML, and SAML commands, if supported). If no parameter is specified, no conclusions can be drawn by the SENDER concerning the maximum message size the receiver will support. If a zero (0) is specified, that indicates that no maximum message size is enforced.

Otherwise, the parameter should be a numeric value, which is the maximum message size in bytes the receiver will accept, for example:

```
S: (opens network connection)

R: 220 guys.com ESMTP Service Ready
S: EHLO hosta.gals.com
R: 250-guys.com
R: 250 SIZE
S: MAIL FROM:<Atoklas@gals.com>
R: 250 OK
```

This indicates that the receiver supports the SIZE extension but gives no information about the maximum size it will accept. On the other hand, the following example indicates that the receiver supports the SIZE extension and will not accept any message larger than 100,000 bytes:

```
S: (opens network connection)

R: 220 guys.com ESMTP Service Ready
S: EHLO hosta.gals.com
R: 250-guys.com
R: 250-SIZE 100000
R: 250 EXPN
S: MAIL FROM:<Atoklas@gals.com>
R: 250 OK
```

The second aspect involves the sender specifying the actual size of the message, in bytes, with a new parameter in the MAIL command, for example:

```
S: (opens network connection)

R: 220 guys.com ESMTP Service Ready
S: EHLO hosta.gals.com
R: 250-guys.com
R: 250-SIZE
R: 250 EXPN
S: MAIL FROM:<Atoklas@gals.com> SIZE=1573
R: 250 OK
```

A receiver specifically cannot assume that the message is exactly that length (say, to look for the final "period" line after exactly that many bytes or to truncate any bytes beyond the declared message size). It is to be considered merely an indication or guideline as to how much storage may be required for the message. A receiver should never respond with a 552 (insufficient storage) code to the DATA transmission once it has accepted a MAIL command with a declared size (assuming the message does not exceed the declared size, or at least not by very much—a receiver should tolerate an error of a few percent).

The receiver can respond to such a MAIL command with three possible responses:

▶ 250 OK;

▶ 452 Insufficient system resources;

▶ 552 Message exceeds maximum allowable message size.

The 250 code is an indication that there probably is room on the server to hold the entire message (but it is not an absolute guarantee).

The 452 code indicates that there currently is not enough room on the server to accept that message, but there might be later, so it should be requeued for another attempt at some time in the future.

The 552 code indicates that a maximum message size limit has been proclaimed, and the declared message size in the MAIL command is greater than that limit. Such a message should not be requeued. It should be returned to sender as undeliverable.

If the sender gets either a 452 or a 552 response, either it should QUIT immediately or, if it wants to try sending other (smaller) messages, it should first do an RSET command.

PIPELINING extension (RFC 1854)

The PIPELINING extension defines a way to increase throughput (in other words, to decrease the time required for a given session) by batching up certain SMTP commands into a single TCP transmission, if the other end can handle it. It is defined in RFC 1854 (Oct. 1995). To announce that it can accept batched (or "pipelined") commands, the receiver advertises the keyword PIPELINING in response to the EHLO command. No new verbs are defined, nor are any new parameters defined for the MAIL command.

If the receiver indicates that it supports the PIPELINING extension, the sender can issue multiple commands without waiting for any response from the receiver. In particular, the RSET, MAIL, SEND, SOML, SAML, and RCPT commands can occur anywhere in a pipelined command group. The EHLO, DATA, VRFY, EXPN, TURN, QUIT, and NOOP commands can occur only as the last command in a pipelined group (the inclusion of NOOP in this group allows it to be used to achieve synchronization). An extended SMTP command can occur only as the last command in a group unless specifically defined otherwise.

The responses to the pipelined commands all will be sent and in the correct order; it is unnecessary to wait for them before sending another acceptable command. No fewer characters are sent (each line is still terminated by both CR and LF). The increase in throughput comes about by eliminating many of the time-consuming delays while awaiting a response from the other end. Various rules are defined in RFC 1854 concerning exactly how to implement this functionality, in both clients and servers.

Following is an example of a session without pipelining; there are nine separate waits for a response:

```
S: (opens network connection)

R: 220 guys.com ESMTP Service Ready
S: EHLO hosta.gals.com
R: 250 guys.com
S: MAIL FROM:<Atoklas@gals.com>
R: 250 sender OK
S: RCPT TO:<Asmith@@guys.com>
R: 250 recipient OK
S: RCPT TO:<Bjones@guys.com>
R: 250 recipient OK
S: RCPT TO:<Cjohnson@guys.com>
R: 250 recipient OK
S: DATA
R: 354 Start mail input, end with <CRLF>.<CRLF>
  . . .
S: .
R: 250 OK
S: QUIT
R: 221 signing off
```

Here is the same session with pipelining, with only four waits for response from the other end:

```
S: (opens network connection)
```

```
R: 220 guys.com ESMTP Service Ready
S: EHLO hosta.gals.com
R: 250-guys.com
R: 250 PIPELINING
S: MAIL FROM:<Atoklas@gals.com>
S: RCPT TO:<Asmith@guys.com>
S: RCPT TO:<Bjones@guys.com>
S: RCPT TO:<Cjohnson@guys.com>
S: DATA
R: 250 sender OK
R: 250 recipient OK
R: 250 recipient OK
R: 250 recipient OK
R: 354 Start mail input, end with <CRLF>.<CRLF>
S: (message content)
S: .
S: QUIT
R: 250 OK message received
R: 221 signing off
```

8bit-MIMEtransport extension (RFC 1652)

The 8bit-MIMEtransport extension originally was defined in RFC 1426 (Feb. 1993) and was refined in RFC 1652 (July 1994). It supports the sending of characters beyond the basic 128 ASCII characters (which require only 7 bits each). Examples are characters from various European languages (such as the umlaut characters in German).

The ESMTP keyword for this extension is 8BITMIME (with no parameters). One new parameter defined for the MAIL command is used to indicate whether the message body contains only 7-bit ASCII character or can also contain extended 8-bit characters. The two possible values are:

```
BODY=7BIT
```

and

```
BODY=8BITMIME
```

No new command verbs are defined.

A message body including some 8-bit characters still is limited to no more than 1,000 characters per line, and so on. It is not for raw binary data. Here is an example of a session using this extension:

```
S: (opens network connection)

R: 220 guys.com ESMTP Service Ready
S: EHLO hosta.gals.com
R: 250-guys.com
R: 250 8BITMIME
S: MAIL FROM:<Atoklas@gals.com> BODY=8BITMIME
R: 250 sender and 8BITMIME OK
S: RCPT TO:<Asmith@guys.com>
R: 250 recipient OK
S: DATA
R: 354 Start 8BITMIME mail input, end with <CRLF>.<CRLF>
S: (message content, including some 8 bit data)
S: .
R: 250 OK message received
S: QUIT
R: 221 signing off
```

Large/binary message extensions (RFC 1830)

This experimental RFC is not widely implemented. Defined in RFC 1830 (Aug. 1995), it is intended to allow efficient transfer of large quantities of (possibly binary) data. There are really two extensions defined here: CHUNKING and BINARYMIME. The CHUNKING extension can be implemented without the BINARYMIME extension and would be useful without it. The BINARYMIME extension, however, works only in conjunction with the CHUNKING extension.

The CHUNKING extension provides an alternative to the DATA command in SMTP (called BDAT) and a more efficient way of sending large messages. It also allows the sender and the receiver to not have to "special-case" lines starting with a period. Before an SMTP sender can use the BDAT method of sending the message content, it must see the CHUNKING keyword during an ESMTP handshake with the SMTP receiver. If it does see that, then when it is time to send the DATA

command the sender optionally can send one or more BDAT blocks
(BDAT command followed by data) instead.

The new scheme for sending the message content is to break the data
into chunks (no minimum or maximum is defined for a chunk, nor do the
chunks have to be the same size). The SMTP sender sends a "BDAT *nnn*"
command (where *nnn* is the number of bytes in the chunk), followed by
exactly nnn bytes of data. The receiver replies with a 250 response when it
is ready for the next chunk. That continues until the final chunk, which is
sent using a "BDAT *nnn* LAST" command and the remaining *nnn* bytes of
data (which also must receive a 250 response), after which normal SMTP
is resumed.

Here is an example of the use of CHUNKING with nonbinary data:

```
S:  (opens network connection)

R: 220 guys.com ESMTP Service Ready
S: EHLO hosta.gals.com
R: 250-guys.com
R: 250 CHUNKING
S: MAIL FROM:<Atoklas@gals.com>
R: 250 sender OK
S: RCPT TO:<Asmith@guys.com>
R: 250 recipient OK
S: BDAT 1000
S: (first 1000 bytes of 7-bit message content)
R: 250 1000 bytes received
S: BDAT 1000
S: (second 1000 bytes of 7-bit message content)
R: 250 1000 bytes received
S: BDAT 345 LAST
S: (final 345 bytes of 7-bit message content)
R: 250 message received, 2345 bytes total
S: QUIT
R: 221 signing off
```

The BINARYMIME extension adds a new parameter (BINARYMIME)
to the BODY keyword of the MAIL FROM: command, as defined in the
8BITMIME extension (RFC 1652). As usual, an SMTP sender cannot use
the BINARYMIME extension unless it sees the BINARYMIME keyword
from the corresponding SMTP receiver during the ESMTP handshake.

Assuming the SMTP receiver has advertised BINARYMIME and CHUNKING extensions in the ESMTP handshake, the SMTP sender then can optionally use the BINARYMIME value in the BODY parameter of the MAIL FROM: command. Use of the BINARYMIME body type requires use of the BDAT mechanism for transmitting the message content (you cannot use the conventional DATA mechanism with a BINARYMIME body type). If the receiver accepts such a MAIL FROM: command, that means it agrees to preserve all bits in every byte passed using the BDAT command (otherwise, the receiver can strip the most significant bit of each byte, making it a 7-bit channel).

Here is an example of chunking of BINARYMIME message content:

```
S: (opens network connection)

R: 220 guys.com ESMTP Service Ready
S: EHLO hosta.gals.com
R: 250-guys.com
R: 250-BINARYMIME
R: 250 CHUNKING
S: MAIL FROM:<Atoklas@gals.com>  BODY=BINARYMIME
R: 250 sender OK
S: RCPT TO:<Asmith@guys.com>
R: 250 recipient OK
S: BDAT 1000
S: (first 1000 bytes of 8-bit message content)
R: 250 1000 bytes received
S: BDAT 1000
S: (second 1000 bytes of 8-bit message content)
R: 250 1000 bytes received
S: BDAT 345 LAST
S: (final 345 bytes of 8-bit message content)
R: 250 message received, 2345 bytes total
S: QUIT
R: 221 signing off
```

If a given SMTP receiver does not support BINARYMIME, but the SMTP sender has a message with binary MIME content, the sender can either indicate a failed delivery or convert the binary content into 7-bit MIME on the fly (using "quoted printable" or "base64" encoding as appropriate) and send the message anyway.

It is possible to combine CHUNKING with PIPELINING (RFC 1854). In that case, the SMTP sender can stream any number of BDAT blocks (BDAT command followed by data) and collect the 250 responses as they come in or at some point in the future. All MAIL FROM: and RCPT TO: commands should be processed prior to the first BDAT block being sent. Unfortunately, there is no way to associate a given 250 response with a given BDAT block, other than the order in which they are received.

Here is an example of using CHUNKING with nonbinary data, using PIPELINING:

```
S: (opens network connection)

R: 220 guys.com ESMTP Service Ready
S: EHLO hosta.gals.com
R: 250-guys.com
R: 250-PIPELINING
R: 250 CHUNKING
S: MAIL FROM:<Atoklas@gals.com>
S: RCPT TO:<Asmith@guys.com
R: 250 sender OK
R: 250 recipient OK
S: BDAT 1000
S: (first 1000 bytes of 7-bit message content)
S: BDAT 1000
S: (second 1000 bytes of 7-bit message content)
S: BDAT 345 LAST
S: (final 345 bytes of 7-bit message content)
R: 250 1000 bytes received
R: 250 1000 bytes received
R: 250 message received, 2345 bytes total
S: QUIT
R: 221 signing off
```

Remote queue processing declaration (RFC1985)

The TURN command in RFC 821 leaves a possibly serious security hole, which basically would allow a rogue server to obtain mail not intended for it. Because of that security problem, not many server implementors support the TURN command. RFC 1985 (Aug. 1996) defines an extended

TURN command (ETRN) that solves those problems. Instead of turning the direction around right then, it requests that another connection (in the opposite direction) be scheduled to take later (possibly even while the first one is going on), complete with all the normal announcing of domain names and opportunities to reject a session. A good implementation of ETRN will both validate the domain name and determine if any mail is queued in the specified direction. It is up to the server implementor whether to allow an ETRN command to request queuing of mail to any domain other than the one that the message containing the command is from. In any case, the server will send mail to the correct domain, regardless of who requested the messages to be queued for transmission. Therefore, the worst that might happen is for deferred mail to be delivered to the correct destination earlier than otherwise might have happened. Any of the following codes may be returned immediately in response to an ETRN command:

```
250 OK, connection queued for <x>
251 OK, no messages waiting for <x>
252 OK, pending messages for <x> queued
253 OK  pending messages for <x> queued
458 Unable to queue messages for <x>
459 ETRN request for  denied: <x>
500 Syntax Error
501 Syntax Error in parameter(s)
```

The keyword advertised in response to an EHLO command is ETRN, with no parameters. One new command verb (also ETRN) is defined, with a single parameter, which is the name of the domain or node for which to schedule a connection.

Here is an example of various uses of ETRN:

```
S: (opens network connection)

R: 220 guys.com ESMTP Service Ready
S: EHLO hosta.gals.com
R: 250-guys.com
R: 250-EXPN
R: 250-HELP
R: 250 ETRN
S: MAIL FROM:<Atoklas@gals.com>
```

```
R: 250 sender and 8BITMIME OK
S: DATA
R: 354 Start mail input, end with <CRLF>.<CRLF>
S: (message content)
S: .
R: 250 OK
S: ETRN uu.net
R: 458 unable to queue messages for uu.net
S: ETRN gals.com
R: 250 OK, messages queued for gals.com
S: ETRN losers.com
R: 251 no messages waiting for losers.com
S: ETRN megacorp.com
R: 253 OK, 153 pending messages for megacorp.com queued
S: QUIT
R: 221 signing off
```

Other SMTP extensions

Two major extensions to the SMTP protocol are so elaborate and important that they are broken out into chapters of their own.

One of those extensions is MIME (Multipurpose Internet Mail Extension), which is primarily an extension of the RFC 822 message syntax and content. One of the simplest new things possible with MIME is to attach binary files (similar to using uuencode but more standardized and better thought out). However, it also supports rich text (text with different fonts, sizes, colors, etc.), multimedia (embedded audio and images), and more complex organizations of a message (multipart and even nested). MIME is widely implemented in current Internet Mail clients (basic SMTP servers can largely ignore it), but typically only a small part of the overall functionality is implemented (little more than what was possible with uuencode). MIME was originally defined in RFCs 1521–1523, but the current specification (as of January 1997) is in RFCs 2045–2049.

The second of those extensions is Delivery Status Notification (DSN), which defines standard ways to track the delivery status of messages (where they are in the system, whether the message has been received at the recipient's post office, etc.). A more complex set of response codes also is defined (for use in status notification messages, not as a replacement for the SMTP three-digit response codes), as well as a standard way for servers to use MIME to handle error conditions (such as undeliverable mail). DSN is defined in RFCs 1891–1894. Very few products so far have implemented much if any of this functionality, but customers migrating to Internet Mail from proprietary systems will create a demand for it, since they have been using analogous features for some time.

Relevant RFCs

Quite a few RFCs are related to SMTP. Following is the complete list (the most important ones are indicated by asterisks).

*1985	PS	J. De Winter, "SMTP Service Extension for Remote Message Queue Starting," 08/14/1996. (Pages=7) (Format=.txt)
*1894	PS	K. Moore, G. Vaudreuil, "An Extensible Message Format for Delivery Status Notifications," 01/15/1996. (Pages=31) (Format=.txt)
*1893	PS	G. Vaudreuil, "Enhanced Mail System Status Codes," 01/15/1996. (Pages=15) (Format=.txt)
*1892	PS	G. Vaudreuil, "The Multipart/Report Content Type for the Reporting of Mail System Administrative Messages," 01/15/1996. (Pages=4) (Format=.txt)
*1870	S	J. Klensin, N. Freed, K. Moore, "SMTP Service Extension for Message Size Declaration," 11/06/1995. (Pages=9) (Format=.txt) (Obsoletes RFC1653) (STD 10)
*1869	S	J. Klensin, N. Freed, M. Rose, E. Stefferud, D. Crocker, "SMTP Service Extensions," 11/06/1995. (Pages=11) (Format=.txt) (Obsoletes RFC1651) (STD 10)
*1854	PS	N. Freed, A. Cargille, "SMTP Service Extension for Command Pipelining," 10/04/1995. (Pages=7) (Format=.txt)
1846	E	A. Durand, F. Dupont, "SMTP 521 reply code," 10/02/1995. (Pages=4) (Format=.txt)
1845	E	D. Crocker, N. Freed, A. Cargille, "SMTP Service Extension for Checkpoint/Restart," 10/02/1995. (Pages=8) (Format=.txt)
*1830	E	G. Vaudreuil, "SMTP Service Extensions for Transmission of Large and Binary MIME Messages," 08/16/1995. (Pages=8) (Format=.txt)

1653	DS	J. Klensin, N. Freed, K. Moore, "SMTP Service Extension for Message Size Declaration," 07/18/1994. (Pages=8) (Format=.txt) (Obsoletes RFC1427) (Obsoleted by RFC1870)
*1652	DS	J. Klensin, N. Freed, M. Rose, E. Stefferud, D. Crocker, "SMTP Service Extension for 8bit-MIMEtransport," 07/18/1994. (Pages=6) (Format=.txt) (Obsoletes RFC1426)
1651	DS	J. Klensin, N. Freed, M. Rose, E. Stefferud, D. Crocker, "SMTP Service Extensions," 07/18/1994. (Pages=11) (Format=.txt) (Obsoletes RFC1425) (Obsoleted by RFC1869)
1495	PS	H. Alvestrand, S. Kille, R. Miles, M. Rose, S. Thompson, "Mapping Between X.400 and RFC-822 Message Bodies," 08/26/1993. (Pages=15) (Format=.txt) (Updates RFC1327)
1428	I	G. Vaudreuil, "Transition of Internet Mail From Just-Send-8 to 8Bit-SMTP/MIME," 02/10/1993. (Pages=6) (Format=.txt)
1427	PS	K. Moore, N. Freed, J. Klensin, "SMTP Service Extension for Message Size Declaration," 02/10/1993. (Pages=8) (Format=.txt) (Obsoleted by RFC1653)
1426	PS	J. Klensin, N. Freed, M. Rose, E. Stefferud, D. Crocker, "SMTP Service Extension for 8bit-MIMEtransport," 02/10/1993. (Pages=6) (Format=.txt) (Obsoleted by RFC1652)
1327	PS	S. Hardcastle-Kille, "Mapping Between X.400(1988) / ISO 10021 and RFC 822," 05/18/1992. (Pages=113) (Format=.txt) (Updates RFC0822) (Obsoletes RFC1148) (Updated by RFC1495)
1148	E	B. Kantor, S. Kille, P. Lapsley, "Mapping Between X.400 (1988) / ISO 10021 and RFC 822," 03/01/1990. (Pages=94) (Format=.txt) (Obsoletes RFC0987) (Obsoleted by RFC1327)
1138	I	S. Kille, "Mapping Between X.400(1988) / ISO 10021 and RFC 822," 12/01/1989. (Pages=92) (Format=.txt) (Updates RFC1026)
1090		R. Ullmann, "SMTP on X.25," 02/01/1989. (Pages=4) (Format=.txt)
1047		C. Partridge, "Duplicate Messages and SMTP," 02/01/1988. (Pages=3) (Format=.txt)
1026	PS	S. Kille, "Addendum to RFC 987: Mapping Between X.400 and RFC-822," 09/01/1987. (Pages=4) (Format=.txt) (Updates RFC0987) (Updated by RFC1138)
0987	PS	S. Kille, "Mapping Between X.400 and RFC 822," 06/01/1986. (Pages=69) (Format=.txt) (Updates RFC0822) (Obsoleted by RFC1148) (Updated by RFC1026)

0876		D. Smallberg, "Survey of SMTP Implementations," 09/01/1983. (Pages=13) (Format=.txt)
*0822	S	D. Crocker, "Standard for the Format of ARPA Internet Text Messages," 08/13/1982. (Pages=47) (Format=.txt) (Obsoletes RFC0733) (STD 11) (Updated by RFC1327, RFC0987)
*0821	S	J. Postel, "Simple Mail Transfer Protocol," 08/01/1982. (Pages=58) (Format=.txt) (Obsoletes RFC0788) (STD 10)
0788		J. Postel, "Simple Mail Transfer Protocol," 11/01/1981. (Pages=62) (Format=.txt) (Obsoletes RFC0780) (Obsoleted by RFC0821)
0780		S. Sluizer, J. Postel, "Mail Transfer Protocol," 05/01/1981. (Pages=43) (Format=.txt) (Obsoletes RFC0772) (Obsoleted by RFC0788)
0733		D. Crocker, J. Vittal, K. Pogran, D. Henderson, "Standard for the Format of ARPA Network Text Messages," 11/21/1977. (Pages=38) (Format=.txt) (Obsoletes RFC0724) (Obsoleted by RFC0822)
0724		D. Crocker, K. Pogran, J. Vittal, D. Henderson, "Proposed Official Standard for the Format of ARPA Network Messages," 05/12/1977. (Pages=0) (Format=.txt) (Obsoleted by RFC0733)

MIME: Multipurpose Internet Mail Extension

The simple "flat" message syntax defined in RFC 822 is adequate for ASCII text-only messages, possibly including binary files that have been translated into additional ASCII text via a scheme like uuencode (see Chapter 3). Today, however, people want to send much more complex mail. MIME allows users to send messages that include any of the following:

▸ Text, including extended character sets (e.g., for European languages);

▸ Rich text messages (multiple typefaces, font sizes, color, etc.);

▸ Attachments with hints that allow the recipient's client to interpret them correctly (i.e., invoke the appropriate application and feed the attached data into it);

▶ Multiple, possibly even nested (hierarchically structured) attachments in a single message (message/rfc822);

▶ A large message automatically split among a number of smaller messages (and automatically stitched back together at the other end) to get around per-message limits (message/partial);

▶ A message whose actual content is to be obtained through some channel external to e-mail, such as FTP, the Web, and so on (message/external);

▶ Multiple messages in a digest format that allows a client to display any subset, possibly even with an automatically generated table of contents (multipart/digest);

▶ Multiple versions of a given text in a single message, so a client can identify and display the most appropriate version for it, for example, simple ASCII text, rich text, and audio (multipart/alternative);

▶ Multiple parts of a compound message that the client is supposed to present simultaneously, for example, a moving video image and an audio soundtrack (multipart/parallel).

Some proprietary e-mail systems (e.g., Microsoft Exchange Server) support some of the above features. However, most, if not all those features are the very ones that have the most problems going between different e-mail systems (e.g., Exchange to Internet). Full support for MIME typically can be done only in an end-to-end native Internet Mail system. Some gateways (e.g., the Exchange Internet Mail Service) can map proprietary features onto some of the MIME features as messages go through a gateway (typically, only simple attachments), but many of the more desirable features (e.g., rich text, audio) are almost impossible for a gateway to convert properly. When it comes to secure e-mail extensions like S/MIME or PGP/MIME, there are essentially unsolvable problems with a gateway being able to map proprietary schemes onto the Internet "open" standard. The widespread use of MIME and the richness of the standard is one of the strongest arguments for pushing Internet Mail all the way out to the desktop, rather than using it just as a universal messaging backbone between proprietary islands.

This complex and rich extension to RFC-822 allows Internet e-mail to equal or exceed many of the most advanced and powerful features of proprietary messaging systems. It does support the attachment of binary files, but it goes well beyond what can be done with uuencode. Few vendors have supported the entire range of features described here. As more do, proprietary messaging systems will have fewer and fewer advantages over Internet e-mail. No doubt future extensions to MIME (e.g., S/MIME) will take Internet e-mail into even further vistas.

History and backward compatibility

MIME started life in a much simpler form in RFC 934 (Jan. 1985) and RFC 1049 (March 1988). It was extended significantly in RFCs 1341 and 1342 (June 1992), which were later refined and expanded in RFCs 1521–1523 (Sept. 1993). Recently (Dec. 1996) those RFCs have been reworked one more time into a group of five RFCs (2045 through 2049). The bulk of that work was done by Ned Freed and Nathaniel Borenstein.

Unfortunately, few vendors have chosen to support MIME in all its glory. Most have used it to do only a slightly cleaner version of what uuencode used to do (simple binary attachments). As you will see from this chapter, full support for MIME goes well beyond what possibly could have been done with uuencode. When vendors support all its features, together with the DSN extensions, Internet Mail will have most of the advanced capabilities of the best proprietary LAN mail systems but in a form that is optimized for, and capable of supporting, worldwide interoperation.

Since RFC 822 said very little about the exact contents or syntax of the message body, a message containing even very complex MIME content is in compliance with RFC 822 (given a few new message headers, which themselves are consistent with the syntax of headers defined in RFC 822). In fact, RFC 822 specifically states that unrecognized headers should be ignored. Therefore, an Internet Mail UA that is RFC 822 compliant but not MIME compliant should be able to receive messages with complex MIME content (although it will not be able to make sense of some or all of it). In fact, if the entire message (headers and all) is saved to a file by such a UA, there are programs that can extract the attachments from such a file (much like the uuencode program).

Unlike uuencode, however, it is not possible to do offline creation of a MIME-compliant message for later transmission with a UA that does not support MIME. That is because part of the MIME content is in the message headers (and there is no way to add new headers to an outgoing message either by cutting and pasting or by an insert-file mechanism; such new content is restricted to the message body).

A MIME-compliant UA can accept and work with an incoming straight RFC 822 format message (sent by a non-MIME UA). Such a message will be missing the top-level MIME headers (e.g., "MIME-Version:"), which tips off the UA to treat the rest of the message as straight RFC 822 format with no MIME content.

The important current RFCs (as of March 1997, when this chapter was written) for MIME are indicated by asterisks in the complete list of relevant RFCs at the end of this chapter.

Basic MIME message structure

MIME essentially allows you to create a composite message with one or more subparts, each of which can contain other subparts, to as many levels as you like (like the little wooden Russian dolls within dolls within dolls). The subparts at a given level are separated with a MIME boundary (a string defined in one of the real message headers, chosen so as to not be likely to occur in the message itself). Each subpart can have its own headers similar to (but not identical to) the real message headers.

Because MIME was defined in such as way as to easily incorporate extensions, there is a central registry to locate information about any official content type. You can view the current list with a Web browser at:

```
ftp://ftp.isi.edu/in-notes/iana/assignments/media-types/
```

The list is maintained by the Internet Assigned Names Authority (IANA). You can find out more about IANA at their Website:

```
http://www.iana.org/iana
```

There are several basic message content types in MIME (specified in the top-level Content-Type: header). There are several possible single-part organizations and several possible multipart organizations. There are three special content types, which allow nested, split, or even external content.

The general form of a straight RFC822 (non-MIME) message is as follows:

```
[message headers]
[blank line]
[message body]
```

For example:

```
To: bjones@guys.com
From: asmith@guys.com (Albert Smith)
Subject: example "straight RFC822" message

This is the message body. It will appear in the "received
message window" of the user agent upon receipt of the
message.
```

The general form for a single-part MIME message is as follows:

```
[message headers]
Mime-Version: 1.0
Content-Type: [content_type/subtype for subpart 0]
[blank line]
[subpart 0 content (primary message body)]
```

For example:

```
To: bjones@guys.com
From: asmith@guys.com (Albert Smith)
Subject: example single part MIME message
Mime-Version: 1.0
Content-Type: plain/text; charset="us-ascii"

This is the message body. It will appear in the "received
message window" of the user agent upon receipt of the
message.
```

The general form for a multipart MIME message (in this case, message body plus two attachments, but there could be any number of attachments) is as follows:

```
[message headers]
Mime-Version: 1.0
Content-Type: multipart/mixed; boundary="<mime_boundary>"
[blank line]
__<mime_boundary>
Content-Type: [content_type/subtype for subpart 0]
[blank line]
[subpart 0 content (primary message body)]
[blank line]
__<mime_boundary>
[MIME headers for subpart 1]
[blank line]
[subpart 1 content (first attachment)]
__<mime_boundary>
[MIME headers for subpart 2]
[blank line]
<subpart 2 content (second attachment)>
__<mime_boundary>--
```

For example:

```
To: bjones@guys.com
From: asmith@guys.com (Albert Smith)
Subject: example multipart MIME message
Mime-Version: 1.0
Content-Type: multipart/mixed; boundary="********"

--********
Content-Type: text/plain; charset="us-ascii"

This is the main message body. There are two attachments
associated with this message. The first one is binary
data, which is encoded in base64, and will be written to
the file demo.dat upon receipt by a MIME User Agent. The
second attachment is simple ASCII text, hence does not
need to be Encoded. It will be written to the file
demo.txt upon receipt. This main message body is all
that will appear in the normal "received message window"
in the User Agent upon receipt of this message. There
may be summary lines that indicate that the attachments
```

were processed, but these are not required by the MIME
standard.

___********
Content-Type: application/octet-stream
Content-Transfer-Encoding: base64
Content-Disposition: attachment; filename="demo.dat"

AAECAwQFBgcICQoLDA0ODxAREhMUFRYXGBkaGxwdHh8gISIjJCUmJygpK
issLS4vMDEyMzQ1Njc4OTo7PD0+P0BBQkNERUZHSElKS0xNTk9QUVJTVF
VWV1hZWltcXV5fYGFiY2RlZmdoaWprbG1ub3BxcnN0dXZ3eHl6e3x9fn+
AgYKDhIWGh4iJiouMjY6PkJGSk5SVlpeYmZqbnJ2en6ChoqOkpaanqKmq
q6ytrq+wsbKztLW2t7i5uru8vb6/wMHCw8TFxsfIycrLzM3Oz9DR0tPU1
dbX2Nna29zd3t/g4eLj5OXm5+jp6uvs7e7v8PHy8/T19vf4+fr7/
P3+/w==
___********
Content-Type: text/plain; charset="us-ascii"
Content-Disposition: attachment; filename="demo.txt"

This is a simple ASCII text file. It contains no extended
characters, hence can be represented with the US-ASCII
character set. It is not a part of the main message
body, but is an attachment just like the first one
(which was binary data).

___********___

Single-part organizations

text/plain; charset=US-ASCII

This is the content type for subparts containing only simple ASCII text.
Typically it is used for the primary message body (that which is entered in
the UA's text editor, as distinct from attachments). This is the default content type for any subpart if none is specified. Specifically, US-ASCII refers
to the character set defined in ANSI X3.4-1986 or the international reference version of ISO 646 (1991).

text/plain; charset=ISO-8859-x

This is the content type for text using one of the extended character sets
defined in ISO-8859 (the *x* after the ISO-8859 should be replaced with the
appropriate number for the specific ISO-8859 character set used in a

given message subpart). Each of these character sets (ISO-8859-1 through ISO-8859-9) includes US-ASCII as the first 128 characters, then various extended characters for specific European languages as the second 128 characters. (See Appendix A for information on those character sets.) If a message really contains only characters in the range 0 to 127 (e.g., characters from the first half of any of the ISO-8859 character sets), then the charset for that message should be specified as US-ASCII, to allow the greatest possible number of UAs to work with it. A good Western world-aware UA should include support for both creating and viewing messages in all the ISO-8859 character sets. True world-aware e-mail should allow subparts using the Unicode standard of 16 bits per character. There are schemes to encode UNICODE content into 7- or 8-bit characters today (UTF-7, UTF-8; see RFC 1641). Although few if any UAs support UNICODE today, Windows NT likely will hasten that evolution. Windows NT uses UNICODE as its internal character set and is already available in native Far Eastern languages, such as Thai, simple and unified Chinese, and Japanese.

text/enriched

RFC 1896 defines MIME enriched text as a set of formatting codes (e.g., <bold>) that can be embedded in otherwise ASCII text, as a simple markup language similar to (but simpler than) Standard Generalized Markup Language (SGML) or HTML. It allows typical word processing effects to be added to any part of your message, such as font (<Font-Family><param>Times</param>);   relative font sizes (e.g., <bigger>, <smaller>; effects (<bold>, <italic><underline>)and even color (<color><param>red</param>). A given formatting code affects text from the character immediately following that code up to a balancing formatting code (e.g., </bold>).

Care should be exercised in sending messages as enriched text, because not all UAs will be able to interpret such messages correctly. At minimum, a current MIME-compliant UA should be able to at least filter out the enriched-text formatting codes, leaving the simple ASCII text. Ideally, it should be able to display the text using all the enriched-text attributes. A good UA should allow you to keep track (in the address book) of which recipients can handle enriched text. Anyone else should have all enriched-text attributes stripped out before a message is sent.

Another possibility would be to send both the simple ASCII and the enriched-text version using multipart/alternative content type (unfortunately, few UAs are able to process this correctly). There is no way for the sending UA to be able to query the capabilities of the receiving UA itself.

Note that MIME enriched-text format is similar in concept, but incompatible in implementation, with Microsoft's Rich Text Format (RTF). If you receive a "rich text message" from Microsoft Exchange via the Internet connector, it will not be in MIME enriched-text format, and your MIME-compliant UA will not be able to correctly interpret the rich text. Fortunately, the gateway extracts the simple text from such a message and sends that as the main body part, with the rich text version of the message sent as a useless (to a MIME client) attachment. That is typical of the problems associated with a proprietary mail system that couples to the Internet via a gateway.

In spite of the (coming) availability of text/html and text/sgml, the text/enriched content type definitely has its place. The full functionality easily can be implemented right in the UA; its capabilities handle the overwhelming majority of the formatting necessary in an e-mail message; and it includes certain features useful for text e-mail that neither text/html nor text/sgml has. It also is easier for a UA that does not support text formatting to filter out the formatting codes of text/enriched than those of text/html or text/sgml.

text/html

With the immense popularity of the World Wide Web and familiarity of its markup language (HTML) and the tools for working with it, text/html soon will become an alternative to text/enriched, as a MIME content type for rich text (complete with images, audio, etc.). The MIME text/html format is specified in RFC 1866 ("Hypertext Markup Language—2.0"). A text/html subpart that contains HTML v3.0 or proprietary Netscape (or Microsoft) HTML content possibly would not be interpreted correctly by a MIME-compliant UA. A UA that supports this content type could either attempt to interpret the HTML itself (a difficult enough task today and becoming more so every year as HTML evolves) or, more likely, bring up a separate browser (e.g., Microsoft Internet Explorer) and feed it the received HTML.

In general, it is difficult to create HTML (other than for the simple rich-text attributes available in text/enriched). Invoking an external HTML browser can be time consuming on some systems. Composing a reply to a complex HTML message is quite difficult. Many people (myself included) consider use of this markup language (or text/sgml) for e-mail to be somewhat like swatting a fly with a thermonuclear warhead.

text/SGML

SGML is an older and more mature markup language than HTML, more suited to high-quality documents than to Web pages. SGML is defined in ISO 8879, and IETF RFC 1874 defines a MIME content type using this standard. SGML allows far better control over the appearance of a document than is possible with text/enriched or with text/html (especially if the message is to be printed). Again, the SGML content could be interpreted right in the UA, but SGML is a complex standard. Unfortunately, SGML viewers are not as common as HTML viewers, but it still might be easier to implement one of those than trying to embed the functionality in the UA itself (of course, such a viewer would be useful from a variety of other applications). Such a viewer also should be able to print high-quality documents directly from the SGML content.

audio/<subtype>

This is the content type for digitized audio (e.g., human speech or a soundtrack). Two subtypes currently are defined, the simplest of which is "basic." This format uses single-channel 8-bit pulse coded modulation (PCM) with a 8,000-Hz sample rate (suitable for voice but not much else). A more sophisticated subtype is defined in RFC 1911 ("32kadpcm"). It is hoped that future RFCs will standardize subtypes suitable for two-channel (or more) music, up to CD quality. Few current UAs support the audio content type. Expect future UAs to allow direct recording and playback of audio subparts within the UA. Those could be integrated with voice mail systems to allow you to "view" (listen to) and archive your voice mail messages right along with your textual and graphic e-mail.

image/<subtype>

This is the content type for still (nonmoving) images. A number of image formats have been registered with the IANA, including those listed in Table 14.1.

Table 14.1
Seven image formats

Format	Relevant RFC	Description
JPEG	RFC 1521	JPEG using JFIF encoding
GIF	RFC 1521	Graphics Interchange Format
IEF	RFC 1314	Image Exchange Format
G3FAX	RFC 1494	Group 3 facsimile
TIFF	—	Tagged Image File Format
CGM	—	Computer Graphics Metafile
PNG	RFC 2083	Portable Network Graphics

GIF is patented by Compuserve, which has tried to enforce royalties for its use. PNG was developed by a working group of the World Wide Web Consortium (W3C) as a public domain alternative to GIF.

As with the other complex content types, those formats might be best handled with an external "viewer" (or "helper") application. Several commercial applications currently are available for viewing and printing images in various formats, including all those listed in Table 14.1. A full-featured UA might include a simple viewer/printer application that covers all the common graphic formats registered with the IANA.

video/<subtype>

This is the content type for moving images. Currently defined subtypes include motion pictures experts group (MPEG) (RFC 1521) and Quicktime. Real-time video requires either a very fast CPU (perhaps an Intel MMX technology Pentium) and/or specialized video hardware. Even with the high compression ratios possible with MPEG, short video messages tend to be fairly large for e-mail (typically megabytes). It will be a few years before it is possible to have Internet Mail provide full-screen 30-frame-per-second, 24-bit-per-pixel video, complete with CD-quality stereo soundtrack, but MIME is ready for it when the rest of the infrastructure catches up. Wide-scale deployment of Asynchronous Transfer Mode (ATM) networks (typically 600 Mbps or more) would help a lot.

application/<subtype>

This is a kind of catch-all content type for data files useful to some application programs (e.g., Microsoft Word and Excel). It is the ideal content type for electronic document interchange (EDI) applications. Two of the subtypes defined in RFC 1521 for this content type are "octet-stream" (a basic stream of raw binary data) and "postscript." Many people do not realize it, but Postscript is actually more than just a markup language (like HTML or enriched text format); it is a general, powerful, stack-oriented programming language. Postscript is used to communicate exactly what to print or draw on Postscript-compatible printers, but it includes control structures (loops) and file I/O statements. Therefore, a UA that can interpret a general Postscript file (received from potentially anyone) is quite a security risk.

Multipart organizations

The multipart content types allow more than one subpart in a single message. Each subpart can have its own MIME headers to specify details about that subpart (e.g., content-type, charset, disposition). The various subparts do not (in general) all have to be of the same content type. These content types are as follows.

multipart/mixed

This content type allows multiple subparts in a single message, with no particular relationship between the parts, and would be used for attaching multiple documents. It is up to the UA (or possibly even the user) to determine what to do with the received subparts (possibly with hints in the MIME headers, e.g., content-disposition).

multipart/alternative

This is similar to multipart/mixed, except that the various subparts are different versions of the same message (e.g., simple ASCII text, enriched text, and SGML). It is up to the UA to select the most appropriate one to display or print (the other subparts would be ignored).

multipart/parallel

This is similar to multipart/mixed, except that all the subparts are intended to be displayed together (e.g., text and a still picture, a moving video image and audio sound track). If a UA is not capable of displaying the subparts in parallel, it is allowed to display them in sequence (perhaps with a note to the user that they were intended to be displayed simultaneously).

multipart/digest

This is similar to multipart/mixed, except that each subpart is an RFC 822 message itself (possibly complete with MIME content). This format would be ideal for use by a mailing list server that supports a digest mode. A UA that supports this content type might display a table of contents listing the sender and the subject for each message, with a way to view any of the messages individually. Microsoft's Outlook Express UA supports this quite well.

Special content types

message/rfc822

A subpart with this content type contains a complete RFC 822-style message, including optional MIME content. It could be used in the multipart/digest organization or for a server to return an undeliverable message.

message/partial

Some e-mail transports or servers might limit the size of a single message. This content type allows a large message to be split into two or more pieces, each of which is sent in a separate MIME message. A good UA would be able to split any outgoing message larger than a given user-defined limit and piece back together a split message received from another UA that supports message/partial.

message/external-body

This content type is perhaps the most flexible, because the subpart in this case is merely a pointer to where the real subpart can be found, which could be a Web page via HTTP, a file via FTP, and so on. Essentially, the UA would invoke the appropriate companion application, feeding it the pointer to where the external-body is located.

Other MIME headers

In addition to the content-type header, there are other MIME headers, as follows:

▶ Content-transfer-encoding, which defines the type of encoding used to represent the data;

▶ Content-ID, which identifies a MIME subpart with a unique string, for reference purposes;

▶ Content-description, which is a way to associate some description with a MIME subpart;

▶ Content-disposition, which is a way to specify what to do with the subpart on receipt by a UA.

Content-transfer-encoding

The concept and details of encoding were covered in Chapter 3. To refresh your memory, encoding is basically the process of representing data in another form for some reason (in this case, encoding binary data into an ASCII format so they will be able to go through a 7-bit e-mail transport). The change in the representation of the data must be completely reversible. Changing the representation of the data back into the original form is called decoding. In this case, it involves recovering the original binary data from the ASCII representation. The data are there in both forms (original and encoded); it is merely the representation that has been changed. For example, the integer resulting from adding 1 + 1 + 1 + 1 is the same whether it is represented with the arabic numeral 4 or the roman numeral IV. This process is analogous to translating binary serial data into audio tones and back again (the function of a telephone modem). In the case of modems, the encoding is done to be able to send the data over a voice grade telephone line that cannot reliably transmit digital waveforms.

Several kinds of encoding have been defined for use with MIME. In the basic RFCs, the following types are defined. (Other RFCs or standards define MIME encoding styles for binhex, which is popular on the Macintosh, and uuencode, the older UNIX standard).

77bit

This is the basic encoding style for simple ASCII text (i.e., no change in the data representation is required). It is the default value for content-transfer-encoding if none is specified in a given message or subpart. The only charset that can be used with 7bit is US-ASCII. Specifically, all data are in the form of short lines of US-ASCII characters.

8bit

This is the same as 7bit encoding (no change in representation) except that some bytes can have the high-order bit set (i.e., values in the range 128 to 255 decimal). It can support use of one of the ISO 8859 character sets. Basic SMTP mail transports (compliant with RFC 821 and RFC 822) cannot support this encoding style. An extension to SMTP (RFC 1426, "8bitMIME") defines how SMTP clients and servers can be modified to allow this encoding style to be used. That standard provides for the automatic conversion of 8-bit data into 7-bit form if a given node in the path of a message does not support the extension. Specifically, all data must be in the form of short lines of characters, some of which may have the most significant bit set.

binary

This encoding is basically the same as 8bit, but the data need not be in the form of lines short enough for SMTP transport. There currently are no Internet Mail transport mechanisms capable of supporting this encoding style.

quoted-printable

A generalization of 7bit encoding to allow 8-bit characters to be represented by a group of 7-bit characters (using a "quoting" mechanism). Most printable ASCII characters (except =) require no such change in representation. Any 8-bit character can be represented by a three-character sequence (an equals sign, then two hex digits that can represent any value from 0 to 255 decimal). The equals sign itself is represented as the three-character sequence "=61" (0x61 is the ASCII code for equals sign). Quoted-printable encoding is more efficient (fewer resulting characters) than base64 encoding for messages that are mostly 7-bit ASCII text, with a few 8-bit codes (e.g., a message using one of the ISO-8859 character sets, for a European language, like French). For general binary data (where values in the range 128 to 255 are just as likely as values in the range 0 to 127), quoted-printable encoding is less efficient (more resulting characters) than using base64.

base64

This is the preferred encoding style for general 8-bit binary data. Each group of three 8-bit bytes is encoded into four 7-bit ASCII characters. See Chapter 3 for details.

Content-ID

This header identifies a MIME subpart with a unique string, for reference purposes. It is required only in the multipart/external content type.

Content-description

This header is a way to associate some description with a MIME subpart, for example, a caption that might be displayed along with an image. The parameter is any arbitrary ASCII text. It is implementation specific what a UA should do with the parameter on receipt.

Some examples would be:

▶ Content-type: image/png;

▶ Content-disposition: inline;

▶ Content-description: My daughter, Bronwen, at her 2-year birthday party.

Content-disposition

This header has semiofficial status, since it is defined in an RFC (1806), but the RFC is in the "experimental" category. This header is widely used in the Qualcomm Eudora clients (in fact, Steve Dorner, the original author of Eudora, is one of the two authors of RFC 1806) and is quite useful, so I am including it here. However, you should be aware that not all UAs will understand this header. Part of the reason it has not achieved standards-track status is that it can include a filename, which tends to be platform specific.

There are three possible disposition types: inline, attachment, and extension token (which means additional types may be defined in the future). The inline disposition type means this subpart should be displayed automatically when the message is viewed. The attachment disposition type means the subpart is logically separate from the message in some sense, and it should be up to the user whether and when to display it (e.g., the user clicks an icon representing the attachment to cause it to be displayed). In the case of an attachment, it is possible to also specify a filename. Assuming that a filename is specified and it is compatible with the platform on which the UA is running, the UA should write the decoded content of that subpart to the specified file. The file typically should be created and written in some user-definable attachment directory; drive names and directory paths should not be specified and may be ignored if they are. It would be acceptable for a UA to mangle the specified name in some way to make it compatible with the available file system. The RFC does not give any hints as to what to do if there is already an existing file with the same name, but Eudora typically appends a digit to the received name to make it unique. For example, if the filename is demo.txt and there is already a demo.txt and demo1.txt, then the new data are written to demo2.txt.

Examples would include the following:

▶ Content-disposition: inline;

▶ Content-disposition: attachment;

▶ Content-disposition: attachment; filename=chap14.doc.

Relevant RFCs

The following RFCs specify or are related to MIME. (The asterisks indicate those RFCs that are more important.)

*2077	PS	S. Nelson, C. Parks, Mitra, "The Model Primary Content Type for Multipurpose Internet Mail Extensions," 01/10/1997. (Pages=13) (Format=.txt)
*2076		Common Internet Message Headers

2068	PS	R. Fielding, J. Gettys, J. Mogul, H. Frystyk, T. Berners-Lee, "Hypertext Transfer Protocol—HTTP/1.1," 01/03/1997. (Pages=162) (Format=.txt)
*2049	DS	N. Freed, N. Borenstein, "Multipurpose Internet Mail Extensions (MIME) Part Five: Conformance Criteria and Examples," 12/02/1996. (Pages=24) (Format=.txt) (Obsoletes RFC1590)
*2048	DS	N. Freed, J. Klensin, J. Postel, "Multipurpose Internet Mail Extensions (MIME) Part Four: Registration Procedures," 12/02/1996. (Pages=21) (Format=.txt) (Obsoletes RFC1590)
*2047	DS	K. Moore, "MIME (Multipurpose Internet Mail Extensions) Part Three: Message Header Extensions for Non-ASCII Text," 12/02/1996. (Pages=15) (Format=.txt) (Obsoletes RFC1590)
*2046	DS	N. Freed, N. Borenstein, "Multipurpose Internet Mail Extensions (MIME) Part Two: Media Types," 12/02/1996. (Pages=44) (Format=.txt) (Obsoletes RFC1590)
*2045	DS	N. Freed, N. Borenstein, "Multipurpose Internet Mail Extensions (MIME) Part One: Format of Internet Message Bodies," 12/02/1996. (Pages=31) (Format=.txt) (Obsoletes RFC1590)
2017	PS	N. Freed, K. Moore, A. Cargille, "Definition of the URL MIME External-Body Access-Type," 10/14/1996. (Pages=5) (Format=.txt)
*2015	PS	M. Elkins, "MIME Security With Pretty Good Privacy (PGP)," 10/14/1996. (Pages=8) (Format=.txt)
1927	I	C. Rogers, "Suggested Additional MIME Types for Associating Documents," 04/01/1996. (Pages=3) (Format=.txt)
*1911	E	G. Vaudreuil, "Voice Profile for Internet Mail," 02/19/1996. (Pages=22) (Format=.txt)
*1896	I	P. Resnick, A. Walker, "The Text/Enriched MIME Content-Type," 02/19/1996. (Pages=21) (Format=.txt, .ps) (Obsoletes RFC1563)
1895	I	E. Levinson, "The Application/CALS-1840 Content-Type," 02/15/1996. (Pages=6) (Format=.txt)
*1892	PS	G. Vaudreuil, "The Multipart/Report Content Type for the Reporting of Mail System Administrative Messages," 01/15/1996. (Pages=4) (Format=.txt)
*1874	E	E. Levinson, "SGML Media Types," 12/26/1995. (Pages=6) (Format=.txt)
1873	E	E. Levinson, J. Clark, "Message/External-Body Content-ID Access Type," 12/26/1995. (Pages=4) (Format=.txt)
1872	E	E. Levinson, "The MIME Multipart/Related Content-Type," 12/26/1995. (Pages=8) (Format=.txt)
1867	E	E. Nebel, L. Masinter, "Form-Based File Upload in HTML," 11/07/1995. (Pages=13) (Format=.txt)

*1866	PS	T. Berners-Lee, D. Connolly, "Hypertext Markup Language—2.0," 11/03/1995. (Pages=77) (Format=.txt)
1848	PS	S. Crocker, N. Freed, J. Galvin, S. Murphy, "MIME Object Security Services," 10/03/1995. (Pages=48) (Format=.txt)
1847	PS	J. Galvin, S. Murphy, S. Crocker, N. Freed, "Security Multiparts for MIME: Multipart/Signed and Multipart/Encrypted," 10/03/1995. (Pages=11) (Format=.txt)
1844	I,	E. Huizer, "Multimedia E-mail (MIME) User Agent Checklist," 08/24/1995. (Pages=8) (Format=.txt) (Obsoletes RFC1820)
1820	I	E. Huizer, "Multimedia E-mail (MIME) User Agent Checklist," 08/22/1995. (Pages=8) (Format=.txt) (Obsoleted by RFC1844)
*1806		Content-Disposition Header (Experimental)
1767	PS	D. Crocker, "MIME Encapsulation of EDI Objects," 03/02/1995. (Pages=7) (Format=.txt)
*1741	I	P. Faltstrom, D. Crocker, E. Fair, "MIME Content Type for BinHex Encoded Files," 12/22/1994. (Pages=6) (Format=.txt)
*1740	PS	P. Faltstrom, D. Crocker, E. Fair, "MIME Encapsulation of Macintosh files—MacMIME," 12/22/1994. (Pages=16) (Format=.txt)
*1641	E	D. Goldsmith, M. Davis, "Using Unicode With MIME," 07/13/1994. (Pages=6) (Format=.txt, .ps)
1563	I	N. Borenstein, "The Text/Enriched MIME Content-Type," 01/10/1994. (Pages=16) (Format=.txt, .ps) (Obsoletes RFC1523) (Obsoleted by RFC1896)
1556	I	H. Nussbacher, "Handling of Bi-Directional Texts in MIME," 12/23/1993. (Pages=3) (Format=.txt)
1523	I	N. Borenstein, "The Text/Enriched MIME Content-Type," 09/23/1993. (Pages=15) (Format=.txt) (Obsoleted by RFC1563)
1522	DS	K. Moore, "MIME (Multipurpose Internet Mail Extensions) Part Two: Message Header Extensions for Non-ASCII Text," 09/23/1993. (Pages=10) (Format=.txt) (Obsoletes RFC1342)
1521	DS	N. Borenstein, N. Freed, "MIME (Multipurpose Internet Mail Extensions) Part One: Mechanisms for Specifying and Describing The Format of Internet Message Bodies," 09/23/1993. (Pages=81) (Format=.txt, .ps) (Obsoletes RFC1341) (Updated by RFC1590)
1505	E	A. Costanzo, D. Robinson, R. Ullmann, "Encoding Header Field for Internet Messages," 08/27/1993. (Pages=36) (Format=.txt) (Obsoletes RFC1154)
*1494	PS	H. Alvestrand, S. Thompson, "Equivalences Between 1988 X.400 and RFC-822 Message Bodies," 08/26/1993. (Pages=26) (Format=.txt)
*1486	E	M. Rose, C. Malamud, "An Experiment in Remote Printing," 07/30/1993. (Pages=14) (Format=.txt) (Obsoleted by RFC1528, RFC1529)

1437	I	N. Borenstein, M. Linimon, "The Extension of MIME Content-Types to a New Medium," 04/01/1993. (Pages=6) (Format=.txt)
1344	I	N. Borenstein, "Implications of MIME for Internet Mail Gateways," 06/11/1992. (Pages=9) (Format=.txt, .ps)
1343	I	N. Borenstein, "A User Agent Configuration Mechanism for Multimedia Mail Format Information," 06/11/1992. (Pages=10) (Format=.txt, .ps)
1342	PS	K. Moore, "Representation of Non-ASCII Text in Internet Message Headers," 06/11/1992. (Pages=7) (Format=.txt) (Obsoleted by RFC1522)
1341	PS	N. Borenstein, N. Freed, "MIME (Multipurpose Internet Mail Extensions): Mechanisms for Specifying and Describing the Format of Internet Message Bodies," 06/11/1992. (Pages=80) (Format=.txt, .ps) (Obsoleted by RFC1521)
*1314	PS	D. Cohen, A. Katz, "A File Format for the Exchange of Images in the Internet," 04/10/1992. (Pages=23) (Format=.txt)
1154	E	R. Ullmann, D. Robinson, "Encoding Header Field for Internet Messages," 04/16/1990. (Pages=7) (Format=.txt) (Updates RFC1049) (Obsoleted by RFC1505)
1049	S	M. Sirbu, "Content-Type header field for Internet messages," 03/01/1988. (Pages=8) (Format=.txt) (STD 11) (Updated by RFC1154)
1036		M. Horton, R. Adams, "Standard for interchange of USENET messages," 12/01/1987. (Pages=19) (Format=.txt) (Obsoletes RFC0850)
934		M. Rose, E. Stefferud, "Proposed standard for message encapsulation," 01/01/1985. (Pages=10) (Format=.txt)

DSN: Delivery Status Notification

DSN is another important extension to SMTP (RFC 821) and the Internet Message Format (RFC 822). Basically, it adds mechanisms so a sender can track the progress of a sent message. The standard appears to be primarily the work of Keith Moore and Gregory Vaudreuil. DSN, together with MIME and IMAP4, gives standards-based Internet Mail much of the functionality of the best proprietary e-mail systems (e.g., Microsoft Exchange Server). One important distinction is that successful delivery only means delivery to the recipient's server mailbox and says nothing about when—or even whether—the recipient has actually retrieved or read the message.

DSN is one of the more powerful, yet still widely unsupported features of Internet e-mail. This feature, once widely supported by both e-mail clients and servers, allows users to track the progress of messages or at least determine how and why they went

277

astray. Without DSN, messages sometimes simply disappear into the ether, and the sender has no way of knowing where the problem might have been. Because of concerns about privacy, DSN intentionally stops short of being able to know if and when a message has actually been read. The most it will tell you is that it was successfully delivered to the destination postoffice (and when). What happens to it from there, if anything, is outside the purview of DSN. As with MIME, DSN helps Internet e-mail compete on equal footing with X.400 and most proprietary messaging systems that provide various levels of message tracking. It can even allow limited interoperation with the tracking facilities of foreign e-mail systems, such as X.400.

DSN is a new standard, and not many vendors support it. DSN must be supported by the sending UA (and all MTAs along the way) for it to work correctly. In comparison, for MIME to work, it must be supported by both the sending and the receiving UAs, but not necessarily any MTAs in between. If the sending UA and all MTAs up to some point support DSN, then the message hits a nonconforming server, the original sender will at least get a "message relayed" notification message that indicates at what point the original message left DSN space. It is not necessary that the recipient's UA support DSN. On the other hand, IMAP4 is entirely concerned with the link between a given UA and the local Internet Mail server (the sender need not have IMAP4 available for the recipient to be able to make full use of it).

Technically, it would be feasible to have an IMAP server or a POP server generate a (hypothetical) "message retrieved" notification message when the recipient actually retrieves the message from the Mail server. That would allow senders to know whether (and when) their messages have been retrieved (but possibly still not read). However, that is specifically disallowed in the DSN standard (and there is strong resistance among e-mail server implementors to provide that final link, presumably for privacy reasons). The result is that delivery notifications with SMTP/DSN are not quite the same as read receipts in typical proprietary e-mail systems. Perhaps in a future revision of DSN, that could be added. There still may be some people who do not want other people to know when (or if) they read their messages. For such people, there may have to be a way to disable such status reports on a per-account basis (similar to the disabling of telephone caller-ID reporting).

DSN is defined in RFCs 1891–1894 (listed in Section 15.6). It defines the various pieces of the overall DSN functionality. In theory, only part of DSN could be implemented, but it really was intended that the functionality described in all four RFCs would be implemented (the pieces all work together).

RFC 1891 defines features that constitute an ESMTP extension, including:

▶ A new DSN reply to the ESMTP EHLO command (to let a connecting SMTP sender know that the SMTP receiver supports DSN);

▶ Two new optional parameters for the RCPT command: NOTIFY (specify conditions under which notification messages should be generated) and ORCPT (to convey the original sender-specified recipient address);

▶ Two new optional parameters for the MAIL command: RET (specify whether "delivery failure" notification messages will return the entire contents or just the headers of the failed message) and ENVID (specify an identifier for this message transmission envelope, which is to be returned in any notification messages relevant to the message).

RFC 1892 defines a new multipart MIME organization (multipart/ report) type that further extends the Internet Message Format (originally RFC 822). The new content type should be used at the top level of a generated notification message. It contains two or three subparts, as follows:

▶ Part 1 (required) is a human-readable message that describes the event that caused the notification message to be generated (e.g., free format, any content-type, charset). Typically, this part would be read by the human recipient of the notification message (e.g., viewed in a window like a normal e-mail message). If the machine parsable information is used to update status indicators in the user interface, this part could be discarded.

▶ Part 2 (required) is a machine-parsable account of the event that caused the DSN to be generated. This part could be used by an intelligent UA to update status indicators in the list of sent messages, discard the original message on successful delivery, retry failed deliveries, and so on. It is entirely up to the UA implementor as to what to do with that information.

▶ Part 3 (optional) is the returned failed message (or just the headers, depending on the value of the RET keyword), which may be useful for diagnosing the problem (ideally if an envelope ID is returned, that should be sufficient for the sender to locate the message that it originally sent).

RFC 1893 defines a new set of mail status response codes similar to (but not compatible with) the RFC 821 three-digit codes (e.g., 250) but used for reporting detailed delivery status. These codes currently are used only in the machine-parsable subpart of a DSN message (they are not intended to replace the response codes in SMTP itself).

RFC 1894 defines another new MIME content type (message/ delivery-status) for use in the new MIME multipart/report organization type (defined in RFC 1892). This RFC defines the syntax of the second (machine-parsable) part of a notification message and covers the details of how it is to be used in various situations.

SMTP service extensions for DSN (RFC 1891)

RFC 822 did require that an SMTP receiver notify the SMTP sender of a delivery failure if it determines that the message cannot be delivered to one or more of the recipients. Typically, that was just an RFC 822 (non-MIME) message sent to the address from the MAIL command in the RFC 821 envelope, with an explanation of the problem and part or all of the failed message.

That is inadequate to diagnose or fix many mail delivery problems. DSN is designed to provide the following capabilities (assuming a DSN was requested by the sender):

▶ Reliable notification of either (1) a failure in the delivery some-where along the way and, if so, when and how it failed or (2) suc-cessful delivery to a destination;

- ▶ Prevention of message looping by ensuring that a failed attempt to deliver a notification message does not generate another notification message;

- ▶ Provision of a mechanism for the DSN requests to be propagated forward through multiple MTAs to either the point of failure or the final recipient MTA (or possibly where it encounters a noncon-forming server and leaves DSN space);

- ▶ Ability of the sender to identify the mail transaction and the recipient address that generated the DSN, even when mail has crossed a gateway to another mail system (e.g., X.400);

- ▶ Preservation of DSNs from non-Internet Mail systems that already provide such notifications (e.g., X.400 or MS Exchange), including at least (1) the ability to specify whether to return the original message content and (2) the ability to specify whether no notification, notification of successful delivery, notification of failed delivery, or notification of any delivery is generated.

A DSN-compliant mail server should indicate that it supports DSN to any client (or other MTA) that connects to it using ESMTP. That is done during the ESMTP negotiation at the start of the connection. An ESMTP sender should not attempt to use the new DSN keywords in the RFC 821 RCPT or MAIL commands unless it has seen the DSN keyword returned during that negotiation. An example follows:

```
S: (opens network connection)

R: 220 guys.com ESMTP Service Ready
S: EHLO hosta.gals.com
R: 250-guys.com
R: 250-SIZE
R: 250-DSN
R: 250 EXPN
```

The RFC 821 MAIL command normally specifies the sender of the mail message (unlike recipients, there is only one sender in a given message). A DSN-compliant SMTP sender optionally can specify either or both of two new keywords in the command (RET and ENVID). The RET

keyword can have one of two values: FULL (return entire message content) or HDRS (return just message headers). The ENVID keyword can
have an arbitrary ASCII string as its value (up to 100 characters long). An
example follows:

```
S: MAIL FROM:<atoklas@gals.com> RET=HDRS ENVID=AA123456
R: 250 OK
```

The RFC 821 RCPT command normally specifies one recipient of a
mail message. There is one RCPT command for each recipient of the message; in theory, different DSN parameters could be specified for each
recipient. In practice, a DSN-conforming UA might allow the DSN options
to be specified only per message (not per recipient), which are then
included on all RCPT commands for that message.

A DSN-conforming SMTP sender optionally can specify either or both
of two new keywords in this command: NOTIFY and ORCPT. The NOTIFY
keyword can have one of four possible values: NEVER (do not send any
notifications for this message), SUCCESS (send notification on delivery
success), FAILURE (send notification on delivery failure), or DELAY (it is
permissible, but not required, to send notification if delivery is delayed
significantly). The default value (if NOTIFY is not specified) can be either
NOTIFY=FAILURE or NOTIFY=FAILURE,DELAY (this is up to the server
implementor). The ORCPT keyword allows specification of the original
recipient (which is useful especially when mail crosses gateways).

```
S: RCPT TO: <bjones@guys.com>NOTIFY=FAILURE,SUCCESS,DELAY
R: 250 OK
```

The DSN request is passed along from one DSN-conforming server to
another DSN-conforming server according to some simple rules:

▶ If a server receives a message with ENVID and/or RET keywords on
the MAIL command, it must include the same keyword(s) (with the
same values) on the MAIL commands it uses to relay the message
onward. If there was no ENVID keyword on a given incoming MAIL
command, the server cannot add one to the corresponding outgoing MAIL command (ditto for RET).

▶ If a server receives a message with NOTIFY and/or ORCPT
keywords on any RCPT command, it must include the same

keyword(s) (with the same values) on the corresponding RCPT commands it uses to relay the message onward. If there was no NOTIFY keyword on a given incoming RCPT command, the server cannot add one to the corresponding outgoing RCPT command. However, if there was no ORCPT keyword on a given incoming RCPT command, it is permissible for the server to add one (using the recipient address from the RCPT command) to the corresponding outgoing RCPT command.

When a DSN-conforming server relays mail to a nonconforming server, however, the rules are a bit different:

- No DSN keywords (ENVID, RET, NOTIFY, ORCPT) may be used on any outgoing MAIL or RCPT commands to the nonconforming SMTP receiver.

- If the MAIL command contained a NOTIFY keyword with SUCCESS among the values, and the message was successfully handed off to the nonconforming SMTP receiver (2xx response code), the SMTP sender should generate a "relayed successfully" notification message to the original sender.

- If the MAIL command contained a NOTIFY keyword with FAILURE among the values, and the message could not be delivered (5xx response code), the SMTP sender should generate a "failed" notification message to the original sender (which is really just normal operation).

- If the MAIL command contained a NOTIFY keyword with the value NEVER, the SMTP sender should not generate a notification message regardless of whether the delivery succeeded or failed (again, normal operation). If the delivery failed, it is permissible to inform the local postmaster of that failure.

The SMTP envelope for a notification message is defined as follows:

- The sender address must be empty (i.e., MAIL FROM: <>).

- The recipient address must be copied from the MAIL command in the message that caused the notification message to be generated (if the original message included MAIL FROM: <asmith@guys.com>, the notification message must include a single RCPT command, e.g., RCPT TO: <asmith@guys.com>).

▸ The RET keyword may not be specified in the MAIL command of a notification message. If the NOTIFY keyword is specified, it must include the value NEVER, to prevent message loops (feedback) from happening.

▸ The ENVID (with a new, unique ID value) and/or the ORCPT keywords may be used if desired.

The notification message itself consists of headers and a MIME multipart/report body part. The first subpart typically is text/plain (but may be other content types). The second subpart is message/delivery-status (see Section 15.5). The third subpart optionally contains part or all of the failed message.

The first subpart (human-readable status) can contain essentially anything, but in general it should be human readable.

The rules for what is returned in the second subpart (machine-parsable status) are:

▸ If the MAIL command of the original message contained an ENVID keyword, the value from that must be returned in an Original-Envelope-ID field. Otherwise, there must not be an Original-Envelope-ID field.

▸ If the RCPT command of the original message contained an ORCPT keyword, the value from that must be returned in an Original-Recipient field. Otherwise, there must not be an Original-Recipient field.

▸ The Reporting-MTA field must be supplied. Assuming the server that generated the notification message knows its own fully qualified Internet domain name (e.g., mailserver.megacorp.com), the MTA-name-type field contains "dns," and the Reporting-MTA field contains that FQDN.

▸ The Final-Recipient field must be supplied, and it contains the recipient's address from the message envelope (RCPT command). If the message was received via SMTP/ESMTP, the address-type is rfc822.

▶ The Action-Field must be supplied.

▶ The Status Field must be supplied, with the extended SMTP response code from RFC 1893.

▶ A few other fields are returned in various situations.

The rules for what is returned in the third subpart (returned message) are:

▶ If the message delivery fails, the RET keyword had the value FULL, and the failed message is smaller than some implementation-specific size limit, the third part should contain the entire failed message, and the content type should be message/rfc822.

▶ If the message delivery fails and the RET keyword had the value HDRS, or if the failed message is larger than the implementation-specific size limit, the third part should contain only the headers of the failed message. In that case, the content type of the third part should be text/rfc822-headers.

▶ If the message delivery succeeds, the third part should contain only the headers of the failed message, regardless of what value the RET keyword had. Again, the content type of the third part should be text/rfc822-headers.

A typical message (complete with ESMTP protocol envelope) that requests a DSN follows:

S: (opens TCP connection to port 25)

```
R: (accepts connection)
R: 220 gals.com ESMTP Server ready
S: EHLO guys.com
R: 250-gals.com
R: 250-DSN
R: 250 SIZE
S: MAIL FROM:<asmith@guys.com> RET=HDRS ENVID=AA123456
R: 250 sender OK
S: RCPT TO: <atoklas@gal.com> NOTIFY=SUCCESS,FAILURE DELAY ORCPT=rfc822;
    atoklas@gals.com
R: 250 recipient OK
S: DATA
R: 354 send data, end with CRLF.CRLF
S: To: atoklas@gals.com (Alice Toklas)
S: From: asmith@guys.com (Albert Smith)
```

```
S: Subject: Important meeting!
S: Content-Type: text/plain; charset=us-ascii
S:
S: It is very important that you come to a meeting tonight at 10pm,
S: at the University Union bldg, room 201.
S:
S:.
R: 250 message received
S: QUIT
R: 221 ESMTP server signing off
S: (closes connection)
R: (closes connection)
```

Multipart/report content type (RFC 1892)

The basic MIME specification defines several content types for entire messages or message subparts. Some of those are multipart content types (e.g., multipart/mixed, multipart/parallel, and multipart/alternative). MIME also makes a provision for new content types to be defined. RFC 1892 defines a new MIME content type (multipart/report), which in theory could be used for a variety of purposes, but it was defined specifically for the top-level content type of returned DSNs, as defined in RFC 1891. Another MIME content type (text/rfc822-headers) also is defined. A third content type (which is intended for the second subpart of a DSN and hence is nested inside a multipart/report content type) called message/delivery-status is defined in RFC 1894.

The syntax of a multipart/report MIME content type is similar to that of multipart/mixed (see Chapter 14). In a DSN message, it must be the top-level content type (i.e., not nested in some other content type). One of the main reasons for defining a new content type is to allow any software that might be receiving a notification message to be able to recognize it quickly as such. As already described, a multipart/report content type message has either two or three components: human-readable status (required), machine-parsable status (required), and returned message or message headers (optional).

Two keywords are required in the multipart/report declaration: report-type (which must have the value "delivery-status") and boundary (as used with multipart/mixed).

An example of a DSN message (that would have been returned on successful delivery of the message in the previous example) follows:

```
To: asmith@guys.com

From: postmaster@gals.com
Subject: Delivery Notification (success) for atoklas@gals.com
Content-type: multipart/report; report-type=delivery-status; Boundary=*****
MIME-Version: 1.0

--*****
Content-type text/plain; charset=us-ascii

Your message (id AA123456) was successfully delivered to atoklas@gals.com,
 on 7 March, 1997 14:17:23

--*****
Content-type: message/delivery-status

Reporting-MTA: dns; mailserver.gals.com
Original-Envelope-ID: AA123456

Original-Recipient: rfc822;atoklas@gals.com
Final-Recipient: rfc822;atoklas@gals.com
Action: delivered
Status: 2.0.0

--*****
Content-type: text/rfc822-headers

To: atoklas@gals.com (Alice Toklas)
From: asmith@guys.com (Albert Smith)
Subject: Important meeting!
Content-Type: text/plain; charset=us-ascii

--*****--
```

Enhanced mail system status codes (RFC 1893)

The discussion in Chapter 13 about RFC 821 examined the three-digit response codes used by an SMTP receiver (e.g., an Internet Mail server) to indicate whether a given command from the SMTP sender (e.g., an Internet Mail UA or another Internet Mail server relaying mail) succeeded or failed. RFC 1893 defines a new set of response codes similar to (but not compatible with) the SMTP response codes for use in the machine-parsable subpart of DSN messages (not in the SMTP protocol itself). One of the main motivations for defining a new set of codes was that the SMTP response codes were inadequate for handling DSNs (and could not be adequately extended). The new codes are intended to be used only for media- and language-independent status reporting (not system- or implementation-specific errors).

While the SMTP response codes are three adjacent digits (e.g., 250), the new mail status codes are three numeric fields separated by periods. The first field is a single digit (specifically, 2, 4, or 5). The second and third fields are values with one to three digits (this particular standard, however, defines only single-digit codes for these fields). A status code can contain no spaces, tabs, or comments and no leading zero digits (e.g., use 7, not 07 or 007).

The first field (exactly like the first digit of the SMTP response codes) can have the values listed in Table 15.1.

New codes in the first field can be defined only by standards-track documents (a special class of RFCs).

The second field classifies the status. The currently defined codes are listed in Table 15.2.

Table 15.3 lists in numerical order all the status codes defined in RFC 1893 (X means "does not matter" or "any digit").

Table 15.1
First-field values of new mail status code

Value	Meaning	Description
2	Success	Indicates a successful delivery action. The subcodes (the second and third fields) can be used to provide notification of transformations required for delivery.
4	Persistent transient (recoverable) failure	Indicates a valid message that could not be delivered successfully at this time (e.g., server rejected message due to lack of space). Resending of exactly the same message at some point in the future may be successful.
5	Permanent (unrecoverable) failure	Indicates a failure that likely will not succeed just by resending the same message (either the message itself or the destination must be changed before it can be successfully delivered).

Table 15.2
Second-field values of new mail status code

0	Other or undefined	No details available, or error is not covered by currently defined codes.
1	Addressing status	Either the origination or the destination address is in error.
2	Mailbox status	There is a problem with the recipient's mailbox (e.g., no such mailbox).
3	Mail system status	There is a problem with the destination system.
4	Network and routing status	There was a problem with the network plumbing along the way (e.g., unexpected timeout).
5	Mail delivery protocol status	There was a problem with the SMTP protocol (e.g., implementation error or corrupt data).
6	Message content or media status	There was a problem with the message MIME structure, encoding, or other content (e.g., unknown content-type or error in specification).
7	Security or policy status	There was a problem with system policy (e.g., mail not accepted from that domain) or security (e.g., cannot obtain necessary encryption key).

Table 15.3
Status codes defined in RFC 1893

Code	Definition
X.0.0	Other or undefined status
X.1.0	Other address status
X.1.1	Bad destination mailbox address
X.1.2	Bad destination system address
X.1.3	Bad destination system address syntax
X.1.4	Destination mailbox address ambiguous
X.1.5	Destination mailbox address valid
X.1.6	Mailbox has moved
X.1.7	Bad sender's mailbox address syntax
X.1.8	Bad sender's system address

Table 15.3 (continued)

X.2.0	Other or undefined mailbox status
X.2.1	Mailbox disabled and not accepting messages
X.2.2	Mailbox full
X.2.3	Message length exceeds administrative limit
X.2.4	Mailing list expansion problem
X.3.0	Other or undefined mail system status
X.3.1	Mail system full
X.3.2	System not accepting network messages
X.3.3	System not capable of selected features
X.3.4	Message too big for system
X.4.0	Other or undefined network or routing status
X.4.1	No answer from host
X.4.2	Bad connection
X.4.3	Routing server failure
X.4.4	Unable to route
X.4.5	Network congestion
X.4.6	Routing loop detected
X.4.7	Delivery time expired
X.5.0	Other or undefined protocol status
X.5.1	Invalid command
X.5.2	Syntax error
X.5.3	Too many recipients
X.5.4	Invalid command arguments
X.5.5	Wrong protocol version
X.6.0	Other or undefined media error
X.6.1	Media not supported
X.6.2	Conversion required but prohibited
X.6.3	Conversion required by not supported
X.6.4	Conversion with loss performed
X.6.5	Conversion failed

Table 15.3 (continued)

X.7.0	Other or undefined security status
X.7.1	Delivery not authorized, message refused
X.7.2	Mailing list expansion prohibited
X.7.3	Security conversion required but not possible
X.7.4	Security features not supported
X.7.5	Cryptographic failure
X.7.6	Cryptographic algorithm not supported
X.7.7	Message integrity failure

Message/delivery-status content type (RFC 1894)

RFC 1894 defines a new MIME content type for use as the machine-parsable subpart of a DSN message (the message/delivery-status content type normally is nested inside a top-level MIME multipart/report content type as defined in RFC 1892).

Basically, this content type consists of two or more groups of short lines of ASCII text (called fields). Each field consists of a field name (e.g., Original-Envelope-ID), a separator (:), then the value associated with that field name (e.g., AA123456). A blank line separates the various groups of fields. A typical field might look like this:

```
Original-Envelope-ID: AA123456
```

The first group contains per-message fields (which apply to all recipients of the message), and each field can occur only once in a given message/delivery-status subpart (hence, only once in a given DSN message).

The remaining group(s) contain per-recipient fields for each of the message recipients (one group per recipient).

Per-message fields

The per-message fields, if present, occur in the following order.

Original-envelope-ID

This optional field is the returned envelope identifier from the ENVID parameter on the MAIL command that requested the DSN be sent. That allows an SMTP sender to unambiguously associate the DSN with a specific message that it sent.

An example is:

```
Original-envelope-ID: AA123456
```

Reporting-MTA

This required field is the name of the MTA that actually sent the DSN message. It could be any MTA along the path of the mail message. A successful delivery message can be sent only from the final recipient's MTA, but a failure could be reported anywhere along the way (including from your own local MTA). The value of this field is the MTA name type (typically "dns" to indicate a normal Internet node), followed by the node name.

An example is:

```
Reporting-MTA: dns; mailserver.gals.com
```

Dsn-gateway

This optional field is the nodename of the gateway or MTA from which the reporting MTA got the (non-Internet) delivery status information used to create this DNS message. It is present only if the information came from a foreign mail system. A foreign mail system to SMTP gateway typically would have "smtp" as the name type.

An example is:

```
Dsn-gateway: smtp; exchange.megacorp.com
```

Received-from-MTA

This optional field is the nodename (and normally IP address) of the MTA from which the message (that this DSN message is about) was received. If coming from an Internet host via SMTP, this field should be the node/domain name from the HELO (or EHLO) command. The IP address of that node should be included as a comment in parentheses.

An example is:

```
Received-from-MTA: smtp; mailserver.guys.com (123.45.67.89)
```

Arrival-date

This optional field is the date and time at which the message arrived at the reporting MTA (see also the Last-Attempt-Date field in the per-recipient fields). The date and time are expressed in RFC 822 syntax, complete with timezone field.

An example is:

```
Arrival-date: Sat, 8 Mar 1997 12:17:00 -0800
```

Per-recipient fields

The per-recipient fields, if present, occur in the following order.

Original-recipient

This optional field is the original recipient to whom the message was addressed, from the ORCPT parameter on the RCPT command used to send the message (that this DSN message is about). The address-type normally is rfc822. This field may differ from the Final-recipient value if the message was altered (e.g., by going through a gateway) while in transit. If there was no ORCPT keyword on the RCPT command of the original message, there should not be an Original-recipient field in the DSN message.

An example is:

```
Original-recipient: rfc822;atoklas@gals.com
```

Final-recipient

This required field is the recipient to whom this set of per-recipient fields is directed. It may have been modified in various ways while in transit from the original specification of the recipient (see Subsection 15.5.2.1).

An example is:

```
Final-recipient: rfc822;atoklas@gals.com
```

Action

This required field is the action performed by the Reporting-MTA while attempting to deliver the message (that this DSN is about). The possible actions are as follows:

▶ Failed. The message could not be delivered to the recipient.

- Delayed. The message has not been delivered yet, but MTA is still trying.

- Delivered. Delivery to the user (or mailing list exploder, etc.) was successful.

- Relayed. The message was forwarded on to a non-DSN compliant MTA.

- Expanded. The message was relayed to a multiple-recipient alias (not a terminal delivery; an additional success or failure notice may be generated from exactly one of the final deliveries).

An example is:

```
Action: delivered
```

Status

This required field is the enhanced mail system status code (see the description of RFC 1893 in Section 15.4) that indicates the exact result of the delivery attempt (e.g., 4.4.1 for "no answer from host, please try again").

An example is:

```
Status: 4.4.1
```

Remote-MTA

This optional field is the nodename of the remote node that reported delivery status to the actual reporting MTA.

An example is:

```
Remote-MTA: smtp;exchange.megacorp.com
```

Diagnostic-code

This optional field is provided primarily for relaying detailed diagnostic code from foreign mail systems.

Last-attempt-date

This optional field is the date and time of the last attempt by the reporting MTA to deliver the message. It uses RFC-822 date-time format.

An example is:

```
Last-attempt-date: Sat, 8 Mar 1997 12:17:00 -0800
```

Will-retry-until

This optional field is for delayed DSN messages and shows the date and time after which the reporting MTA is no longer planning to attempt delivery. It should not be used in any other type of DSN message.

An example is:

```
Will-retry-until: Sat, 15 Mar 1997 12:17:00 -0800
```

Relevant RFCs

The following RFCs are relevant to DSN.

1893	PS	G. Vaudreuil, "Enhanced Mail System Status Codes," 01/15/1996. (Pages=15) (Format=.txt)
1892	PS	G. Vaudreuil, "The Multipart/Report Content Type for the Reporting of Mail System Administrative Messages," 01/15/1996. (Pages=4) (Format=.txt)
1891	PS	K. Moore, "SMTP Service Extension for Delivery Status Notifications," 01/15/1996. (Pages=31) (Format=.txt)

POP: Post Office Protocol

As of the writing of this book, POP3 is far and away the most popular way to retrieve messages from an Internet e-mail postoffice. Over the next few years, you will see the tide rapidly turn toward IMAP for that function. However, POP3 will continue to be supported as a legacy protocol for many years to come. It is likely that ISPs will start to offer IMAP as a premium service in the near future (as an alternative to POP3).

POP has been through quite an evolution. The original version, titled just Post Office Protocol with no version number (it should be considered POP version 1), is described in RFC 918 (Oct. 1984).

POP2, which made some major changes to the protocol and was the first widely deployed, is described in RFC 937 (Feb. 1985).

POP3 had a major change in direction and is really a new protocol. Numerous RFCs define POP version 3, each RFC obsoleting the previous one. RFC 1081 (Nov. 1988) is a major redefinition of POP to v3; RFC 1225 (May 1991) contains clarifications; RFC 1460 (June 1993) removes RPOP and adds APOP; RFC 1725 (Nov. 1994) removes LAST and adds UIDL; and RFC 1939 (May 1996) examines security and implementation issues.

Another RFC (1734) extends POP3 to allow cryptographic authentication and privacy.

Like SMTP, POP is a command/response protocol implemented over a reliable transport (TCP). A POP server sends a greeting in response to a new connection, then the client repeats a cycle of sending a command and waiting for the response. Each command and response is a short sequence of ASCII characters, followed by CR and LF. Each command starts with a four-character keyword, followed by optional arguments. The response is simpler than in SMTP, consisting of a line starting with a success/failure indicator (+, -, =, or # in POP2; +OK or –ERR in POP3), followed by returned data in some cases and human-readable text in others. A few responses are multiline (terminated by a line with only a single period in the first byte).

POP2

Version 2 of the POP protocol is defined in RFC 937. It has not evolved anywhere near as much as version 3 and is not as widely implemented. POP2 is implemented over a connection-oriented, reliable transport (TCP) and uses socket port 109. Like IMAP (and unlike POP3), POP2 assumes that the mailbox resides on the Internet Mail server, not the workstation. POP2 does support connection to various mailboxes but has no mechanism for creating or deleting them (mailbox management is handled outside POP2, e.g., by a system or mail administrator).

A given connection starts with the client attempting to connect to the server via port 109. The server accepts the connection and responds with a greeting ("+ OK POP2 server Ready"). Then the client sends a request, which the server processes, after which it sends a response. The response can start with various characters that characterize the type of response:

- A plus sign (+) is a positive response; the request was processed with no problems.

- A minus sign (-) is an error response; the request failed in some way.

- An equal sign (=) is followed by a number indicating the size of a message.

- A number (or hatch) symbol (#) is followed by a number indicating a message count.

This command/response cycle can be repeated any number of times, until the QUIT command is transmitted or the server unilaterally aborts the connection due to some error.

POP2 server states

From the server's viewpoint, a given connection starts (on accepting the connection and sending the greeting) in the AUTH (authentication) state and progresses through various states as commands are received. After AUTH, the states are MBOX, ITEM, NEXT, and DONE.

AUTH state

Client command	Server response

HELO <user> <pass>	If valid user name and password, then
	    Set current mailbox = default for user
	    Open current mailbox R/W, exclusive access
	    Set N = 1
	    Send "#<num messages in current mailbox>"
	    Go to MBOX state
	
	Else
	    Send "- Error"
	    Disconnect

QUIT	Send "+ Bye" Disconnect
Any other command	Send "- Error" Disconnect

MBOX state

Client command	Server response

FOLD <mailbox>	Delete marked messages in current mailbox
	Close current mailbox
	If valid mailbox specified, then
	    Set current mailbox to specified mailbox
	    Open current mailbox R/W, exclusive access
	    Set N = 1
	    Send "# msgs in current mailbox"
	
	Else
	    Set N = 1
	    Send "#0"
	    Go to MBOX state

READ	Send "=<message size>" Go to ITEM state
QUIT	Send "+ Bye" Go to DONE state
Any other command	Send "- Error" Disconnect

ITEM state

Client command	Server response

FOLD <mailbox>	Delete marked messages in current mailbox
	Close current mailbox
	If valid mailbox specified, then
	    Set current mailbox to specified mailbox
	    Open current mailbox R/W, exclusive access
	    Set N = 1
	    Send "#<num msgs in current mailbox>"
	
	Else
	    Set N = 1
	    Send "#0"
	    Go to MBOX state

READ	If there are messages left then
	    Send "=<message size>"
	
	Else
	    Send "=0"
	    Go to ITEM state

ITEM state (continued)

READ<nnn>	Set N = nnn
	If there is a message nnn then
	Send "=e of message N
	
	Else
	  Send "=0"
	Go to ITEM state

RETR	If at least 1 message left, then
	    Send data for message N
	    Go to NEXT state
	
	Else
	    Disconnect

QUIT	Send "+ Bye" Go to DONE state
Any other command	Send "- Error" Disconnect

NEXT state

Client command	Server response
ACKS	Leave message N in mailbox Set N = N+1 Go to ITEM state
ACKD	Mark message N for deletion Set N = N+1 Send "=<size of Message N>" Go to ITEM state
NACK	Leave message N in mailbox Send "=<size of Message N>"
Any other command	Send "- Error" Disconnect

DONE state

Client command	Server response
Any command	Send "- Error" Disconnect
(close connection)	(close connection)

POP2 commands

HELO username password

This command authenticates the user. If the user has an account on this server, and the password is the correct one for that user, it opens the default mailbox for that user in Read/Write mode, with exclusive access until the end of session or until a new folder is opened.

Possible responses are:

```
#nnn       nnn = number of messages in default mailbox for user
- error    if user is not a known account, or password is wrong
```

An example follows (again, C stands for "client" and S for "server"):

```
S:   + OK POP2 Server ready
C:   HELO asmith swordfish
S:   #2
```

FOLD mailbox

This command deletes any messages marked for deletion in the current mailbox. It closes the current mailbox and relinquishes exclusive access. If the specified mailbox exists, this command opens it in Read/Write mode with exclusive access and makes it the current mailbox. It returns the number of messages in the new current mailbox. If the specified mailbox does not exist, the command returns the number of messages as 0.

Possible responses are:

```
#nnn       nnn = number of messages in specified mailbox
```

An example is:

```
C:   FOLD /usr/mail/asmith
S:   #3
```

READ

This command prepares to read the next message in sequence. It returns the number of bytes in the message. If no more messages remain, it returns a message size of 0.

Possible responses are:

```
=ccc ccc = message length in bytes (0 if no more messages)
```

An example is:

```
C:   READ
S:   =120
```

READ message_number

This command prepares to read the specified message and makes it the current message. It returns the number of bytes in the message. If the specified message does not exist, the command returns a message size of 0.

Possible responses are:

```
=ccc ccc is the message length in bytes (0 if no such message)
```

An example is:

```
C:   READ 2
S:   =200
```

RETR

This command requests that the server send the message just selected with a READ command. The response is the message in rfc822 format. The client must send ACKS, ACKD, or NACK as the next command. If the previous READ command returned a 0, RETR will cause the connection to terminate.

Possible responses are message text (message headers, null line, and message body) or a closing of the connection (if an attempt is made to read a nonexistent message).

An example is:

```
C:   RETR
S:   first line of message
S:   . . .
S:   last line of message
S:   .
```

ACKS

This command acknowledges correct receipt of a message as initiated by the last RETR command and saves it on the server. It is an increment current message indicator. It returns the number of bytes in the next message or 0 if there are no more messages.

Possible responses are:

```
=cccccc is the message length in bytes (0 if no more
messages)
```

Example:

```
C:   ACKS
S:   =200
```

ACKD

This command acknowledges correct receipt of a message as initiated by the last RETR command and marks it for deletion from the server. It is an increment current message indicator. It returns the number of bytes in the next message or 0 if there are no more messages.

Possible responses are:

```
=cccccc is message length in bytes (0 if no more mes-
sages)
```

Example:

```
C:   ACKD
S:   =200
```

NACK

This command is a negative acknowledgment of receipt of a message as initiated by the last RETR command (i.e., something went wrong, like insufficient disk space to store the message). The command leaves the message on the server, does not increment current message indicator, and returns the number of bytes in the same message.

Possible responses are:

```
=ccc        ccc = message length in bytes (0 if no more
messages)
```

An example is:

```
C:    NACK
S:    =200
```

QUIT

This command terminates the session. Any messages marked for deletion are deleted. The current mailbox is closed, and exclusive access to it is relinquished. Server sends + OK and then closes the connection.

Possible responses are:

```
+ OK
```

An example is:

```
C:    QUIT
S     + OK POP2 server signing off
```

POP3

POP3 is the most commonly used scheme for retrieving messages from an Internet Mail server. It likely will be phased out slowly and replaced with IMAP over the next few years. IMAP can do everything POP3 can do, plus quite a few new and useful tricks. POP3 is not just a newer version of POP2; the commands are quite different, even the port is different. It is entirely possible for a given Internet Mail server to support both POP2 and POP3 (and possibly even IMAP4). POP3 is implemented over a connection-oriented, reliable transport (TCP) and uses socket port 110.

POP3 is never used between Internet Mail servers. In fact, it primarily exists only within a given LAN within the Internet, between individual users' computers and their local Internet Mail server. It does not place any requirements on the syntax of the messages it retrieves (although they typically are RFC 822 compliant, optionally with MIME subparts). Also, use of POP3 by a given recipient to retrieve e-mail messages does not impose any burden on anyone sending message to them to do anything beyond normal RFC 821/RFC 822 messaging (it is not an end-to-end protocol). The sender would do things exactly the same (and send exactly the same set of bits) whether the recipient is going to use POP3 or IMAP (or something completely different) to retrieve messages from the destination Internet Mail server.

Any processing of MIME subparts by an SMTP/POP3 UA is an implementation detail and has nothing to do with the protocol used to retrieve the message from the POP3 server. In comparison, IMAP allows access to the MIME message structure and individual MIME subparts of a message.

A POP3 server must be able to handle a large number of simultaneous connections from UAs. Each connection requires exclusive (Read/Write) access to the mailbox corresponding to the logged-in user for the duration of the connection. A typical connection involves authenticating, checking for (new) messages, downloading messages, and deleting messages from the server. Each user has only a single mailbox, and mail is assumed to be passing through the mail server (as opposed to IMAP, in which the server has multiple mailboxes per user and is a permanent, centralized message store).

POP3 authentication mechanisms

On accepting a TCP socket connection from a client (on port 110), a POP3 server responds with this type of greeting:

```
+OK <human-readable text>
```

An example is:

```
+OK POP3 server ready
```

At this point, here are three possible ways for a client to authenticate itself. The first (USER/PASS) involves sending a plaintext user name and password. The second (APOP) is optional and uses a somewhat more sophisticated scheme that does not expose the password in plaintext on the network. The third (AUTH) also is optional (defined separately in RFC 1734) and uses the most sophisticated scheme, using one of several cryptographic authentication schemes as defined for IMAP4 (Kerberos v4, GSSAPI, S/KEY, etc.).

The most common authentication method uses the USER and PASS commands. Almost all POP servers support at least this authentication scheme (but RFC 1939 for the first time makes them optional). First, a plaintext user name (also called the POP3 login name, or the POP3 account name) is sent in a USER command, followed immediately by a plaintext password in a PASS command. Many Internet Mail servers maintain their own account/password database, independent of the operating system's account/password database. Some servers (e.g., Software.com's Post.Office) optionally allow linkage of e-mail accounts to operating system accounts (so the password required to authenticate a POP3 session is the same as the operating system login password). As of RFC 1939, the rules for using USER and PASS were tightened up quite a bit, to make it more difficult to hack into a POP server (see descriptions of those commands for details). For example, if "asmith" is a valid POP3 login account, and "swordfish" is the password for that account, the following sequence would authenticate that user (with the USER/PASS scheme):

```
C:   (open TCP connection to server on port 110)
S:   +OK POP3 Server Ready
C:   USER asmith
S:   +OK valid user
C:   PASS swordfish
S:   +OK user authenticated
```

There are two main problems with that scheme. First, the user name and the password are specified in two separate commands, which means, in theory, that a hacker could determine valid account names by trying a list of likely names—any positive response is a hit. RFC 1939 addresses that problem in two ways. First, a USER command is valid only immediately after the POP3 greeting or after a failed USER/PASS command sequence (which might take a few seconds). Second, it is specifically allowed to return a positive response (+OK) for any user name but then return a negative response (-ERR) to the subsequent PASS command if the user name was not a valid one (in effect, that puts the user name and the password into a single command). However, that still does not address the possibility of a hacker obtaining user names and passwords by snooping on the network. That is what the APOP and AUTH commands are for.

The APOP authentication command is optional (and not commonly implemented). However, if a POP3 server supports it, the greeting (in response to a new connection) must include a timestamp, as defined in RFC 822. Therefore, if the greeting from a given POP3 server includes a string between the left and right angle brackets, that server likely supports APOP. If not, it definitely does not (and the client should not try to use APOP).

The basic idea of the APOP command is for the client to take the timestamp from the greeting, append a "shared secret" (a string known only to the server and the client), produce a digest of the resulting string with MD5 (a well-known message digest algorithm, defined in RFC 1321), and send the user's POP account name and the digest back to the server as the arguments to an APOP command. The server will append the shared secret to the timestamp it sent, produce a digest of the resulting string with MD5, and compare that with the received digest from the APOP command. If the received and generated hashes match, the client must have known the shared secret.

The shared secret is not defined in the RFC, other than that it must be a string known only to the POP server and the POP clients using that server, and the string should be considerably longer than the eight-character example used. That means the details of the shared secret, including how it is made known to both the server and the client, are completely up to the implementor (not really in the spirit of open standards). It could be the same shared secret for everyone, or since the user's POP account name is sent as part of the APOP command (prior to the server having to know the shared secret), the shared secret could be different for each POP account. The server also has knowledge of the IP address from which the client connected, so the shared secret could be a function of which node the client is on.

If the same shared secret is used for all POP accounts, there must be a secure way for that secret to be obtained by the server and all clients. One way would be to set it manually in the server and all clients at installation time (difficult to do in a truly secure way, unless one person does all such installations for the entire site). Another method would be for the server and all clients to obtain the shared secret either from a commonly accessible file or via a network connection (outside the POP protocol). Any of those schemes requires use of a server and clients that support the scheme used for sharing the secret (which means that not just any client can be used, at least not for APOP authentication).

If the shared secret is different for each POP3 account, it could be a derived function of the account name itself (e.g., the account name mangled with some algorithm known to both server and clients that is not known to third parties trying to hack into your system). It also could be an arbitrary string that the server can look up in a table (one for each POP account). That is basically the same thing as a POP account password. Unfortunately, that precludes the use of the user's operating system logon password (as the POP account password) on NT, because it is not possible to obtain a user's login password itself to append to the timestamp (you can determine whether a particular password is correct or not only by attempting to do an internal logon).

If the shared secret is a function of the IP address from which the client is accessing the server, that could be done by consistently mangling an IP address into a string (in some way known to the server and all clients but not to third parties attempting to hack into your system) or by a table lookup of a string that would be unique for each IP address. That would involve setting the correct string at client install time, coordinated with a list of such strings on the server (which is difficult to do at all, let alone securely). It also means a client could not change its IP address (which precludes use of DHCP).

The best bet is probably for the POP server to maintain a password (not linked to any operating system account) for each POP account and use that as the shared secret. The server and the clients still would need to know to use that scheme. It would be a bad idea to use a null string for the shared secret; in that case, *any* client could always authenticate. Problems with implementing and managing the shared secret are the main reason that APOP is not widely implemented.

Take the following example:

```
C:    (open TCP connection to server on port 110)
S:    +OK POP3 Server Ready <1234.697170952@guys.com>
C:    APOP asmith c9dcb935ce1d21d02afdebcbd9cb1d5a
S:    +OK user authenticated
```

In that example, asmith's client extracts the timestamp from the greeting:

```
<1234.697170952@guys.com>
```

It then appends asmith's POP account password to the timestamp:

```
<1234.697170952@guys.com>swordfish
```

Running that string through MD5 yields this digest:

```
c9dcb935ce1d21d02afdebcbd9cb1d5a
```

So, asmith's client sends an APOP command with his POP3 account name and that digest:

```
APOP asmith c9dcb935ce1d21d02afdebcbd9cb1d5a
```

The server remembers the timestamp it sent to asmith:

```
<1234.697170952@guys.com>
```

It then gets the account name from the first parameter of the APOP command:

```
asmith
```

The server then looks up the password for this POP3 account and appends it to the timestamp:

```
<1234.697170952@guys.com>swordfish
```

Running that through MD5 yields the following digest:

```
c9dcb935ce1d21d02afdebcbd9cb1d5a
```

Because that matches the digest sent by asmith as the second parameter in the APOP command, asmith must have known the shared secret and therefore is authenticated. Note that at no time is the shared secret itself (in this case, asmith's POP password) sent over the net in plaintext. MD5 is a one-way algorithm, so there is no way for a snooping hacker to recover asmith's password. However, asmith's account name is still in plaintext, which gives a hacker a known valid account name. For that reason, it may not be difficult for hackers to guess or figure out the password for that account (assuming they know that the shared secret is the POP account password). It would be better to send neither the account name nor the password in plaintext and also to use a stronger algorithm (MD5 has recently been cryptanalyzed). That is what the AUTH command is for.

With the AUTH command (which is exactly like the one in IMAP4), no change is required to the greeting from the POP server. That greeting may or may not include a timestamp, depending on whether the POP server also supports the APOP command. However, the AUTH command itself specifies only the desired cryptographic authentication algorithm you want to use. If a client gets a negative response (-ERR) to all the algorithms it supports (e.g., KERBEROS_V4, KERBEROS_V5, and GSS_API), you can assume the server does not support the AUTH command (or at least not with any algorithm you know how to do, which amounts to the same thing). In that case, the client should drop back to the less secure APOP command (if supported) or even the totally unsecure USER/PASS commands (if not supported).

Assuming a given POP3 server supports the AUTH command, using the KERBEROS_V4 algorithm, an authentication might look like the following:

```
C:  (open TCP connection to server on port 110)
S:  +OK POP3 Server ready
C:  AUTH KERBEROS_V4
S:  + AmFYig==
C:  BAcAQU5EUkVXLkNNVS5FRFUAOCAsho84kLN3/
    IJmrMG+25a4DT+nZImJjnTNHJUtxAA+oOKPKfHEcAF
    s9a3CL5Oebe/ydHJUwYFdWwuQ1MWiy6IesKvjL5rL9
    WjXUb9MwT9bpObYLGOKi1Qh
S:  + or//EoAADZI=
C:  DiAF5A4gA+oOIALuBkAAmw==
S:  +OK Kerberos V4 authentication successful
```

For details on how authentication algorithms such as Kerberos work, see Chapter 5.

In general, a client that supports all three authentication mechanisms should try the AUTH command first, then the APOP command (if the greeting indicates the server supports it), and fall back to USER/PASS only as a last resort. A server always should accept any method it supports (although a really paranoid site that supports either or both of the cryptographic mechanisms actually could reject USER/PASS or even APOP, if desired, even if they are valid, possibly just for particularly sensitive accounts). If advanced authentication is required, all affected users must use appropriate mail clients (ones that support the accepted authentication mechanisms) and be instructed in how to configure those mail clients properly.

POP3 commands

Like SMTP, a given session can be in one of several states:

- ▶ Authorization state: Connection has been accepted; client must authenticate.

- ▶ Transaction: Client has authenticated.

- ▶ Update: Client has sent QUIT command.

Most commands are permitted in only one state.

Authorization-state

Every connection starts in the authorization state, on the server accepting the connection (there is no preauthentication mechanism, as in IMAP).

In the authorization state, only five commands are legal: USER, PASS, APOP, AUTH, and QUIT (QUIT is documented in the transaction state section). Also, this is the *only* state in which those commands, except QUIT, are legal. All the authentication commands are optional, but at least one of the mechanisms must be supported (USER/PASS, APOP, and/or AUTH). Attempts to issue any other command while in the authorization state or the four authentication commands in any other state will receive a negative response (-ERR). The AUTH command is defined in a separate standard (RFC 1734).

USER username

This command is optional (but if not supported, APOP and/or AUTH must be supported). The user name (also called the POP3 login name or POP3 account name) is a name that the POP3 server uses to locate the appropriate mailbox in the mail server's message store (where the MTA deposits incoming messages). The name can be (but is not required to be) the same as the user's operating system account name and also can be (but is not required to be) related to the user's e-mail address. The mailbox name cannot include any spaces.

The USER command is valid only in response to the server greeting or immediately after a failed authentication command (USER, PASS, APOP, or AUTH).

A server may respond with +OK to any mailbox name (valid or not), if desired. It may respond with -ERR for an invalid mailbox name, if desired (in which case, another USER command may be issued immediately). It also may respond with -ERR for a valid mailbox name that does not permit authentication via USER/PASS.

An example is:

```
C:    (open TCP connection to server on port 110)
S:    +OK POP3 Server Ready
C:    USER asmith
S:    +OK
```

PASS password

This command is optional (supported if and only if the USER command is supported). The password is the POP3 password for the specified mailbox name in the associated USER command (this may or may not be the same as any other password, e.g., the user's NT login password).

A PASS command is valid only immediately after a successful USER command.

A positive response (+OK) indicates that all the following are true:

▶ Both the user name and password are valid.

▶ This account can be authenticated with USER/PASS.

▶ The mailbox is available for this mail client to use (exclusively).

A negative response (-ERR) may indicate that one or more of the following are true. The human-readable text following the -ERR may indicate the specific problem, but that is not required by the RFC. In general, if any text is returned, it should be displayed to the user of the client.

▶ The password was incorrect for the user specified in the USER command.

▶ The user specified in the USER command was invalid (even if the USER command received a positive response).

▶ The user specified in the USER command is valid, and the password may even be correct, but this mailbox will not accept authentication with the USER/PASS authentication mechanism (APOP or AUTH must be used).

▶ The user's mailbox currently is unavailable (e.g., it is busy because another mail client has locked it, the user's mailbox is being backed up, other maintenance activity is underway, the user's mailbox has expired, or the user's mailbox is disabled because the ISP bill is unpaid).

APOP username digest

This command performs POP account authentication without sending a password in plaintext.

An APOP command is valid only in response to the server greeting, or immediately after a failed authentication command (USER, PASS, APOP, or AUTH).

A positive response (+OK) indicates that all the following are true:

▶ The client knew the correct shared secret for this server.

▶ This account can be authenticated with APOP.

▶ The mailbox is available for this mail client to use (exclusively).

A negative response (-ERR) may indicate that any one or more of the following are true. The human-readable text following the -ERR may indicate the specific problem, but that is not required by the RFC. In general, if any text is returned, it should be displayed to the user of the client.

▶ The client did not know the correct shared secret for this server.

▶ The user specified in the APOP command was invalid.

▶ The user specified in the APOP command is valid, and the password may even be correct, but this user's mailbox will not accept authentication with the APOP authentication mechanism.

▶ The user currently is unavailable (e.g., it is busy because another mail client has locked it, the mailbox is being backed up, other maintenance activity is underway, the user's mailbox has expired, or the user's mailbox is disabled because the ISP bill is unpaid).

An example is:

```
C:    (open TCP connection to server on port 110)
S:    +OK POP3 Server Ready i.697170952@guys.com
C:    APOP asmith c9dcb935ce1d21d02afdebcbd9cb1d5a
S:    +OK user authenticated
```

AUTH authentication_mechanism

This command performs POP account authentication using an advanced cryptographic authentication mechanism, such as Kerberos. The following mechanisms are some that might be supported (see RFC 1731), although particular implementations may include some or all of these or even other schemes (e.g., KERBEROS_V5):

▶ KERBEROS_V4

▶ GSSAPI

▶ SKEY

There is no way for a POP client to know which authentication mechanisms are supported, other than by trying them with the AUTH command (there is no equivalent to the IMAP4 CAPABILITY command).

An AUTH command is valid only in response to the server greeting or immediately after a failed authentication command (USER, PASS, APOP, or AUTH).

It is possible for the authentication exchange to optionally enable a protection mechanism for the following session (symmetric key encryption using the exchanged session key). If a protection mechanism is selected, encryption begins immediately following the CR,LF at the end of the positive acknowledgment and continues for the rest of the session. Encrypted messages are sent in binary, with a 4-byte message length (network byte order, MSByte first), followed by the encrypted message.

An immediate negative response (-ERR) indicates that the requested authentication mechanism is not supported.

A server challenge (a plus sign followed by a space followed by a base64 string) indicates that the requested authentication mechanism is supported and a cryptographic authentication exchange is underway. At the end of that exchange, the final response will be sent.

A positive response (+OK) after the authentication exchange has begun indicates that all the following are true:

▶ Authentication was completed and was successful.

▶ This account can be authenticated with AUTH.

▶ The mailbox is available for this mail client to use (exclusively).

A negative response (-ERR) after the authentication exchange has begun may indicate that one or more of the following are true. The human-readable text following the -ERR may indicate the specific problem, but that is not required by the RFC. In general, if any text is returned, it should be displayed to the user of the client.

▶ The mailbox identified during the authentication exchange was invalid.

▶ The mailbox identified during the authentication exchange currently is unavailable (e.g., it is busy because another mail client has locked it, the mailbox is being backed up, other maintenance activity is underway, the mailbox has expired, or the mailbox is disabled because the ISP bill is unpaid).

▶ The mailbox identified during the authentication exchange was valid, but authentication failed.

▶ The mailbox identified during the authentication exchange is valid, and authentication may have completed successfully, but this mailbox will not accept authentication with the specific authentication mechanism selected.

Transaction state

Some POP commands are available only in the transaction state, which means the connection must have advanced successfully from the authorization state (by one of the USER/PASS, APOP, or AUTH commands). Once in the transaction state, the authorization-state commands no longer are accepted (except for QUIT). Being in the transaction state implies the user's mailbox is available for the exclusive use of this connection ("open and locked"). Required transaction state commands include STAT, LIST, RETR, DELE, NOOP, RSET, and QUIT. Optional transaction state commands are TOP and UIDL. All the commands leave the connection in the transaction state, except QUIT, which advances the connection to the update state.

STAT

The STAT command, which is required, requests the status of the mailbox. The response consists of +OK, followed by one space, followed by the number of messages in the mailbox, one more space, then the size of the mailbox in bytes. This command can be used at the start of a session to determine how many messages are currently on the server for the authenticated user and what the total size of those messages is.

An example is:

```
C:   STAT
S:   +OK 2 320
```

LIST [message_number]

This required command can be specified with no argument, which results in a multiline response with information on all messages in the mailbox (that are not currently marked for deletion). If an argument is specified, it should be a message number (of a message not currently marked for deletion), in which case the response is a single line with information on the requested message. The information returned for each message consists of the message number, followed by one space, followed by the message size in bytes.

This command typically is used at the start of a session to determine what messages exist, so that RETR commands can be done for each of them.

If an argument is specified and there is no corresponding message number (or that message is marked for deletion), a negative response (-ERR) is sent.

If the mailbox is empty, the response to a LIST command with no arguments is +OK 0 0.

An example with an argument is:

```
C:   LIST 2
S:   +OK 2 200
```

An example with no argument is:

```
C:   LIST
S:   +OK 2 messages (320 octets)
S:   1 120
S:   2 200
S:   .
```

UIDL [message_number]

This optional command is similar to the LIST command but returns a unique message identifier for the specified message (or for all messages if no argument is specified). Each response contains the message number, followed by one space, followed by the unique message identifier, followed by end of line.

A unique message identifier can be any string, from 1 to 70 characters long (each character being a printable ASCII character in the range 0x21 to 0x7E). Each identifier must be unique within the mailbox (not necessarily for the entire message store) and be persistent across sessions (if a given message has a particular UID in one session, it must have the same ID in all sessions). Once used, a given UID can never be reused (even if the corresponding message is deleted), as long as that mailbox exists. If the specified message does not exist or has been marked for deletion, a negative response (-ERR) is sent.

This command is used to implement a poor man's server-based message store (not as good as that provided in IMAP). Basically, a client does not automatically delete messages after each RETR; then at the start of each session, it does a UIDL (with no argument) to obtain a list of UIDs for all messages currently on the server. Any UIDs not currently known by

the client are retrieved. That of course requires the client to keep track of
UIDs for all retrieved messages.

An example with an argument is:

```
C:   UIDL 2
S:   +OK 2 QhdPYR:00WBw1Ph7x7
```

An example with no argument is:

```
C:   UIDL
S:   +OK
S:   1 whqtswO00WBw418f9t5JxYwZ
S:   2 QhdPYR:00WBw1Ph7x7
S:   .
```

RETR message_number

This required command retrieves a single message. The argument must be
a valid message number (and that message must not currently be marked
for deletion). If the specified message exists, a positive response is sent,
with the message size in bytes. Following that, a multiline response is sent
that is the rfc822 message (headers and body). Like any multiline
response, the final line from the server is a period on a line by itself. If the
specified message does not exist, a negative response is sent.

Examples are:

```
C:   RETR 1
S:   +OK 120 octets
S:   <first line of message>
S:   . . .
S:   <final line of message>
S:   .
C:   RETR 3
S:   -ERR no such message
```

TOP message_number number_of_lines

This optional command is similar to the RETR command but sends all
header lines, the null line terminating the headers, and the specified
number of lines of the message body as a multiline response. If the

specified number of lines (which must be a positive integer) is larger than the number of lines in the specified message, the entire message is sent. This mechanism can be used to provide a simple selective retrieval scheme in a POP client (weak compared to equivalent functionality in IMAP). The only way a client can determine if a given POP server supports the TOP command is to try using it to retrieve the headers of a message that is known to be retrievable with RETR. In that case, if TOP is not supported, a negative response (-ERR) is sent.

Examples are:

```
C:    TOP 1 5
S:    +OK 120 octets
S:    <all header lines>
S:    <null line following headers>
S:    <first 5 lines of message body>
S:    .
C:    TOP 1 0
S:    +OK 120 octets
S:    <all header lines>
S:    <null line following headers>
S:    .
C:    TOP 3 100
S:    -ERR no such message
```

DELE message_number

This required command marks a specified message for deletion (it will not actually be deleted until the update state). If the message exists (and is not already marked for deletion), a positive response (+OK) is sent. Otherwise, a negative response (-ERR) is sent.

An example is:

```
C:    DELE 1
S:    +OK message 1 deleted
```

```
C:    DELE 2
S:    -ERR message 2 already deleted
```

NOOP

This required command does nothing and always sends a positive response (+OK). It can be used to prevent an autologout timer from expiring or to determine if the server is still functioning.

An example is:

```
C:    NOOP
S:    +OK
```

RSET

This required command unmarks any messages previously marked for deletion, then sends a positive response (+OK). It is used if some error occurs (e.g., the client did not have room to store downloaded messages) and you want to return the mailbox to the state it was in at the beginning of the session.

An example is:

```
C:    RSET
S:    +OK
```

QUIT

This required command ends the transaction state and enters the update state.

If some error prevents the server from deleting all messages marked for deletion, a negative response (-ERR) is sent. Otherwise, a positive response (+OK) is sent. Note that e-mail clients from Netscape are notorious for not sending the required QUIT command at the end of a session, which wreaks havoc with some e-mail servers. The RFCs go to some length to make it clear that the QUIT command is *not* optional.

An example is:

```
C:    QUIT
S:    +OK POP3 server signing off
```

Update-state

Once the QUIT command has been sent, the server goes through and actually (quietly) deletes all messages that were marked for deletion. The mailbox is closed, and the connection's exclusive access to it is relinquished (the mailbox is "unlocked"). If the client terminates the connection without a valid QUIT command, the server should not enter the update state (the equivalent of an automatic RSET should be done). Finally, the network socket connection to the client is closed.

Accessing one mailbox from multiple POP3 clients

Some POP3-based Internet Mail clients (e.g., Qualcomm's Eudora) have the option of leaving mail on the server. That allows a single POP server mailbox to be accessed from more than one client (e.g., one at work and one at home). If configured as described here, you will be able to retrieve all messages at both sites (regardless of which one retrieves messages first); however, messages will not continue to accumulate on the server indefinitely.

The POP server must support the optional UIDL command.

The clients both should be configured as follows:

▶ Leave mail on server: selected configuration;

▶ Delete mail from server after n days: selected configuration (set n to value of 7 to 30);

▶ Determine first unread message: first message not read by this machine;

▶ Delete from server when emptied from trash: deselected configuration.

The number of days to leave mail on server should be set long enough to ensure that both clients have a chance to download all messages. The shorter that time, the less total mail will be on the server at any given time (n days worth). That can be set by experience, but something in the 7- to 30-day range should suffice. Note that this scheme extends to any number of clients accessing the central store from multiple locations. Unlike IMAP, one local message store is created for each client.

Examples of entire sessions

The first example is of a POP2 client doing a simple retrieval from a POP2 server.

```
C:   (initiate TCP connection on port 109)
S:   (accept TCP connection)
S:   + OK POP2 server ready
C:   HELO asmith swordfish
S:   #2
C:   READ
S:   =120
C:   RETR
S:   (contents of message 1, headers, null line &
     body)
S:   .
C:   ACKD
S:   =200
C:   READ
S:   =200
C:   RETR
S:   (contents of message 2, headers, null line &
     body)
S:   .
C:   ACKD
S:   =0
C:   QUIT
S:   + OK POP2 server signing off
C:   (client drops its end of connection
S:   (server drops its end of connection)
```

The second example is of a simple POP3 server that supports only plaintext (USER/PASS) authentication, with the client deleting messages immediately upon retrieving them (i.e., the "keep messages on server" option is not selected).

```
C:    (initiate TCP connection on port 110)
S:    (accept TCP connection)
S:    +OK POP3 server ready
C:    USER asmith
S:    +OK
C:    PASS swordfish
S:    +OK user authenticated
C:    LIST
S:    +OK 2 messages (320 octets)
S:    1 120
S:    2 200
S:    .
C:    RETR 1
S:    +OK 120 octets
S:    (contents of message 1, headers, null line &
      body)
S:    .
C:    DELE 1
S:    +OK message 1 deleted
C:    RETR 2
S:    +OK 200 octets
S:    (contents of message2, headers, null line &
      body)
S:    .
C:    DELE 2
S:    +OK message 2 deleted
C:    QUIT
S:    +OK POP3 server signing off
S:    (close server end of TCP connection)
C:    (close client end of TCP connection)
```

The third example is of a POP3 server and client that support a POP authentication, with the client leaving messages on the server. Assume that the first message (with UID XXX001) has been retrieved previously (which means the client has UID XXX001 in a locally maintained list of previously retrieved messages, but not the UID for message 2, which is

XXX079). However, the client has determined (from local information it has, not from information obtained via POP in this session) that the message associated with UID XXX001 is over 30 days old (and the client has been configured to delete any such messages from the server).

```
C:  (initiate TCP connection on port 110)
S:  (accept TCP connection)
S:  +OK POP3 Server Ready <1234.697170952@guys.com>
C:  APOP asmith c9dcb935ce1d21d02afdebcbd9cb1d5a
S:  +OK user authenticated
C:  UIDL
S:  +OK 2 messages (320 octets)
S:  1 XXX001
S:  2 XXX079
S:  .
C:  RETR 2
S:  +OK 200 octets
S:  (contents of message 2, headers, null line &
    body)
S:  .
C:  DELE 1
S:  +OK message 1 deleted
C:  QUIT
S:  +OK POP3 server signing off
S:  (close server end of TCP connection)
C:  (close client end of TCP connection)
```

At the end of the last example, the server will be holding the message with UID XXX079 (originally message 2, now called message 1 if another LIST or UIDL is issued). The message with UID XXX001 (originally message 1) is no longer on the server.

Relevant RFCs

The more important RFCs in the following list are indicated by asterisks.

1957	I	R. Nelson, "Some Observations on Implementations of the Post Office Protocol (POP3)," 06/06/1996. (Pages=2) (Format=.txt)
*1939	S	J. Myers, M. Rose, "Post Office Protocol—Version 3," 05/14/1996. (Pages=23) (Format=.txt) (Obsoletes RFC1725) (STD 53)
*1734	PS	J. Myers, "POP3 AUTHentication Command," 12/20/1994. (Pages=5) (Format=.txt)
1731	PS	J. Myers, "IMAP4 Authentication Mechanisms," 12/20/1994. (Pages=6) (Format=.txt)
1725	DS	J. Myers, M. Rose, "Post Office Protocol—Version 3," 11/23/1994. Pages=18) (Format=.txt) (Obsoletes RFC1460) (Obsoleted by RFC1939)
1460	DS	M. Rose, "Post Office Protocol—Version 3," 06/16/1993. (Pages=19) (Format=.txt) (Obsoletes RFC1225) (Obsoleted by RFC1725)
1225	DS	M. Rose, "Post Office Protocol—Version 3," 05/14/1991. (Pages=16) (Format=.txt) (Obsoletes RFC1081) (Obsoleted by RFC1460)
*1082	H	M. Rose, "Post Office Protocol—Version 3: Extended Service Offerings," 11/01/1988. (Pages=11) (Format=.txt)
1081	PS	M. Rose, "Post Office Protocol—Version 3," 11/01/1988. (Pages=16) (Format=.txt) (Obsoleted by RFC1225)
*0937	H	M. Butler, D. Chase, J. Goldberger, J. Postel, J. Reynolds, "Post Office Protocol—Version 2," 02/01/1985. (Pages=24) (Format=.txt) (Obsoletes RFC0918)
0918		J. Reynolds, "Post Office Protocol," 10/01/1984. (Pages=5) (Format=.txt) (Obsoleted by RFC0937)

IMAP: Internet Message Access Protocol

IMAP4 is a powerful new alternative to POP3. It can be used to do everything that POP3 does, if desired. It also supports keeping messages in multiple mailboxes on the central server (and managing those mailboxes) for access from any number of sites. It allows far more sophisticated and selective retrieval of information than does POP3. It also supports server-based searching, public folders, and various other advanced features not possible with POP3. It places far more demands on the server and the underlying message store than does POP. It likely will replace POP3 as the primary mechanism for retrieving message from Internet postoffices over the next few years.

Although POP3 has served reasonably well as a way to retrieve messages from the message store of an Internet Mail postoffice, it does tend to force you into downloading all messages into your local message store. If you always access your mail from the same system and have a very high bandwidth channel to access the mail server (e.g., a 10-Mbps LAN connection), that is acceptable. However, if you want to access your mail from a number of systems or have a limited bandwidth connection between your UA and the message store, a few things definitely could be improved, for example, keeping mail on the server and being able to do selective retrieval and management of the messages kept on the server.

Some POP3-based UAs (e.g., Eudora Pro v3.0) provide an option to leave mail on the server (which is accomplished by not issuing the DELE command after retrieving each message and issuing a UIDL command at the start of every connection to determine which messages on the server previously have been downloaded to a given UA's local message store). However, those options have several shortcomings:

▶ There are limited ways to manage the mail left on a server (a typical scheme might delete all messages older than n days. A value of 30 should give you time to retrieve them from all your UAs but still keep the messages stored on the server from building up indefinitely.

▶ If your UA somehow loses its list of UIDLs for messages already retrieved (which seems to happen all too often), your only choice is to retrieve all messages still in your server's message store, whether they previously have been retrieved or not. That can result in a very long download (for most people, 30 days' worth of messages, including attachments, is a very large amount of data) and in duplicate copies of messages in your local message store.

▶ The minimum amount of information you can retrieve (the "granularity" of the retrieval mechanism) is still one entire message. That does not allow you to retrieve only a few of the headers (e.g., sender and subject) to determine if you want to download the entire message or perhaps delete it from the server's message store.

IMAP was developed as an alternative to POP to address exactly those issues. The initial work was based on POP2, not POP3. Like SMTP and POP, IMAP is connection oriented and assumes a reliable transport with built-in flow control, so it typically is implemented on TCP (using port 143).

Like POP, IMAP is entirely concerned with the retrieval of messages by a UA from your local mail server's message store (as opposed to sending messages to your local MTA or between MTAs). There is no change to the protocols or message syntax of messages going between servers due to IMAP. As long as the recipient's UA and the local mail server support compatible versions of IMAP, it can be used by that recipient. Therefore, the decision of which message access protocol (POP or IMAP) a mail recipient uses is completely independent of what UA or message access protocol the sender might use. In contrast, extensions to the SMTP message syntax (RFC 822 issues) require both the sender's and the receiver's UAs (and any MTAs in the path) to support (or at least tolerate) the extensions. Extensions to the SMTP protocol itself (RFC 821 issues) require both the sender's UA and any MTAs involved to support (or at least tolerate) the extensions.

In practice, a single message store that supports IMAP typically also provides POP as a parallel access mechanism (POP will continue to be used widely for some time after IMAP becomes popular). Some users of a given mail server might use POP-compliant clients, and others might use IMAP-compatible clients against the same server message store. On the other hand, it is likely that some problems would arise (e.g., messages mysteriously vanishing from the server message store) if both POP- and IMAP-based UAs are used against a single user's mailbox. If a person decides to use an IMAP client to retrieve messages from his or her server mailbox, he or she should use IMAP consistently, regardless of how or from where the retrievals are done (at work, elsewhere in the company, at home, or from a notebook computer). In an emergency, a POP-based UA could be used, but if it does not support the "leave messages on server" option, messages will be deleted from the server mailbox that IMAP would expect to still be there; either way, the IMAP server would not be aware that those messages had been read.

In addition to better management of messages on the server mailbox and selective retrieval of parts of a message (as opposed to only entire messages), IMAP also provides a way to search a user's server mailbox (right on the server) on a variety of criteria, which could leverage the speed or power of a high-end central server or database engine to do sophisticated searching without increasing the complexity of the UA or requiring the user's computer to be very high performance. Only the search command itself and the results of that search need be transmitted over the network connection, which would minimize network traffic for a centrally located mailbox. Of course, searching a local mailbox would not generate network traffic, but then every UA a given person uses has to maintain an up-to-date local copy of the message store and must itself include complex message-searching functionality. The downside is that the searching might be a significant (or even the major) work load on the computer running the IMAP server. Certainly that requires much greater CPU (and disk I/O) capabilities than a server running POP would require (or correspondingly, a given hardware platform would be able to support fewer IMAP users than POP users, all other things being equal).

Another major change is the ability to maintain multiple mailboxes per user on the server (similar to the way that some POP-based UAs manage their local message store). Such mailboxes can be created and managed under control of the users (via their IMAP4 UA) and organized in any way (including hierarchical) that makes sense to a given user. One advantage of doing that on the server is that the same mailbox organization will be seen from any IMAP client (for a given user), regardless of where it is being used (so long as the same IMAP server is being used).

IMAP maintains a good deal more information about the state of a message (whether it has been read, replied to, etc.). That allows an IMAP client to provide more of the status information the user of a proprietary e-mail system is used to having available. Combined with the DSN extensions to SMTP, that should provide most such functionality. However, the additional information, together with the message attributes that make some of IMAP's functionality possible (or more efficient), makes the message store for an Internet Mail server that supports IMAP more complex (and slower) than one that supports just POP.

One interesting change is the standardized support for authentication through cryptographic schemes like Kerberos, instead of the use of a simple plaintext user name and password. While a POP3 extension to support authentication exists, it is not widely implemented or used (it is, in fact, a retrofit of IMAP concepts to POP3). With authentication being a part of the IMAP4 standard, it hopefully will be implemented far more widely than the POP3 authentication scheme. IMAP even provides a standard way to encrypt all messages in both directions after the authentication (called a protection mechanism).

Finally, IMAP is designed to work well in an asynchronous, or pipe-lined implementation in which many messages can be sent as fast as the underlying transport will allow, with responses being received and handled later in the transaction.

Evolution of IMAP

IMAP evolved from work originally done at the University of Washington's Networks and Distributed Computing department. The first IETF RFC on the protocol was for version 2 (RFC 1064, July 1988, in which IMAP stood for Internet Mail Access Protocol). The initial server and client implementations were done by Mark Crispin, who also wrote the RFC. Various other people contributed to the protocol definition and wrote early clients for it. The next IMAP standard, called IMAP4 (in which the name of the protocol was changed to Internet *Message* Access Protocol) was RFCs 1730-1733 (Dec. 1994). The latest IETF standard as of the writing of this book is IMAP 4rev1, RFCs 2060-2062 (Dec. 1996).

This evolution was driven by feedback from actual implementation and deployment of IMAP servers and clients, plus changes in the available platforms, access channels, and other Internet Mail standards. There has been some effort to provide backward compatibility, but there may be some problems with mixing products based on the various versions of the standard. Some commands, responses, and even data formats used by IMAP2 clients and servers have been rendered obsolete (see RFC 2062). As long as your server and clients are compliant with IMAP4 or later, they should interoperate correctly (even if not all the latest functionality is available to everyone in a mixed system).

A complementary standard called Application Configuration Access Protocol (ACAP), which is still an IETF draft proposed standard as of January 1997, extends the concepts to a very large system with multiple servers running on different computers. This standard was originally known as Internet Message Support Protocol (IMSP). Especially for smaller systems (say, an IMAP server running on a single computer), ACAP is not required. ACAP is not covered in this book due to its very early stage of development.

IMAP characteristics

Like POP, IMAP clients and servers exchange messages consisting of ASCII characters followed by a CR,LF. Unlike POP, IMAP also allows some messages to be a sequence of bytes (not necessarily ASCII) of a fixed, known length, which is then followed by a line.

Also, like POP, IMAP clients send a request message to the server and receive a response from the server with the status of the processing of that request (possibly including information resulting from the request, such as a retrieved header). Unlike POP, an IMAP client must be ready to accept multiple responses from the server (including unsolicited responses, not associated with any request the client sent). Also unlike POP, all messages start with a tag (a short identifier unique within that connection, such as A0001, A0002) that helps associate a given response to a particular request (the response to a given request includes the same tag). Unsolicited messages from the server to the client use an asterisk (*) as the special tag.

In a few situations, a message is continued from a previous message, in which case the special tag + is used. That tag scheme (together with the requirement that clients must always be ready to accept messages from the server) is primarily to allow the possibility of an asynchronous design (for both clients and server) in which commands can be sent as fast as the transport allows, and responses can be received once the function is complete (possibly even out of order, unlike pipelined SMTP). That optimizes throughput by eliminating needless turnaround delays (twiddling your thumbs while you await the response to a sent command before you send the next command). There are certain situations in which ambiguity can result from not waiting for responses. An IMAP client must await responses from certain commands to avoid such ambiguities (see RFC 2060 for details).

It is possible for a client to be IMAP4 compliant yet only be ready to receive responses to commands it sent (e.g., send a command, then receive an arbitrary number of unsolicited responses, then receive the tagged response to the sent command). However, even as this chapter is being written, an extension is being worked out to allow a client to inform a server that it is *always* watching for untagged messages from the server. If a server is aware that the client has such capability, it can send messages out of the blue to inform the client that new mail has arrived, and so on.

That capability could significantly reduce client/server network traffic (no need for a client to poll every few minutes to see if any new mail has arrived).

One possible client design capable of such operation would have the main execution thread fire off requests as fast as the transport (TCP) allows and keep a list of outstanding requests. A separate execution thread would watch for messages from the server; when one appears, it would check to see if a message is a response to an outstanding request. If it is, the thread would then perform appropriate actions based on whether the response indicated success or failure. Unsolicited messages could be processed immediately on receipt. If an outstanding request does not receive a corresponding response within a certain amount of time, it could be assumed to have been lost (and appropriate error handling done). In Windows NT (and even Windows 95), it is possible for client software, as well as server software, to be multithreaded.

The three possible specific responses by a server to a client's request message are:

▶ OK: Command was valid and completed successfully.

▶ NO: Command was valid, but failed in some way, e.g., message not found.

▶ BAD: Command was invalid.

Numerous other unsolicited (untagged) server response messages might be sent before a specific response is sent, with a number of possible syntaxes. For example, when you open a mailbox, you might get the following messages, only the last of which is a "specific" response to your request to open the mailbox. (As in previous chapters, C stands for "client" and S for "server.")

```
C:   a002 select inbox
S:   * 18 EXISTS
S:   * FLAGS (\Answered \Flagged \Deleted \Seen\Draft)
S:   * 2 RECENT
S:   * OK [UNSEEN 17]
S:   * OK [UIDVALIDITY 3857529045]
S:   a002 OK [READ-WRITE] SELECT completed
```

Unique message identifiers

IMAP introduces a 32-bit unique message number and a 32-bit unique identifier validity value. Together, those values form a 64-bit value that is permanently associated with a given message (even across sessions) and refers to no other message in that mailbox. The message number is assigned in a monotonically ascending (but not necessarily contiguous) sequence. The identity validity value (sent as an unsolicited message when a given mailbox is selected) is used to uniquely identify a particular mailbox, and a good choice might be a 32-bit representation of the date and time that mailbox was created. That would help detect that a particular mailbox had been deleted and a new one with the same name created since the last access.

The unique identifier must be strictly ascending in the messages in a given mailbox at all times. If for some reason the message store is reordered by some other agent, the unique identifiers must be regenerated to be strictly ascending. If for some reason, the unique identifiers (or the strictly ascending sequence) cannot be maintained over sessions, then the unique identifier validity value sent each time a mailbox is selected must be larger than the value sent in any previous session. (Hence, the 64-bit value created by putting the unique identifier validity value in the most significant 32 bits and the unique identifier in the least significant 32 bits is guaranteed to be unique and monotonically increasing even over multiple sessions.)

Message flags

IMAP maintains flags for each message. A given message can have any number of flags associated with it. There are two types of flags: system flags and keyword flags. Both kinds can be valid only during a given session (which means they could be kept in memory) or persistent across sessions (which means they would have to be stored in the mailbox).

The system flags all start with a backslash (\) and are listed in Table 17.1.

The keyword flags are defined by a given server implementation (and do not begin with a backslash). It is even possible for a given implementation to allow user-defined keyword flags.

Table 17.1
IMAP system flags

Flag	Description
\Seen	Message has been read.
\Answered	Message has been replied to.
\Flagged	Message has been marked for urgent/special attention.
\Deleted	Message has been marked deleted for later EXPUNGE.
\Draft	Message is still being composed.
\Recent	Message is new in this session.

Internal date message attribute

The internal date message attribute defines a date and time to be permanently associated with each message (and hence stored in the mailbox). This is not the date and time from the RFC 822 Date: header (which is the time the message was sent).

Messages received via SMTP should be assigned the date and time of the final delivery of the message into the mailbox by SMTP (which should agree with the latest date and time from the RFC 822 Received: headers in the message).

Messages received via the IMAP4 COPY command should be assigned the internal date and time of the source message.

Messages received via the IMAP4 APPEND command should be the date and time specified in the APPEND command.

Other message attributes

The other message attributes that must be stored in the mailbox for each message are as follows:

▶ Message size, which is the number of bytes in the message as expressed in rfc822 format;

- Envelope structure, which is a parsed representation of the rfc822 message headers;

- Body structure, which is a parsed representation of the MIME body structure.

Retrieval granularity

In POP, the smallest unit you can retrieve (the granularity) is one whole message. In IMAP, the granularity of retrieval is finer. Specifically, the following can be retrieved independently (in addition to the entire message):

- The entire rfc822 message header;

- The entire rfc822 message body;

- The MIME message structure;

- Any MIME body part;

- Any MIME header.

Normal mailboxes in IMAP

A traditional mail server that supports only POP maintains a single storage area on the server for a given user (normally used only for temporary storage from the receipt of messages via SMTP until they are retrieved and deleted via POP). The details of how that is kept (locally, flat file, database, or even on a remote message store server) is implementation dependent. However, there is only one storage area per user.

On a mail server that supports IMAP, there can be multiple storage areas for each user, organized into one or more mailboxes, each of which is similar to the single storage area on a POP-only mail server. Every user gets a default mailbox (INBOX) but can create additional mailboxes from the UA. Each mailbox has a name. In an IMAP session, one of the mailboxes for the user making the connection must be selected, after which messages are stored in or retrieved from that mailbox, until another mailbox is selected. Conventions for naming mailboxes are implementation dependent, except for the special name (regardless of case) INBOX, which means "the primary mailbox for this user on this server."

If a hierarchical organization of mailboxes is desired, mailbox names must be left-to-right hierarchical using a single delimiter character (e.g., a period) to separate levels of hierarchy. The same delimiter must be used throughout a given mailbox name. The first hierarchical element of a given mailbox name that begins with a number sign (or hatch symbol) identifies which namespace the mailbox is for (e.g., #news.comp.mail.misc could refer to a mailbox linked to a newsnet group). IMAP defines a scheme for use of international (e.g., Chinese) characters in mailbox names (see RFC 2060, Section 5.1.3 for details).

Shared mailboxes in IMAP

IMAP4, unlike POP, includes the concept of shared mailboxes, which are conceptually more like newsnet newsgroups than normal mailboxes. A shared mailbox might have a primary owner (who might even play the role of proctor to enforce rules about what can be posted). It is possible that the mailbox could be set up for anyone to post to it (deposit messages) or only authorized individuals. It is also possible that you could allow anyone to subscribe to a given mailbox or restrict access to specific individuals. It is even possible for the contents of such folders to be automatically replicated over a number of servers (or perhaps only specifically designated ones), although such replication is not covered in the standard. Existing services, like real Internet usenet newsgroups (e.g., rec.travel) or broadcast news services, could be coupled to specific shared mailboxes so that any incoming messages from the newsgroup (messages posted to that newsgroup from anywhere in the Internet) could magically appear as new messages in the corresponding shared mailbox. Messages sent to the shared mailbox from inside the domain could be posted automatically to the corresponding newsnet newsgroup. That would require an NNTP gateway between the real newsnet newsgroup feed and the IMAP public folders.

An IMAP client should support the ability to view a list of shared mailboxes the user is allowed to subscribe to, as well as allowing the user to subscribe or unsubscribe to any of them. There are several implications of that. First, no separate news reader (NNTP) client is required. Second, shared folders and newsgroups are available via the single-user interface, and cross-forwarding would be simple. Third, attachments could be sent and retrieved easily through the shared folders, exactly the same as with normal person-to-person e-mail.

If you think about it for a second, you will see that the IMAP shared folder scheme provides many of the capabilities of a mailing list (e.g., Majordomo), with a much simpler administration scheme, and much more efficient distribution mechanism. With a real mailing list, every message is sent to (and downloaded by) every member of the list. With shared mailboxes, all subscribers can view a list of posted messages and decide (based on subject, message size, and even message structure) which messages to download. They also are automatically organized into the appropriate mailbox, with no rules processing required.

Server-based rules

While not yet a part of the IMAP standard, a few current server implementations allow the client to set up per-user server-based rules (sometimes called filters) that can be used to sort incoming messages into specific folders on a variety of criteria (e.g., all messages containing "security" in the subject go into the "info-sec" mailbox). A client would have access to a list of all mailboxes and should somehow inform the user (e.g., by changing the color of the mailbox name in the GUI or putting it in bold type) that a given mailbox has new mail (otherwise, it would be up to the user to check all mailboxes for new items on a regular basis). While management of such rules might be considered an implementation detail from the viewpoint of the server administrator (and in fact could be managed via a Web-form administration facility of the server, completely independent of the client), it would be most natural to do this from within the IMAP client, which would require an extension to the protocol.

Disconnected mode

One of the exciting areas still evolving in the IMAP standard is its ability to operate in connected, disconnected, or "sometimes connected, sometimes not" modes. These modes refer to whether the client has continual access to the server. A home computer that you connect only occasionally via a dial-in modem would be an example of a disconnected-mode client. A notebook computer that you sometimes use in the office and sometimes take with you on the road would be an example of the third mode.

The basic idea is to use the server to keep the "official" and most recent set of messages (allowing access from any client computer and simple, reliable backup) but to update automatically a local cache of messages on the client computer while it is connected. Reading and searching could be done on the local cache when the client is in the disconnected state.

A preliminary discussion of these concepts is covered in RFC 1733.

Interaction with other mail standards

Some of the other mail standards (SMTP) are pretty much independent of IMAP. Some of them (e.g., MIME) are intertwined with IMAP. Others (DSN) are quite new and not yet widely deployed, and not all details of their interaction with IMAP are completely understood at this point. The ACAP standard in particular is intimately related to IMAP, and the two probably will continue to evolve in parallel. A few (S/MIME, PGP/MIME) have definite areas of conflict with IMAP that will have to be resolved (interaction with IMAP is one of the primary issues currently holding up release of those standards).

Server states

A particular connection to an IMAP server is in one of four possible states at any given time. Most commands are valid in only one state. Commands issued when the server is in the wrong state will be rejected (with BAD or NO responses, as appropriate).

The four states illustrated in Figure 17.1 are as follows:

▶ Nonauthenticated state is the state that each connection starts in (except for preauthenticated connections). The client must present acceptable authentication credentials (valid user/password or Kerberos authentication) before almost any other commands will be accepted.

▶ Authenticated state is the state a normal connection enters on acceptance of the authentication credentials (or starts in for preauthenticated connections) or returns to after a mailbox selection error.

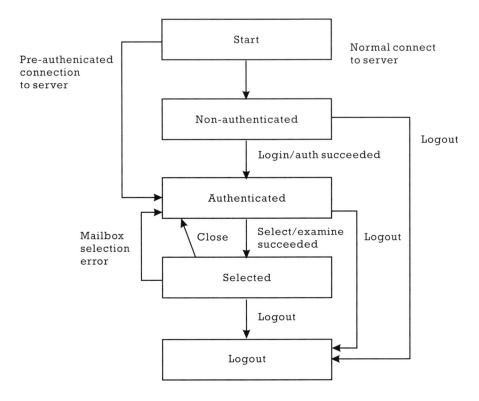

Figure 17.1   IMAP state diagram.

▶ Selected state is the state a connection enters on selection of a valid mailbox.

▶ Logout is termination of the connection. It can be entered either by a client command or by a unilateral decision of the server.

Data representation

The representation of data in messages exchanged between IMAP4 clients and servers (not to be confused with RFC 822 or MIME message syntax) is precisely defined. A data item can be a number, a string, or a parenthesized list.

A number is one or more digit characters, representing a numeric value.

A string can be either a literal string or a quoted string. A literal string consists of a prefix and data. The prefix is an open brace ({), then the length of the data in bytes, then a close brace (}) and a CR,LF. The data can contain any ASCII characters, including control codes (even CR and LF). In messages from server to client, the data themselves are sent immediately after the prefix. In messages from client to server, the prefix is sent first, then the data are sent in response to a continuation request, for example:

{	3	}	CR	LF	A	B	C
0x7B	0x33	0x7D	0x0D	0x0A	0x41	0x42	0x43

A quoted string consists of zero or more 7-bit ASCII characters (not including CR or LF), with double quotes at the start and the end, for example:

"	A	B	C	"
0x22	0x41	0x42	0x43	0x22

Limited 8-bit or multibyte characters can be including in literal strings, but only if the character set is identified (see "Assigned Numbers," IETF STD 2/RFC 1700). Raw binary data should be encoded into an ASCII form, such as base64.

Data structures are represented in a parenthesized list, which is a sequence that starts with an open parenthesis, then zero or more data items separated by spaces, then a close parenthesis, for example:

(item1 item2 item3)

It is possible for parenthesized lists to be nested for representation of more complex data structures, for example:

(item1 (item2a item2b item2c) (item3a (item3b1 item3b2)))

The special atom NIL represents the nonexistence of a particular data item normally represented as a string or a parenthesized list.

IMAP client commands

The commands in this section are organized by the state in which they can be used. Some commands can be used in more than one state, in which case they are listed in the lowest of the four possible states.

In the examples, "C" stands for client and "S" for server.

Client commands in any state

CAPABILITY

Responses: Required untagged response: CAPABILITY
Result: OK; capability enumeration complete

This command requests a list of server capabilities. The server must send at least a single untagged response with IMAP4rev1 as one of the enumerated capabilities. The final response is a tagged OK response.

A capability starting with AUTH= defines a supported authentication mechanism, for example, AUTH=KERBEROS_V4.

An example is:

```
C:   A0123 CAPABILITY
S:   * CAPABILITY IMAP4rev1 AUTH=KERBEROS_V4
S:   A0123 OK
```

NOOP

Responses: No specified response
Result: OK, noop complete

This command does nothing, but it can be used to trigger a server into sending untagged messages if you happen to have time to process some. It also will reset any inactivity autologout timer the server may support.

An example with no untagged messages waiting on a server is:

```
C:   a002 NOOP
S:   a002 OK
```

An example with untagged messages waiting on a server is:

```
C:   a047 NOOP
S:   * 22 EXPUNGE
S:   * 23 EXISTS
S:   * 3 RECENT
S:   14 FETCH (FLAGS (\Seen \Deleted))
S:   a047 OK
```

LOGOUT

Responses: Required untagged response: BYE
Result: OK, logout complete

The LOGOUT command terminates the session. The server must send an untagged BYE response, then a tagged OK response, then close the network connection.

An example is:

```
C:   A023 LOGOUT
S:   * BYE IMAP4rev1 Server logging out
S:   A023 OK
(connection closed)
```

Client commands in nonauthenticated state

A connection that is not preauthenticated must use either the AUTHENTICATE (e.g., Kerberos) or LOGIN (username/password) command to advance to the authenticated state. An implementation optionally can provide nonauthenticated access to certain mailboxes. That is done by doing a LOGIN with the name "anonymous" (a password is required, but details are implementation specific). Once authenticated, it is not possible to reenter the authenticated state.

Available commands in this state are the universal commands (CAPABILITY, NOOP, and LOGOUT), plus AUTHENTICATE and LOGIN.

AUTHENTICATE authentication_mechanism

Responses: continuation data can be requested
Result: OK, authentication complete; state = authenticated
NO   authentication failed; state = nonauthenticated
BAD   syntax error in parameters

This command allows users to authenticate themselves using any of a number of mechanisms defined in RFC 1731, including Kerberos_V4, GSSAPI (defined in RFC 1508), or S/Key.

An authentication exchange consists of a series of challenges by the server and replies by the client (as defined by the specific mechanism). A server challenge consists of a line containing a plus sign (+) followed by base64-encoded data. The client reply consists of a line containing base64 encoded data. If the client wants to cancel an exchange, it sends a line with a single asterisk (*), which causes the server to reject the AUTHENTICATE command with a tagged BAD response.

A protection mechanism provides integrity and privacy to the connection by encrypting all information (after the authentication). Once a protection mechanism is engaged, all messages are encrypted into ciphertext and sent as a byte stream with a 4-byte field (in network order, MSByte first) containing the number of data bytes, followed by the encrypted data in binary.

Both authentication and protection mechanisms are optional. An authentication mechanism can be implemented without a protection mechanism. If one AUTHENTICATE request fails with a NO response, the client can try another AUTHENTICATE command or shift to LOGIN.

An example is:

```
C:   (open connection)
S:   * OK IMAP4rev1 Server
C:   A001 AUTHENTICATE KERREROS_V4
C:   +amFYig==
C:   BAcAQU5EUkVXLkNNVS5FRFUAOC
     Asho84kLN3/IJmrMG+25a4DT+nZImJjnTNHJUtxAA+oOKPKfHE
     cAFs9a3CL5Oebe/ydHJUwYFdWwuQ1MWiy6Ie
     sKvjL5rL9WjXUb9MwT9bpObYLGOKilQh
```

```
S:    + or //EoAADZI=
C:    DiAF5A4gA+oOIALuBKAAmw==
S:    AO01 OK Kerberos V4 authentication successful
```

Note that in a real exchange, there would be no line breaks in the first client response (it would all be one line).

LOGIN username password

Responses:	No specific response
Result:	OK, login complete; state = Authenticated
	NO, login failure; state = Non-authenticated
	BAD, error in parameters

This command allows a client to authenticate itself using a plaintext user name and password (which could be a security risk). The authentication command is considerably more secure.

An example is:

```
C:    (open connection)
S:    * OK IMAP4rev1 Server
C:    a001 LOGIN ASMITH SWORDFISH
S:    a001 OK - AUTHENTICATED
```

Client commands in nonauthenticated state

Once authenticated (either via a preauthenticated connection or by a valid LOGIN or AUTHENTICATE command in a normal connection), commands that manipulate entire mailboxes (as opposed to messages within mailboxes) are permitted. Of those commands, the SELECT and EXAMINE commands enter the selected state. Only one mailbox can be selected at a time in a connection. Simultaneous access to multiple mailboxes requires multiple connections. A SELECT command automatically deselects any currently selected mailbox. If a SELECT command fails, no mailbox is selected (the connection is returned to the authenticated state).

Available commands in the nonauthenticated state are the universal commands (CAPABILITY, NOOP, and LOGOUT), plus SELECT, EXAMINE, CREATE, DELETE, RENAME, SUBSCRIBE, UNSUBSCRIBE, LIST, LSUB, STATUS, and APPEND.

SELECT mailbox

Responses: Required untagged responses:
             FLAGS, EXISTS, RECENT
             Optional OK untagged responses:
             UNSEEN, PERMANENTFLAGS
Results: OK, select completed; state = selected
             NO, select failed; state = authenticated
             BAD, syntax error in parameters

The SELECT command selects a mailbox so that messages can be retrieved from or added to it. The server must respond to a valid select with the following untagged responses, then a tagged OK response:

- FLAGS: defined flags in the mailbox;

- $<n>$ EXISTS: number of messages in the mailbox;

- $<n>$ RECENT: number of messages with the \Recent flag;

- OK [UIDVALIDITY $<n>$]: unique identifier validity value.

The server optionally can also send an UNSEEN response code in an OK untagged response, indicating the sequence number of the first unseen message in the mailbox.

If the user cannot change the permanent state of at least one of the flags listed in the FLAGS untagged response, the server should send a PERMANENTFLAGS response in another untagged response, listing the flags that the user can change permanently.

If the user is permitted to modify the mailbox, the tagged OK response should include [READ-WRITE]; otherwise, it should include [READ-ONLY].

An example is:

```
C:   A142 SELECT INBOX
S:   * 172 EXISTS
S:   * 1 RECENT
S:   * OK [UNSEEN 12]
S:   * OK [UIDVALIDITY 3857529045]
S:   * FLAGS (\Answered \Flagged \ Deleted \Seen\Draft)
S:   * OK [PERMANENTFLAGS (\Deleted \Seen \*)]
S:   A142 OK [READ-WRITE] SELECT completed
```

EXAMINE mailbox

Responses: Required untagged responses:
            FLAGS, EXISTS, RECENT
            Optional OK untagged responses:
            UNSEEN, PERMANENTFLAGS
Result: OK, examine complete; state = selected
            NO, examine failed; state = authenticated
            BAD, syntax error in parameters

The EXAMINE command essentially is identical to the SELECT command except that the mailbox is read-only (no changes to the permanent state of the mailbox, including per-user state, are permitted).

The text of the tagged OK response to the EXAMINE command must begin with the [READ-ONLY] response code.

An example is:

```
C:   A142 EXAMINE PERSONAL
S:   * 17 EXISTS
S:   * 2 RECENT
S:   * OK [UNSEEN 8]
S:   * OK [UIDVALIDITY 3857528123]
S:   * FLAGS (\Answered \Flagged \ Deleted \Seen\Draft)
S:   * OK [PERMANENTFLAGS ()]
S:   A142 OK [READ-ONLY] EXAMINE Completed
```

\

CREATE mailbox

Responses: No specific responses
Result: OK, mailbox created
             NO, mailbox not created
             BAD, syntax error in parameter

The CREATE command creates a new mailbox with the specified name. An attempt to create a mailbox with a name already in use by this user will fail (i.e., will result in a NO response). To create a hierarchy of mailboxes, use the hierarchy separator character returned by the LIST command, as shown in the following example (which assumes that the character is /).

```
C:   A003 CREATE Personal/
S:   A003 OK CREATE completed
C:   A004 CREATE Personal/Chess
S:   A004 OK Create completed
C:   A005 CREATE INBOX
S:   A005 NO - mailbox name already in use
```

DELETE mailbox

Responses: No specific responses
Result: OK, delete succeeded
             NO, delete failed
             BAD, syntax error in parameters

The DELETE command permanently removes the specified mailbox. Any attempt to delete INBOX or a nonexistent mailbox will result in a NO response. It is also an error to delete any name in a hierarchical structure if it has names under it and it has the \Noselect attribute (you must delete the subsidiary names first).

An example is:

```
C:   A002 DELETE Personal/
S:   A002 NO - mailbox has inferior hierarchical names
C:   A003 DELETE Personal/Chess
```

```
S:    A003 OK DELETE completed
C:    A004 DELETE Personal/
S:    A004 OK DELETE completed
C:    A005 DELETE Germany
S:    A005 NO - mailbox not found
C:    A006 DELETE INBOX
S:    A006 NO - INBOX cannot be deleted
```

RENAME existing_mailbox_name new_mailbox_name

Responses: No specific responses
Result: OK, rename completed
             NO, rename failed
             BAD, syntax error in parameters

The RENAME command is used to rename a mailbox. An attempt to rename a mailbox that does not exist or to rename a mailbox to a name already in use will result in a NO response. An attempt to rename INBOX will create a new mailbox with the new name and copy all messages from INBOX to the new mailbox, then leave INBOX empty.

An example is:

```
C:    A002 RENAME Chess Games
S:    A002 OK Rename completed
C:    A003 RENAME Chess Leisure
S:    A003 NO - no such mailbox
C:    A004 RENAME Checkers Games
S:    A004 NO - name already in use
C:    A005 RENAME INBOX January97
S:    A005 OK - messages in INBOX moved to new mailbox
```

SUBSCRIBE mailbox

Responses: No specific responses
Result: OK, subscription completed
             NO, subscription failed
             BAD, syntax error in parameter

This command adds the specified mailbox to the server's set of active or subscribed mailboxes as returned by the LSUB command. An attempt to subscribe to a nonexistent mailbox will result in a NO response. It is also possible that a server might allow or deny permission to a mailbox by user.

```
C:   A002 SUBSCRIBE #news.comp.mail.mime
S:   A002 OK subscribed
C:   A003 SUBSCRIBE #news.alt.sillywalks
S:   A003 NO - no such mailbox
C:   A004 SUBSCRIBE #news.alt.sex.raunchy
S:   A004 NO - access denied
```

UNSUBSCRIBE mailbox

Responses: No specific responses
Result: OK, unsubscription completed
            NO, unsubscription failed
            BAD, syntax error in parameter

This command removes the specified mailbox from the server's set of active or subscribed mailboxes as returned by the LSUB command. An attempt to unsubscribe a nonactive mailbox will result in a NO response.
An example is:

```
C:   A002 UNSUBSCRIBE #news.comp.mail.mime
S:   A002 OK unsubscribed
C:   A003 SUBSCRIBE #news.alt.sillywalks
S:   A003 NO - not currently subscribed
```

LIST reference_names mailbox

Responses: Untagged responses: LIST
Result: OK, list completed
            NO, list failed
            BAD, syntax error in parameters

This command returns a subset (possibly all) of the names from the set of all names available to the user. Zero or more untagged LIST replies are returned, containing the name attributes, hierarchy delimiter, and name.

An example is:

```
C:   A101 LIST "" ""
S:   * LIST (\Noselect) "/" ""
S:   A101 OK LIST Complete
C:   A102 LIST #news.comp.mail.misc ""
S:   * LIST (\Noselect) "." #news.
S:   A102 OK LIST Complete
```

LSUB *reference_names mailbox*

Responses: Untagged responses: LSUB
Result: OK, LSUB completed
            NO, LSUB failed
            BAD, syntax error in parameters

This command returns a subset (possibly all) of the names from the set of names that the user has declared to be active or subscribed. Zero or more untagged LSUB replies are returned.

An example is:

```
C:   A101 LSUB "#news." "comp.mail.*"
S:   * LIST () "." #news.comp.mail.mime
S:   * LIST () "." #news.comp.mail.misc
S:   A101 OK LSUB Complete
```

STATUS *mailbox status_data_items*

Responses: Untagged response: STATUS
Result: OK, status completed
            NO, status failed; no status for that name
            BAD, syntax error in parameters

This command returns the request status of the specified mailbox (it does not change the mailbox in any way, including the \Recent flag). It also does not deselect the current mailbox. Unlike LIST, which can return incomplete information if complete information will take a while, STATUS will take as long as it needs; no wildcards are allowed.

The following status items may be requested:

▶ MESSAGES: number of messages in the mailbox;

▶ RECENT: number of messages in mailbox with \Recent flag;

▶ UIDNEXT: next available UID value for mailbox;

▶ UIDVALIDITY: unique identifier validity value for mailbox;

▶ UNSEEN: number of messages in mailbox without \Seen flag.

An example is:

```
C:   A042 STATUS Personal (UIDNEXT MESSAGES)
S:   * STATUS Personal (MESSAGES 231 UIDNEXT 44292)
S:   A042 OK STATUS Complete
```

APPEND mailbox [flag_parenthesized_list] [date_time]

Responses: No specific response
Result: OK, append completed
            NO, append error; cannot append; error in parameters
            BAD, syntax error in parameters

The APPEND command allows you to append a message (that follows the APPEND command) to the end of the specified mailbox. The message should be in rfc822 format (including 8-bit characters). If flags are specified, they will be set for the new message; otherwise, no flags will be set. If a date/time is specified, the internal date of the new message will be set to that; otherwise, the date will be set to the current date and time.

This command does not create a new mailbox (the specified mailbox must already exist). If the specified mailbox is currently selected, the server should return an untagged EXISTS response (if not, it may be requested with NOOP).

An example is:

```
C:    A003 APPEND Archive (\Seen) (310)
C:    Date: Tue, 21 Jan 1997 19:15:26 -0800 (PST)
C:    From: Albert Smith <asmith@guys.com>
C:    Subject: Howdy from Hawaii
C:    To: Alice B. Toklas <atoklas@gals.com>
C:
C:    Wish you were here
C:
S:    * STATUS Personal (MESSAGES 231 UIDNEXT 44292)
C:    A003 OK APPEND Complete
```

Client commands in selected state

Once a connection has entered the selected state (via the SELECT or EXAMINE command issued in the authenticated state), commands that manipulate messages in a mailbox are available, including the following:

- Universal commands: CAPABILITY, NOOP, and LOGOUT

- Authenticated-state commands: SELECT, EXAMINE, CREATE, DELETE, RENAME, SUBSCRIBE, UNSUBSCRIBE, LIST, LSUB, STATUS, and APPEND.

- In addition, the following available commands: CHECK, CLOSE, EXPUNGE, SEARCH, FETCH, STORE, COPY, and UID.

CHECK

Responses: No specific responses
Result: OK, check complete

The CHECK command requests the checkpoint of the currently selected mailbox. It is implementation dependent but could involve resolving the server's in-memory state with the state on disk, for this mailbox.

An example is:

```
C:   A042 CHECK
S:   A042 OK CHECK Complete
```

CLOSE

Responses:	No specific responses
Result:	OK, close completed, state = authenticated
	NO, close failed; no mailbox selected

The CLOSE command actually removes all messages marked for deletion (those that have the \Deleted flag set), then returns to authenticated state (deselect current mailbox). No untagged EXPUNGE responses are sent (as opposed to the EXPUNGE command, which goes on about every message it removes).

The following commands do an implicit close of the currently selected mailbox before they themselves are done:

SELECT, EXAMINE, LOGOUT

EXPUNGE

Responses:	No specific responses
Result:	OK, expunge completed
	NO, close failed; permission denied

The EXPUNGE permanently removes all messages with \Deleted flag set. Each deleted message results in an untagged EXPUNGE response. The message sequence numbers in the untagged responses reflect the newly deleted messages. For example, messages 3, 4, 7, and 11 had been marked for deletion in the following example.

```
C:   A202 EXPUNGE
S:   * 3 EXPUNGE
S:   * 3 EXPUNGE
S:   * 5 EXPUNGE
```

```
S:    * 8 EXPUNGE
S:    A202 OK EXPUNGE Complete
```

SEARCH [CHARSET_specification] search_criteria... search_criteria

Responses: Required untagged response: SEARCH

Result: OK, search completed

NO, cannot search on specified criteria

BAD, syntax error in parameters

This command searches the currently selected mailbox for messages that match the specified criteria. The untagged SEARCH response lists the messages that match the specified criteria. Specifying multiple criteria results in finding messages that meet ALL the criteria (intersection). Servers optionally may exclude MIME body parts with content types other than TEXT and MESSAGE from the search.

If the CHARSET keyword is specified, the parameter that follows is a character set identifier (e.g., US-ASCII or ISO-8859). Comparisons during the search use the specified character set. US-ASCII is the only character set that must be supported by server implementations and is the default if no CHARSET is specified.

In search keys that use strings, a message matches the key if the specified string is a substring of the field (using case-insensitive compares).

The untagged SEARCH response is a list of message sequence numbers that meet the search criteria, separated by spaces.

An example is:

```
C:    A282 SEARCH FLAGGED SINCE 1-Jan-1997 NOT FROM "Smith"
S:    * SEARCH 2 84 882
S:    A282 OK SEARCH Complete
```

FETCH message_set message_data_item_names

Responses: Untagged response: FETCH

Result: OK, fetch completed

NO, fetch error: cannot fetch specified data

BAD, syntax error in parameters

The FETCH command retrieves part or all of one or more messages in the currently selected mailbox. The message set is a list of one or more message sequence numbers or a range of numbers (e.g., 2:4 means message sequence numbers from 2 to 4, inclusive). A separate untagged FETCH response will be returned for each message, starting with the message sequence number, then the keyword FETCH, then the returned data.

An example is:

```
C:   A654 FETCH 2:4 (FLAGS BODY[HEADER.FIELDS (DATE FROM)])
S:   * 2 FETCH . . .
S:   * 3 FETCH . . .
S:   * 4 FETCH . . .
S:   A654 OK FETCH Complete
```

STORE message_set message_data_item_name value_for_message_data_item

Responses: Untagged FETCH
Result: OK, store completed
             NO, store failed; cannot store those data
             BAD, syntax error in parameters

This command alters data associated with a message in the mailbox. Normally, you will get untagged FETCH responses with the updated values (that can be prevented by adding a .SILENT suffix into the data item name). You will get a FETCH response regardless, if a change to a message's flags from an external source is observed.

Table 17.2 lists what flags can be set with the STORE command.

For example, assume that the following messages have the current flags:

2:   \Seen
3:   None
4:   \Deleted \Flagged \Seen

The STORE command would result in this exchange:

Table 17.2
STORE command flags

Flag	Description
FLAGS <flag list>	Sets flags to new values and returns FETCH responses with new settings
FLAGS.SILENT<flag list>	Sets flags to new values
+FLAGS<flag list>	Adds the argument to the flags for the message; returns FETCH responses
+FLAGS.SILENT<flag list>	Adds the argument to the flags for the message
FLAGS<flag list>	Removes argument from the flags for the message, with FETCH responses
-FLAGS.SILENT<flag list>	Removes argument from the flags for the message

```
C:   A003 STORE 2:4 +FLAGS (\Deleted)
S:   * 2 FETCH FLAGS (\Deleted \Seen)
S:   * 3 FETCH FLAGS (\Deleted)
S:   * 4 FETCH FLAGS (\Deleted \Flagged \Seen)
S:   A003 OK STORE Complete
```

COPY message_set mailbox

Responses: No specific response
Result: OK, copy completed
            NO, copy failed; cannot copy those messages or to that
            mailbox
            BAD, syntax error in parameters

This command copies the specified message(s) to the end of the specified mailbox. The flags and the internal date of the message(s) should be preserved. The destination mailbox must exist (a COPY command will not create a new mailbox). If it fails, the mailbox will be returned to the state prior to the COPY.

An example is:

```
C:    A003 COPY 2:4 Personal
S:    A003 OK COPY Complete
```

UID command command_arguments

Responses: Untagged responses: FETCH, SEARCH
Result: OK, UID command completed
             NO, UID command failed
             BAD, syntax error in parameters

The first form is UID followed by COPY, FETCH, or STORE, with normal parameters following. They perform the normal functions, except that the message set is specified as unique identifiers rather than message sequence numbers.

The second form is UID followed by SEARCH, with normal SEARCH parameters, except that the numbers returned in a SEARCH response are unique identifiers instead of message sequence numbers.

The number after the asterisk in an untagged FETCH response is always a message sequence number, not a unique identifier, even for a UID command response. However, the UID will be returned later in the response, after the keyword UID.

An example is:

```
C:    A999 UID FETCH 4827313:4828442 FLAGS
S:    * 23 FETCH (FLAGS (\Seen) UID 4827313)
S:    * 24 FETCH (FLAGS (\Seen) UID 4827943)
S:    * 25 FETCH (FLAGS (\Seen) UID 4828442)
S:    A999 OK UID FETCH Complete
```

Specifying search criteria

Search keys are listed in Table 17.3. If multiple criteria are specified (by just listing them separated by spaces), messages must match *all* the specified criteria (i.e., the result is the intersection of the messages that meet the individual criteria, which is equivalent to the AND logical function). For example, the search criteria:

NEW FROM Smith NOT SUBJECT Money

Table 17.3
Components from which search criteria can be built

message_set	Messages with the specified message sequence numbers
ALL	All messages in the mailbox
ANSWERED	Messages with an \Answered flag
BCC *string*	Messages whose Bcc: field contains the specified string
BEFORE *date_time*	Messages whose internal date/time is prior to the specified date/time
BODY *string*	Messages whose body contains the specified string
CC *string*	Messages whose Cc: field contains the specified string
DELETED	Messages with a \Deleted flag
DRAFT	Messages with a \Draft flag
FLAGGED	Messages with a \Flagged flag
FROM *string*	Messages whose From: field contains the specified string
HEADER *field_name string*	Messages that have a header with the specified field name (From, To, etc.) whose corresponding field contains the specified string
KEYWORD *flag*	Messages with the specified (implementation specific) keyword
LARGER *n*	Messages whose size is larger than the specified number of bytes
NEW	Messages that have a \Recent flag but not a \Seen flag, equivalent to RECENT UNSEEN
NOT *search_key*	Messages that do not match the specified search key (e.g., NOT ON <date-time>)
OLD	Messages that do not have a \Recent flag, equivalent to NOT RECENT, rather than NOT NEW
ON *date_time*	Messages whose internal date is within the specified date
OR *search_key_1 search_key_2*	Messages that match either (or both) search key(s) (inclusive OR)
RECENT	Messages that have a \Recent flag
SEEN	Messages that have a \Seen flag
SENTBEFORE *date_time*	Messages whose Date: field is prior to the specified date/time
SENTON *date_time*	Messages whose Date: field is within the specified date/time
SENTSINCE *date_time*	Messages whose Date: field is within or later than the specified date/time

Table 17.3 (continued)

SMALLER *n*	Messages whose size is less than the specified number of bytes
SUBJECT *string*	Messages whose Subject: field contains the specified string
TEXT *string*	Messages in which any header field or the body contains the specified string
TO *string*	Messages whose To: field contains the specified string
UID *message_set*	Messages with the specified unique identifiers
UNANSWERED	Messages that do not have an \Answered flag
UNDELETED	Messages that do not have a \Deleted flag
UNDRAFT	Messages that do not have a \Draft flag
UNFLAGGED	Messages that do not have a \Flagged flag
UNKEYWORD	Messages that do not have the specified (implementation specific) keyword
UNSEEN	Messages that do not have a \Seen flag

would match all messages from anyone with Smith in his or her name (even Smithers) that have a /Recent flag (but not a /Seen flag) in which the subject does not contain the word Money.

Specifying what part of a message to fetch

The FETCH command allows the UA to retrieve all or any part of a message (or the same part of each of a group of messages). The way in which the desired part is specified is fairly complicated, due to the complexities of MIME. The FETCH command allows specification of any one of the following parts or a combination of two or more of them by including them in a parenthesized list. The parts that may be specified are listed in Table 17.4.

BODY [section] [partial]

In the parameters, *section* is replaced by a (possibly empty) list of part specifiers, separated by periods. The *partial* parameter is optional (without it, the entire specified body part is returned). If it is present, *partial* is

Table 17.4
Specifications for the FETCH command

ALL	Shorthand for (FLAGS INTERNALDATE RFC822.SIZE ENVELOPE)
BODY	Equivalent to BODYSTRUCTURE
BODY [*section*] <*partial*>	The text of a particular body section.
BODY.PEEK *section* <*partial*>	Equivalent to BODY [*section*] <*partial*>, except that the \Seen flag is not set
BODYSTRUCTURE	Returns the MIME body structure of the message
ENVELOPE	Returns the parsed RFC822 headers, defaulting various fields as required
FAST	Equivalent to (FLAGS INTERNALDATE RFC822.SIZE)
FLAGS	Returns the flags that are set for this message
FULL	Equivalent to (FLAGS INTERNALDATE RFC822.SIZE ENVELOPE BODY)
INTERNALDATE	Returns the internal date of the message
RFC822	Equivalent to BODY[], except for the keyword in the resulting untagged FETCH response (which will be RFC822)
RFC822.HEADER	Equivalent to BODY.PEEK[HEADER], except for the keyword in the resulting untagged FETCH response (which will be RFC822.HEADER).
RFC822.SIZE	Returns the size of the entire message, in bytes.
RFC822.TEXT	Equivalent to BODY[TEXT], except for the keyword in the resulting untagged FETCH response (which will be RFC822 TEXT)
UID	Returns the unique identifier for the message

used to return just a substring of the specified message part. The substring is specified as <*n.m*>, where *n* is the starting byte position, and *m* is the number of bytes to return. The first byte position is 0, so <0.6> would be the first 6 bytes, and <6.3> would be the next three bytes. If the starting byte position is greater than the length of the body part, an empty string is returned. If the ending byte position (start + length) is greater than the length of the body part, the returned string will stop at the end of the body part.

An empty *section* parameter (e.g., BODY[]) returns the entire message.

A part specifier can be a part number or a part name.

With part numbers, the message body (which is the only part of a non-MIME message or of a non-multipart MIME message) is part number 1. The additional parts (typically attachments) of a MIME multipart message are parts 2 through *n*. If one of those MIME multipart components (say, part 5) is itself of type "message" or "multipart" (e.g., an embedded MIME message), the individual components of that embedded message would be parts 5.1 through 5.*n*. This scheme can be applied to arbitrarily deeply nested messages, for example, part 5.1.3.

Part names, which are listed in Table 17.5, can be specified by themselves or as the final part of a complex part number (e.g., 1.2.HEADER). Any request that returns headers will return the selected headers, followed by the null line that separates the headers and the message body. This command causes the \Seen flag to be set.

Table 17.5
Part names for BODY [*section*] [*partial*]

HEADER	All the RFC822 headers of the message.
HEADER.FIELDS	Only the specified field(s) of the message. The HEADER.FIELDS keyword is followed by a list of desired RFC822 header names.
HEADER.FIELDS.NOT	All the RFC822 headers of the message, *except* those listed.
MIME	The MIME header for this message part.
TEXT	The message body (with no headers). See the example in on the next page.
HEADER	All RFC822 headers of the main message (example, lines 1–8).
HEADER.FIELDS (TO FROM)	Just the To: and From: headers of the main message (lines 3, 4 and 8).
HEADER.FIELDS.NOT (Received)	All headers of the main message except Received: (lines 2–8).

Table 17.5 (continued)

TEXT	The entire message body (lines 9–32).
1	First part of the message (lines 12, 13).
2	Second part of the message (lines 15–22).
3	Third part of the message (lines 26–31).
3.MIME	The MIME headers of the third message part (lines 24, 25).
3.HEADER	The RFC822 headers of the third message part (lines 26–29).
3.HEADER.FIELDS (SUBJECT)	The Subject: header of the third message part (lines 28, 29).
3.TEXT	The message body of the third message part (lines 30, 31).
BODY.PEEK section [<partial>]	Equivalent to above command, except that the |Senn flag is not set.
BODYSTRUCTURE	Returns the MIME body structure of the message.
ENVELOPE	Returns the parsed RFC-822 headers, defaulting various fields as required.
FAST	Equivalent to (FLAGS INTERNALDATE RFC822.SIZE)
FLAGS	Returns the flags that are set for this message.
FULL	Equivalent to (FLAGS INTERNALDATE RFC822.SIZE ENVELOPE BODY)
INTERNALDATE	Returns the internal date of the message.
RFC822	Equivalent to BODY [], except for the keyword in the resluting untagged FETCH response (which will be RFC822).
RFC822.HEADER	Equivalent to BODY.PEEK[HEADER], except for the keywprd in the resulting untagged FETCH response (which will be RFC822.HEADER).
RFC822.SIZE	Returns the size of the entire message, in bytes.
RFC822.TXT	Equivalent to BODY[TEXT], except for the keyword in the resulting untagged FETCH response (which will be RFC822.TEXT)
UID	Returns the unique identifier for the message.

Example

Assume the message is as follows (with line numbers added):

```
 1: Received: from mail.gals.com; Thu, 19 Dec 1996 09:50:26 -0800
 2: Date: Thu, 19 Dec 1996 09:50:26 -0800
 3: To: asmith@guys.com
 4: From: atoklas@gals.com (Alice B. Toklas)
 5: Subject: Hello from Hawaii
 6: Mime-Version: 1.0
 7: Content-Type: multipart/mixed; boundary = 4*mime*boundary*
 8:
 9: —mime*boundary*
10: Content-Type: text/plain; charset="us-ascii"
11:
12: Wish you were here. I'm attaching two messages from Barney.
13:
14: —*mime*boundary*
15: Content-Type: message/RFC822;
16:
17: From:<bjones@guy.com>
18: To:<atoklas@gals.com>
19: Subject: Let's get together when you get back
20:
21: When you get home lets you me and Albert all get together.
22:
23: —*mime*boundary*
24: Content-Type: message/RFC822;
25:
26: From:<bjones@guys.com>
27: To:<atoklas@gals.com>
28: Subject: time and place

29:
30: Let's make it 7pm Saturday and Joe's Bar and Grill
31:
32: —*mime*boundary*—
```

Sample retrieval session

The following session shows all the IMAP commands and server responses to do the equivalent to the simple POP session in Chapter 16. The lines starting with C: are commands from the client. (There are three types of lines in this sample: C indicates a line from client to server; S indicates an expected (tagged) response from server to client; and s indicates an unsolicited (untagged) response from server to client.)

```
S: * OK service Simeon IMAP4 v1.4.3 server ready
C: A000001 LOGIN asmith swordfish
S: A000001 OK valid password
C: A000002 SELECT INBOX
```

```
s: * FLAGS (\Answered \Flagged \Draft \Deleted \Seen)
s: * OK [PERMANENTFLAGS (\Answered \Flagged \Draft
   \Deleted \Seen \*)]
s: * 2 EXISTS
s: * 0 RECENT
s: * OK [UIDVALIDITY 828462971]
S: A000002 OK [READ-WRITE] Select completed
C: A000003 FETCH 1 RFC822
s: (text of message one - several lines)
S: A000003 OK Fetch completed
C: A000004 STORE 1 +FLAGS \Deleted
S: A000004 OK Store completed
C: A000005 FETCH 2 RFC822
s: (text of message two - several lines)
S: A000005 OK Fetch completed
C: A000006 STORE 2 +FLAGS \Deleted
S: A000006 OK Store completed
C: A000007 CLOSE
S: A000007 OK Close completed
C: A000008 LOGOUT
s: * BYE Server terminating connection
S: A000008 OK Logout completed
```

The above session should be interpreted as follows:

1. The client opens a connection and receives the IMAP4 greeting.

2. The client logs in using the LOGIN authentication mechanism, which the server accepts.

3. The client SELECTs (opens) the INBOX mailbox. This acquires exclusive access to that mailbox ("locks it"). The client receives a number of unsolicited responses from the server, containing information such as the number of messages in INBOX. Finally, the tagged response for the SELECT command is received.

4. The client fetches the entire text ("RFC822") of message 1.

5. The client marks message 1 for deletion (STORE 1 +FLAGS \Deleted).

6. The client fetches the entire text ("RFC822") of message 2

7. The client marks message 2 for deletion (STORE 2 +FLAGS \Deleted)

8. The client closes the current mailbox (INBOX), which actually quietly deletes all messages marked for deletion, then unlocks any exclusive access to the mailbox.

9. The client then logs out, which causes the connection to be terminated.

Note that the sequence is not necessarily the way a real IMAP client would work, but it is intended to illustrate many of the common commands and typical interaction. Specifically, it is intended to perform exactly the same function as the sample complete session in Chapter 16 but using IMAP. A real IMAP client typically would leave messages on the server rather than immediately delete them, as POP usually does.

Server response codes

Server responses can be tagged or untagged and can contain the status of commands (e.g., OK, NO, BAD), or information related to a command (e.g., LIST). For details on the possible server responses and their syntax and semantics, see the command descriptions in the preceding sections or RFC 2060.

IMAP extensions

Even though IMAP4 is quite new, it was designed to be extended, and there are already four extensions to it in the works. At the time of this writing, none of them has graduated to the status of RFC, but some already are being incorporated in various client or server implementations. The extensions are covered in IETF drafts and are subject to change without notice. They almost certainly will make it to RFC status at some time. The current extensions (and the name of the IETF drafts that this information was drawn from) are as follows:

▶ IMAP4 access control list (ACL) extension (draft-myers-imap-acl-03);

▶ IMAP4 nonsynchronizing literals (draft-myers-imap-literal-01);

▶ IMAP4 optimize-1 extension (draft-myers-imap-optimize-01);

▶ IMAP4 quota extension (draft-myers-imap-quota-01).

With the understanding the following sections are based on preliminary information, here are the current details (as of March 1997) on those four extensions.

ACL extension

The ACL extension defines a scheme for restricting access to mailboxes on an IMAP4 server by ACLs (similar to the mechanism used to restrict access to objects in Windows NT, e.g., files on an NTFS drive). In general, an ACL is a group of access control entries (ACEs), each of which specifies a particular account and a set of permissions (or restrictions) associated with that account.

The ACLs may map onto existing permission mechanisms in the underlying operating system on which the server runs, or new functionality could be implemented in the server itself. They are defined in such a flexible manner that they could be mapped onto either UNIX- or NT-style file system permissions. It is possible for certain rights to be linked in such a way that the rights in a linked group can be granted or denied only in concert (not individually).

An IMAP4 server that supports the ACL extension must return the keyword ACL in response to a CAPABILITY command. The specified account in a given ACL can be either of the following:

▶ Any user name string accepted by the LOGIN and AUTHENTICATE commands;

▶ The special name "anyone" to mean all possible accounts.

A specific right is designated by a single character. Multiple rights can be specified with a string consisting of multiple characters, one per right. If the string starts with a plus sign, the rights are granted (the default if there is no sign). If the string starts with a minus sign, the rights are removed. For example, the string +rw or rw means the Read and Write rights are added to any existing rights. The string -l means the lookup right is removed.

The standard rights that can be granted or denied are listed in Table 17.6.

Five new IMAP4 commands (SETACL, DELETEACL, GETACL, LISTRIGHTS, and MYRIGHTS) are defined in the ACL draft.

SETACL mailbox account rights

Responses:	No specific response
Result:	OK, setacl completed
	NO, setacl failed; cannot set ACL
	BAD, syntax error in parameters

This command grants or denies the specified rights for the specified account on the specified mailbox.

An example is:

```
C:   A003 SETACL Personal anyone +rw
S:   A003 OK SETACL complete
```

Table 17.6
Standard rights

l (lookup)	Mailbox visible to LIST/LSUB commands
r (read)	SELECT the mailbox, perform CHECK, FETCH, PARTIAL, or SEARCH on, or COPY from the mailbox
s (seen/unseen flag permanence)	Keep seen/unseen information across sessions (STORE \Seen FLAG)
w (write)	STORE flags other than \Seen and \Deleted
i (insert)	Perform APPEND, COPY into mailbox
p (post)	Send mail to submission address for mailbox; not enforced by IMAP4 itself
c (create)	CREATE new submailboxes in any implementation-defined hierarchy
d (delete)	STORE \Deleted flag, perform EXPUNGE
a (administer)	Perform SETACL

DELETEACL mailbox account

Responses: No specific response
Result: OK, deleteacl completed
           NO, deleteacl failed: cannot delete acl
           BAD, syntax error in parameters

This command removes all access control entries for the specified account on the specified mailbox.
An example is:

```
C:   A003 DELETEACL Personal anyone
S:   A003 OK DELETEACL complete
```

GETACL mailbox

Responses: Untagged response: ACL
Result: OK, getacl completed
           NO, getacl failed: cannot get ACL
           BAD, syntax error in parameters

This command retrieves entire access control list for the specified mailbox in an untagged ACL response.
An example is:

```
C:   A003 GETACL Personal
S:   * ACL Personal anyone rw asmith rwipslda
S:   A003 OK GETACL complete
```

LISTRIGHTS mailbox account

Responses: Untagged response: LISTRIGHTS
Result: OK, listrights completed
           NO, listrights failed; cannot list rights
           BAD, syntax error in parameters

This command lists what rights may be granted to the specified account for the specified mailbox. The first string in the response is the set of rights the account always will be granted to the mailbox (it could be empty). Following that are zero or more strings, each containing a set of rights that the account may be granted to the mailbox.

An example is:

```
C:    A003 LISTRIGHTS Personal anyone
S:    * LISTRIGHTS Personal anyone "" l r s w i p c d a
S:    A003 OK LISTRIGHTS complete
```

MYRIGHTS mailbox

Responses: Untagged response: MYRIGHTS
Result: OK, myrights completed
            NO, myrights failed; cannot obtain rights
            BAD, syntax error in parameters

This command obtains the rights the logged-in user has on the specified mailbox.

An example is:

```
C:    A003 MYRIGHTS INBOX
S:    * MYRIGHTS INBOX rwipslda
S:    A003 OK MYRIGHTS complete
```

Nonsynchronizing literals extension

This extension defines an alternative way to specify literals that do not incur a so-called round-trip penalty (as does the way specified in the base IMAP4 protocol).

A server that supports this extension must include the string LITERAL+ in the response to a CAPABILITY command.

The form of literal specified in the base protocol (known as a synchronizing literal) sends a byte count in braces, for example, {10}, followed by that number of bytes of data. If a client-to-server message does not include the entire literal, the server will send a continuation message to request that the client continue sending the next portion of the literal.

This new form of literal (nonsynchronizing) adds a plus sign between the byte count and the closing brace, for example, {10+}. A compliant server must check the end of every received line for such a "byte count with plus." If the server finds one, it is the byte count of a nonsynchronizing literal, and the server must treat the specified number of following bytes and the following line as part of the same command.

An example is:

```
C:   A001 LOGIN {6+}
C:   ASMITH {10+}
C:   sword fish
S:   A001 OK LOGIN complete
```

Optimize-1 extension

The optimize-1 extension is intended to reduce the time and the resources required to perform some client operations. In particular, it is intended primarily for disconnected-use clients. No new functionality is specified; clients can do exactly the same operations defined in this extension with existing IMAP commands, but it will take more time to do so.

A server that supports this extension must include the string OPTIMIZE-1 in the response to a CAPABILITY command. Two new commands are defined: GETUIDS and UID EXPUNGE.

GETUIDS starting_UID

Responses: Untagged response: GETUIDS
Result: OK, getuids completed
             BAD, syntax error in parameters

This command obtains information about the UIDs contained in the mailbox. The response is returned in a single untagged response.

An example is:

```
C:   A003 GETUIDS 3475
S:   * GETUIDS 17 3509:3519,2535,3590:3599
S:   A003 OK GETUIDS complete
```

UID EXPUNGE message_set

> Responses: Untagged responses: EXPUNGE
> Result: OK, expunge completed
>   NO, expunge failed, e.g., permission denied
>   BAD, syntax error in parameters

This command permanently removes from the currently selected mailbox all messages that both have been marked for deletion (have \Deleted flag set) and have a UID in the specified message set. If a given message either has not been marked for deletion or is not in the specified message set, it is not removed.

An example is:

```
C:   A003 UID EXPUNGE 3000:3002
S:   * EXPUNGE 3
S:   * EXPUNGE 3
S:   * EXPUNGE 3
S:   A003 OK UID EXPUNGE complete
```

Other Optimize-1 extension commands

In addition, some existing IMAP4 commands (APPEND, COPY, and COPY UID) have had their responses enhanced.

A successful APPEND command now includes an APPENDUID response code in the tagged OK response. It contains as arguments the UIDVALIDITY of the destination mailbox and the UID assigned to the appended message. For example:

```
S:   A003 OK [APPENDUID 38505 3955] APPEND complete
```

A successful COPY or UID COPY command now includes a COPYUID response code in the untagged OK response whenever at least one message was copied. It contains the UIDVALIDITY of the appended-to mailbox, a message set containing the UIDs of the messages copied (in the order they were copied), and the UIDs assigned to the copied message (in the order they were copied). For example:

```
S:    A003 OK [COPYUID 38505 304,319:320 3956:3958]Done
```

Quota extension

The quota extension permits administrative limits on the resource usage to be manipulated via the IMAP protocol. Three new commands are introduced: SETQUOTA, GETQUOTA, and GETQUOTAROOT.

Various server resources can be managed via this mechanism. Two possible resources are:

▶ STORAGE, which is the sum of messages' RFC822.SIZE, in units of 1,024 bytes;

▶ MESSAGE, which is the number of messages.

Each mailbox has zero or more quota roots. Each quota root has zero or more resource limits. All mailboxes that share the same named quota root share the resource limits of (i.e., are controlled by) the shared quota root. The quota root names may—but do not have to—match the names of existing mailboxes. The special quota root "" means all mailboxes.

SETQUOTA quota_root resource_limits_list

Responses:	Untagged responses: QUOTA
Result:	OK, setquota completed
	NO, setquota failed; cannot set the specified quota
	BAD, syntax error in parameters

Apply the specified resource limit to the specified quota root (if there are any previous limits, the new ones supersede them). This affects all mailboxes whose resources are controlled by the specified quota root. The list of resource limits contains pairs of "resource name" and "resource limit." The response contains the quota root and triplets of "resource name," "current usage," and "resource limit."

An example is:

```
C:    A003 SETQUOTA "" (STORAGE 512)
S:    * QUOTA "" (STORAGE 10 512)
S:    A003 OK SETQUOTA complete
```

GETQUOTA quota_root

Responses: Untagged responses: QUOTA
Result: OK, getquota completed
             NO, getquota failed; no such quota root, permission
             denied
             BAD, syntax error in parameters

This command retrieves the current resource usage and limits of the specified quota root. The response contains triplets of "resource name," "current usage," and "resource limit."

An example is:

```
C:    A003 GETQUOTA ""
S:    * QUOTA "" (STORAGE 10 512)
S:    A003 OK GETQUOTA complete
```

GETQUOTAROOT mailbox

Responses: Untagged responses: QUOTAROOT, QUOTA
Result: OK, getquotaroot completed
             NO, getquotaroot failed; no such mailbox, permission
             denied
             BAD, syntax error in parameters

This command retrieves the list of quota roots for the specified mailbox (in an untagged QUOTAROOT response). For each listed QUOTAROOT, a QUOTA response returns triplets of "resource name," "current usage," and "resource limit" of that quota root.

An example is:

```
C:    A003 GETQUOTAROOT INBOX
S:    * QUOTAROOT INBOX ""
```

```
S:   * QUOTA "" (STORAGE 10 512)
S:   A003 OK GETQUOTAROOT complete
```

Relevant RFCs

The following RFCs define the IMAP protocol. Several white papers included in the RFC section on IMAP clarify certain issues or compare it with other protocols.

2062	I	M. Crispin, "Internet Message Access Protocol—Obsolete Syntax," 12/04/1996. (Pages=8) (Format=.txt)
2061	I	M. Crispin, "IMAP4 Compatibility With IMAP2BIS," 12/05/1996. (Pages=3) (Format=.txt)
2060	PS	M. Crispin, "Internet Message Access Protocol—Version 4rev1," 12/04/1996. (Pages=82) (Format=.txt) (Obsoletes RFC1730)
1733	I	M. Crispin, "Distributed Electronic Mail Models in IMAP4," 12/20/1994. (Pages=3) (Format=.txt)
1732	I	M. Crispin, "IMAP4 Compatibility With IMAP2 and IMAP2BIS," 12/20/1994. (Pages=5) (Format=.txt)
1731	PS	J. Myers, "IMAP4 Authentication Mechanisms," 12/20/1994. (Pages=6) (Format=.txt)
1730	PS	M. Crispin, "Internet Message Access Protocol—Version 4," 12/20/1994. (Pages=77) (Format=.txt) (Obsoleted by RFC2060)
1203	H	J. Rice, "Interactive Mail Access Protocol—Version 3," 02/08/1991. (Pages=49) (Format=.txt) (Obsoletes RFC1064)
1176	E	M. Crispin, "Interactive Mail Access Protocol—Version 2," 08/20/1990. (Pages=30) (Format=.txt) (Obsoletes RFC1064)
1064	H	M. Crispin, "Interactive Mail Access Protocol: Version 2," 07/01/1988. (Pages=26) (Format=.txt) (Obsoleted by RFC1203, RFC1176)

NNTP: Network News Transfer Protocol

Most users of the Internet are familiar with the World Wide Web and e-mail services and infrastructure. Somewhat fewer are familiar with another service and infrastructure that is just as powerful. It is the source of an incredible wealth of information and expertise. It is also the source of some of the most serious problems (e.g., neo-Nazis and other hate groups, child pornography) on the Internet today, leading to attempts to control or regulate the entire Internet. That service is known by various names—Usenet, NewsNet, Network News—but basically, it is a set of up to 40,000 special-interest groups (newsgroups) organized into the world's largest BBS (bulletin board system).

There are newsgroups on essentially any subject you can imagine (anyone can create a new newsgroup, and some truly bizarre ones show up on a regular basis). A newsgroup is one of the best ways to contact some of the top experts on any given subject (especially for technical and computer-related topics). If you post a serious (and appropriate) question to a newsgroup, chances are excellent that a knowledgeable person will reply within a few days, either directly to you via e-mail or to the group.

Unfortunately, sometimes even the most effective and useful newsgroups can be rendered completely useless by inane, obvious, or malicious postings. You should always read a group's policy messages (typically reposted on a regular basis) before posting. Some groups are moderated, which means only a moderator can actually post to the newsgroup; proposed postings must be e-mailed to that person for consideration. (You should *never* waste Usenet bandwidth with flames; if a flame is richly deserved, send it directly to the poster via e-mail.)

Like the Web and e-mail services, a protocol (NNTP) is associated with this newsgroup service. There are also various server components (news servers), clients (news readers) and a worldwide infrastructure (the Usenet newsgroups and newsfeeds).

Traditional e-mail is primarily a one-to-one scheme or a one-to-a-few scheme (multiple recipients are specified or a mailing list manager is used). For you to receive a message by e-mail, it must be sent specifically to your e-mail address. There is no simple way for a large number of people to read a single copy of a "public" message or to post public replies to such a message. Attempts to do that with mailing lists tend to be difficult to set up and administer and inefficient in terms of network traffic and storage. It also tends to intrude on all the recipients and clutter up the recipients' e-mail inboxes.

A BBS (of which the Network News system is an example) allows a message to be posted once and read by any number of people (but only the ones interested in that particular message). Posted messages can be organized by topic (there are as many as 40,000 such topics on Usenet, e.g., rec.travel.domestic). Just as with e-mail messages, it is possible to reply to such a message, either privately to the author via e-mail or publicly via another posted message. Since there can be a very large number of readers of each message (any of whom can reply to it), there is a way to keep track of conversation "threads" (chains of messages, replies, replies to replies, etc.).

Unlike a simple (isolated) dialup bulletin board system accessed by modem, the Usenet newsgroups are a true distributed system. Messages posted to any Network News server propagate to all other Network News servers on the Internet (at least all such servers that subscribe to the newsgroup to which the message was posted). There is a worldwide infrastructure of newsfeeds into which anyone can connect a news server.

A given news server can subscribe to any or all the newsgroups relayed over that newsfeed. It also can provide a newsfeed to yet other news servers. A news reader (NNTP client program) can connect to any Network News server and allow its users to subscribe to any of the newsgroups to which the server subscribes. Users can retrieve and search through the headers of all messages posted to any subscribed newsgroup (sender, subject, message size, etc.) They then can retrieve, read, and optionally reply to any message that looks interesting.

Two standards are related to network news: RFC 850 ("Networks News Message Syntax"), which is equivalent to RFC 822 for e-mail, and RFC 977 ("Network News Transfer Protocol"), which is equivalent to RFC 821 for e-mail.

A number of Network News servers are available, for both UNIX and Windows NT. There also are a number of news readers (NNTP clients) available for essentially all workstation platforms (DOS, Windows 3.1, Windows 95/NT, Macintosh, various UNIX flavors, etc.). They run the gamut from simple command-line text interface to sophisticated GUI applications similar to the best e-mail UAs (e.g., the news reader part of Microsoft Mail and News Reader or Outlook Express; both are available free from Microsoft).

However, it is becoming more popular to retrieve all your messages (e-mail, newsgroups, faxes, even voice messages) with a single client. Leaving fax and voicemail alone for the moment, there are two approaches to accessing e-mail and newsgroups from a single client. The first scheme is to create a single client that supports both SMTP/POP3 (or SMTP/IMAP4) and NNTP protocols. You would configure it to point to both an Internet Mail server and an Internet News server (which could be on the same physical machine or on two different machines). The second scheme requires an e-mail server that supports "public folders" (e.g., Microsoft Exchange or IMAP4) and coupling some of those public folders to specific Usenet newsgroups via an NNTP gateway. That approach allows the client software to use a single protocol (e.g., SMTP/IMAP4) to send and retrieve both e-mail and Usenet newsgroup messages (and makes it simple to cross-post mail messages to newsgroups or news messages to mail recipients). The second scheme is more likely the way most people will interact with Usenet newsgroups in the future.

Network news message
syntax (RFC 850)

RFC 850 ("Standard for Interchange of USENET Messages") is the equivalent to RFC 822 in the e-mail world. It is very close to the RFC 822 format but differs in several important respects. There is an effort underway to harmonize the two formats to expedite the convergence of e-mail and News. There also are two earlier formats that might be used by legacy systems; they will not be covered in this chapter, but actual software should be able to accept messages using those formats.

Like e-mail messages, a News message consists of two parts: a group of header lines and a body part, separated by a null (zero length) line. The headers consist of a keyword (e.g., From) followed by a colon, then the associated value, for example:

From: asmith@guys.com (Albert Smith)

Some headers are required, and some are optional. Both client and server news software should ignore any unrecognized headers. Required headers are Relay-Version, Posting-Version, From, Date, Newsgroups, Subject, Message-ID, and Path. Optional headers are Followup-To, Date-Received, Expires, Reply-To, Sender, References, Control, Distribution, and Organization.

Required headers

The following headers must be included in any generated message, and all servers should be able to understand and use them correctly.

Relay-Version version_id; site_id

This header identifies the version of the program and the site sending this news message. It must be the first header in a news message. The associated value contains two fields separated by a semicolon. The first field identifies the version of the sending software (or its protocol); the second field identifies the sending site. Each time the message is relayed, this header is updated to identify the new sender. An example is:

Relay-Version: version B 2.10 2/13/83; site cbosgd.UUCP

Posting-Version version_id; site_id

This header identifies the version of the program that originated the news message. The syntax is the same as for Relay-Version, but this header is not updated at each relay point. An example is:

Posting-Version: version B 2.10 2/13/83; site eagle.UUCP

From: mailing_address

This header identifies the originator of the news message with that person's Internet e-mail address, suitable for responding to with any Internet compatible UA. An example is:

From: asmith@guys.com (Albert Smith)

Date: date_time

This header (the archaic form is Posted:) identifies the time and date that this news message was originally posted. An example is:

Date: Saturday, March-22-97 18:00:09 EST

Newsgroups: newsgroup_list

This header identifies the newsgroup(s) to which the message was posted. The list contains one or more newsgroup names (e.g., rec.travel.domestic) separated by commas. It corresponds to the To: header of e-mail messages. An example is:

Newsgroups: rec.travel.domestic, rec.travel.international

Subject: message_subject

This header (the archaic form is Title:) identifies the subject of the news message. Messages responding to a previous posting should start with Re: then include the subject of the original message (avoiding so-called Re:

buildup). Such followup messages should include a reference header as well. An example is:

> Newsgroups: rec.travel.domestic, rec.travel.international

Message-ID: message_id

This header (the archaic form is Article-ID:) identifies the message with a unique message ID code. A message ID is a string in angle brackets that is unique in the Usenet system. It can be used to list the message to clients, for clients to retrieve the message, and so on. An example is:

> Message-ID: <4105@guys.com>

Path: message_path

This header records the path a news message traverses through the news infrastructure. Each time it is relayed, the server doing the relaying prepends its name to the path. Originally, it could have been used to reply to the sender with e-mail, but today we use domain e-mail addresses, not "bang paths." This header has been kept around because it is still hand for debugging and collecting statistics about the news system. An example is:

> Path: news.megacorp.com!beast.megacorp.com!news.guys.com

Optional headers

The following headers may optionally be included in any generated message, and all servers should be able to understand and use them correctly.

Reply-To: email_address

This header is similar to (and has same syntax) as the From: header. If present, it should be used as the address to reply to the author with e-mail (if not, use the address in the From: header).
An example is:

> Reply-To: asmith@guys.com (Albert Smith)

Sender: email-address

This header is similar to (and has same syntax) as the From: header. If present, it indicates the sender of the message (if not, the sender is assumed to be the user specified in the From: header). An example is:

Sender: bjones@guys.com (Barney Jones)

Followup-To: newsgroups

This header is similar to (and has the same syntax as) the Newsgroups: header. If present, it overrides which group(s) to post news messages in reply to this message (if not, reply messages are posted to all newsgroups in the newsgroup header). An example is:

Followup-To: alt.flames

Date-Received: date_time

This header (the archaic form is Received:) specifies the date and time a particular news message was received. It is solely for local use. If a server receives an incoming message with a Date-Received header, it can discard it or update it with the current date and time. An example is:

Date-Received: Saturday, March-22-97 18:00:09 EST

Expires: date_time

If present, this header specifies a date and time that the message expires. Unless there is a natural reason to expire a message (e.g., to expire a message about a conference after the conference is over), this header should not be used (all servers should have their own default procedures for expiring messages). An example is:

Expires: Saturday, March-22-97 18:00:09 EST

References: message_id_list

This header specifies a list of message-ids in response to which this message was posted. It allows a client to display message chains or conversation threads. An example is:

References: <4105@guys.com>,<4106@guys.com>

Control: control_message

This header allows servers to exchange system control messages with each other. Several defined control messages are specified as the parameter of a Control: header. The ones defined in RFC 977 are:

- ▶ "cancel *message_id*" deletes the message with the specified message id;

- ▶ "ihave *message_id_list remote_system*" is used to inform another server that you have a particular message;

- ▶ "sendme *message_id_list remote_system*" is used to request a message to be sent to you by another server;

- ▶ "newgroup *newsgroup*" is used to create a new newsgroup;

- ▶ "rmgroup *newsgroup*" is used to delete an existing newsgroup;

- ▶ "sendsys" requests a copy of the "sys" file to be sent;

- ▶ "senduuname" requests the uuname program be run and results sent back;

- ▶ "version" requests the version number of the software to be sent. An example is:

Control: cancel <4105@guys.com>

NNTP: Network News Transfer Protocol (RFC 977)

NNTP requires a reliable, connection-oriented transport with flow control, so it typically is implemented over TCP. The well-known port for NNTP is 119. The syntax is similar to SMTP in design. Messages are short sequences of printable ASCII characters (up to 510 per line) terminated by CR,LF. Messages from client to server consist of a command keyword followed by parameters. Replies from server to client consist of three-digit response codes (and optional human-readable explanations). Some commands result in multiline responses with data (e.g., a news message) following the first line containing the three-digit response code.

On receipt of a connection from a client (or another server acting as client), an NNTP server will respond with one of the following:

```
S:   200 server ready - posting allowed
```

or

```
S:   201 server ready - no posting allowed
```

The NNTP commands are ARTICLE, BODY, HEAD, STAT, GROUP, HELP, IHAVE, LAST, LIST, NEWGROUPS, NEWNEWS, NEXT, POST, QUIT, and SLAVE.

Table 18.1 lists the three-digit response codes for NNTP (along with recommended human-readable text).

Table 18.1
NNTP response codes

100	help text follows
199	debug output
200	server ready – posting allowed
201	server ready – no posting allowed
202	slave status noted
205	closing connection – goodbye!
211	n f l s group selected
215	list of newsgroups follows
220	n <a> article retrieved – head and body follows
221	n <a> article retrieved – head follows
222	n <a> article retrieved – body follows
223	n <a> article retrieved – request text separately
230	list of new articles by message-id follows
231	list of new newsgroups follows

Table 18.1 (continued)

235	article transferred OK
240	article posted OK
335	send article to be transferred. End with <CR-LF>.<CR-LF>
340	send article to be posted. End with <CR-LF>.<CR-LF>
400	service discontinued
411	no such news group
412	no newsgroup has been selected
420	no current article has been selected
421	no next article in this group
422	no previous article in this group
423	no such article number in this group
430	no such article found
435	article not wanted – do not send it
436	transfer failed – try again later
437	article rejected – do not try again
440	posting not allowed
441	posting failed
500	command not recognized
501	command syntax error
502	access restriction or permission denied
503	program fault – command not performed

Required commands

The following commands must be present in any server implementation and are fair game for any client to use.

GROUP newsgroup

This command selects an existing newsgroup (e.g., rec.travel.domestic) from the list of newsgroups returned by a previous LIST command. If a valid newsgroup is specified, the current article pointer is set to the first article, and the following four items are returned (following the 211 response code):

- The estimated number of articles in the group;
- The first article number in the group;
- The last article number in the group;
- The name of the group.

If an invalid newsgroup is specified, a 411 response code is returned. Examples are:

```
C:   GROUP rec.travel.domestic
S:   211 10 400 409 rec.travel.domestic group selected
C:   GROUP rec.nosuchgroup
S:   411 no such news group
```

ARTICLE message_id

This command returns the header, a blank line, then the body of the news message with the specified message ID (as returned by a previous NEWNEWS command, the References header from another message, etc.). A message ID can be distinguished from a message number by the surrounding angle brackets. Do not update the current article pointer. If no newsgroup has been selected, a 412 will be returned.

Examples are:

```
C:   ARTICLE <4105@guys.com>
S:   220 402 <4105@guys.com> article retrieved, head
     and body follows
S:   (article headers, blank line, article body)
S:   .
C:   ARTICLE <4105@guys.com>
S:   430 no such article found
```

ARTICLE [message_number]

This command returns the header, a blank line, then the body of the news message with the specified (optional) message number (within the range of message numbers returned when the newsgroup was selected with the

GROUP command). Set the current article pointer to the specified message. If no newsgroup has been selected, a 412 will be returned.

Examples are:

```
C:   ARTICLE 402
S:   220 402 <4105@guys.com> article retrieved, head
       and body follows
S:   (article headers, blank line, article body)
S:   .
C:   ARTICLE 403
S:   423 no such article number in this group
```

HEAD message_id

This command returns the header of the news message with the specified message ID (as returned by a previous NEWNEWS command, the References header from another message, etc.). A message ID can be distinguished from a message number by the surrounding angle brackets. Do not update the current article pointer. If no newsgroup has been selected, a 412 will be returned.

Examples are:

```
C:   HEAD <4105@guys.com>
S:   221 402 <4105@guys.com> article retrieved,
       head follows
S:   (article headers)
```

HEAD message_number

Return the header of the news message with the specified (optional) message number (within the range of message numbers returned when the newsgroup was selected with the GROUP command). Set the "current article pointer" to the specified message. If no newsgroup has been selected, a 412 will be returned.

Examples are:

```
C:   HEAD <4105@guys.com>
S:   221 402 <4105@guys.com> article retrieved,
       head follows
```

```
S:   .
C:   HEAD <4105@guys.com>
S:   430 no such article number in group.
```

BODY [message_id]

Return the body of the news message with the specified message id (as returned by a previous NEWNEWS command, Reference header from another message, etc.). Do not update the "current article pointer." A message id can be distinguished from a message number by the surrounding angle brackets. If no newsgroup has been selected, a 412 will be returned.

Examples are:

```
C:   BODY <4105@guys.com>
S:   221 402 <4105@guys.com> article retrieved,
     body follows
S:   (article body)
S:   .
C:   BODY <4105@guys.com>
S:   430 no such article found
```

BODY [message_number]

Return the body of the news message with the specified (optional) message number (within the range of message numbers returned when the newsgroup was selected with the GROUP command). Set the "current article pointer" to the specified message. If no newsgroup has been selected, a 412 will be returned.

Examples are:

```
C:   BODY 402
S:   222 402 <4105@guys.com> article retrieved,
     body follows
S:   (article headers, blank line, article body)
S:   .
C:   BODY 403
S:   423 no such article number in group
```

STAT message_id

This command returns the statistics of the news message with the specified message ID (as returned by a previous NEWNEWS command, the References header from another message, etc.). Do not update the current article pointer. A message ID can be distinguished from a message number by the surrounding angle brackets. If no newsgroup has been selected, a 412 will be returned. This particular command is of little value, other than being able to obtain the message number for a specified message.

Examples are:

```
C:  STAT <4105@guys.com>
S:  223 402 <4105@guys.com> article retrieved, request
    text separately
C:  STAT <4105@guys.com>
S:  430 no such article found
```

STAT [message_number]

This command returns the statistics of the news message with the specified message number (within the range of message numbers returned when the newsgroup was selected with the GROUP command). Set the current article pointer to the specified message. If no newsgroup has been selected, a 412 will be returned.

Examples are:

```
C:  STAT 402
S:  223 402 <4105@guys.com> article retrieved, request
    text separately
C:  STAT 403
S:  423 no such article number in this group
```

LAST

This command moves the current article pointer to the previous message. If the pointer already is at the first message of the newsgroup, a 422 response is returned. If no newsgroup has been selected, a 412 will be returned. Otherwise, the pointer is set to the previous message, and a 223 response is returned with the message number and the message ID of that message.

Examples are:

```
C:   LAST
S:   223 402 <4105@guys.com> request text separately
C:   LAST
S:   422 no previous article in this group
```

NEXT

This command moves the current article pointer to the next message. If the pointer already is at the last message of the newsgroup, a 421 response is returned. If no newsgroup has been selected, a 412 will be returned. Otherwise, the pointer is set to the next message, and a 223 response is returned with the message number and message id of that message.

Examples are:

```
C:   NEXT
S:   223 403 <4105@guys.com> request text separately
C:   NEXT
S:   421 no next article in this group
```

HELP

This command returns a short summary of the commands that are understood by this server.

An example is:

```
C:   HELP
S:   (help message)
S:   .
```

IHAVE message_id

This command informs the server that the sender has in its possession a message with the specified message ID. A message ID can be distinguished from a message number by the surrounding angle brackets. The server will reply with 335 if it wants that article or with 435, 436, or 437 if it does not.

Examples are:

```
C:   IHAVE <4105@guys.com>
S:   335 send article to be transferred. End with
     <CR-LF>.<CR-LF>
C:   (message head, blank line, message body)
C:   .
S:   235 article transferred OK
C:   IHAVE <4105@guys.com>
S:   435 article not wanted, do not send it
```

LIST

This command returns a list of all newsgroups on the server and the following information on each: newsgroup name, number of last article, number of first article, and whether posting is allowed on this specific newsgroup (Y or N).

An example is:

```
C:   LIST
S:   215 list of newsgroups follows
S:   rec.travel.domestic 1456 1421 Y
S:   rec.travel.international 2314 2135 Y
S:   .
```

NEWGROUPS date time [GMT] [distributions]

This command returns a list of new newsgroups that have been created (within Usenet, not on the server handling the command) since a specific date and time (using the same format as LIST). The date is specified as YYMMDD (YY maps onto the nearest century). The time is specified as HHMMSS in the server's timezone (unless GMT is specified). If "distributions" is specified, that is a list of distribution groups, which allows narrowing the response to newsgroups within a specific subtree of the Usenet newsgroups (e.g., just those under rec, or rec.travel).

```
C:   NEWSGROUPS 800101 000000
S:   231 list of new newsgroups follows
S:   rec.travel.domestic 1456 1421 Y
S:   rec.travel.international 2314 2135 Y
S:   .
```

NEWNEWS newsgroups date time [GMT] [distributions]

This command is similar to the NEWGROUPS command except that it returns a list of message IDs of articles posted to the specified newsgroup (one message ID per line). It is valid to specify any part of the name as an asterisk, which will map any name in that position (e.g., rec.*.international). Multiple newsgroups can be specified, separated by commas (e.g. rec.travel.domestic,rec.travel.international). A leading exclamation point (a "bang") can be used to remove newsgroups from a list that already has been started (e.g., rec.travel.*,!rec.travel.international). The response does not indicate which group a given message ID is from.

An example is:

```
C:   NEWNEWS rec.travel.* 970321 090000
S:   230 list of new articles by message-id follows
S:   <4105@guys.com>
S:   <4106@guys.com>
S:   <4107@guys.com>
S:   <4108@guys.com>
S:   .
```

POST

If posting is allowed on the server (and by this poster if authentication was user), a 340 response code is returned, and the message should be sent. A 440 response code indicates that this server does not accept posting. For syntax of a news message, see the information on RFC 850. On receipt of a message, a further response code will be sent (240 for success, 441 for failure).

Examples are:

```
C:   POST
S:   340 send article to be posted. End with
       <CR-LF>.<CR-LF>
C:   (message head, blank line, message body)
S:   240 article posted ok
S:   .

C:   POST
S:   340 send article to be posted. End with
       <CR-LF>.<CR-LF>
```

```
C:   (message head, blank line, message body,
     for a closed newsgroup)
S:   441 posting failed
S:   .
```

```
C:   POST
S:   440 posting not allowed
```

QUIT

This command terminates the connection.
   An example is:

```
C:   QUIT
S:   205 closing connection - goodbye!
```

SLAVE

This command informs the server that this is another server operating in
"slave" mode, rather than a news reader UA. It can be used to assign priority to such a connection, and so on. The details of the functionality are
entirely implementation dependent.
   An example is:

```
C:   SLAVE
S:   202 slave status noted
```

Example of a complete session

The following example is a typical session in which a news reader UA is
used to read new messages from a single newsgroup.

```
S:   (listens for connection at TCP port 119)
C:   (connects to port 119)
S:   200 NNTP server ready - posting allowed
```

```
C:   NEWNEWS rec.travel.domestic 970321 091803
S:   230 list of new articles by message-id follows
S:   <4105@guys.com>
```

```
S:    <4106@guys.com>
S:    .
C:    GROUP rec.travel.domestic
S:    211 50 10011 10125 rec.travel.domestic group
selected
C:    HEAD <4105@guys.com>
S:    221 10124 <4105@guys.com> article retrieved -
      head follows
S:    (message head for first new message)
S:    .
S:    HEAD <4106@guys.com>
S:    221 10125 <4106@guys.com> article retrieved -
      head follows
S:    (message head for second new message)
S:    .
C:    ARTICLE <4105@guys.com>
S:    220 10124 <4105@guys.com> article retrieved -
      head and body
      follow
S:    (message head, blank line and message body)
S:    .

:     POST
S:    340 send article to be posted. End with
      .
C:    (message head, blank line and message body
      of new message)
C:    .
S:    240 article posted OK

C:    QUIT
S:    205 closing connection - goodbye!
```

In the first group, the client connects to its local news server on TCP port 119. The news server accepts the connection and sends a greeting, informing the client that posting is OK.

In the second group, the user has selected the newsgroup rec.travel.domestic, so the client checks for new news messages in that group. The server returns the message IDs of two new messages. Since new messages are available, the client then selects the group and retrieves the message headers for the two new messages, so it can update its list of message headers in that group (perhaps displayed in a window).

In the third group, the user has requested to read the first of the two new messages (perhaps by double-clicking on the corresponding line in the list of message headers). The client retrieves the entire text of the message (headers and body). It might display the result in a window.

In the fourth group, the user has posted a new message (possibly in reply to the message just read). Because the server allows posting and this particular newsgroup also allows posting, the posting succeeds.

In the fifth group, the user terminates the news reader program, which causes it to inform the server to terminate the connection (QUIT). The server signs off (205) and breaks the connection. The client can then terminate.

S/MIME: Secure MIME

There have been a number of attempts to add cryptography to Internet e-mail. All but two have failed to build any real momentum. The also-rans include PEM (RFCs 1421, 1422, 1423, and 1616), MOSS (RFC 1848), and PGP (RFCs 1991 and 2015).

One standard has built momentum but only within the U.S. government, and there only because it has been specified as a requirement in many situations. That standard is known as Message Security Protocol (MSP) or P42 and is part of the Defense Messaging System (DMS). It also has found some support among people who need to interact with government agencies. Microsoft has created a version of Exchange Server that supports MSP, in addition to all the other aspects of DMS. It has capabilities roughly equivalent to those of S/MIME or X.400 secure e-mail but is not directly interoperable with either.

The clear winner overall is S/MIME. It has had a difficult path receiving the IETF's blessing due to its use of patented domain technology and algorithms (the public key algorithms from RSA). Historically, all standards approved by the IETF have used only public domain technology and algorithms, so anyone can implement them without having to pay royalties. Unfortunately, RSA's patents are so broad as to cover essentially any public key algorithm. The good news is that their patents are due to expire in the very near future (the El Gamal public key algorithms already are in the public domain). Basically, S/MIME involves adding several forms of digital envelopes and digital signatures, based on RSA's Public Key Cryptography Standards (PKCS).

Although the PKCS standards themselves are freely available (and are included on the accompanying CD-ROM), any developers who use the algorithms described therein, whether they use RSA's toolkits or not, are required to pay RSA royalties, *at least for products created and distributed in the United States*. It is unusual for any one company to corner the market in such a manner on an entire technology, and it has held back the development of public key-based products by more than a decade. Expect to see an explosion of cryptography products as the RSA patents finally expire.

Outside the United States, the development and marketing of public key cryptography are a different story. One enterprising fellow, Eric Young (eay@cryptsoft.com) has implemented the important RSA public key algorithms (without access to any of their source code), and placed that implementation in the public domain. Young and another developer, Tim Hudson (tjh@cryptsoft.com) have further used this library to implement SSL versions 2.0 and 3.0. Products built using those tools can be distributed outside the United States (at least in most countries) without payment of any royalties to anyone. Within the United States, you still have to pay royalties to RSA (as long as their patents are in force). If the products are developed in the United States, they are subject to U.S. cryptography export controls (e.g., 40-bit keys).

There are 12 PKCS standards all together, but some have been phased out or replaced. The first 11 are well defined as of the writing of this book, but the twelfth is still in draft. The important ones are the following:

- 1, RSA Encryption Standard;

- 3, Diffie-Hellman Key-Agreement Standard;

- 5, Password-Based Encryption Standard;

- ◗ 7, Cryptographic Message Syntax Standard;

- ◗ 8, Private-Key Information Syntax Standard;

- ◗ 9, Selected Attribute Types;

- ◗ 10, Certification Request Syntax Standard;

- ◗ 11, Cryptographic Token Interface Standard;

- ◗ 12, Personal Information Exchange Syntax.

S/MIME is based heavily on PKCS 1, 7, and 10, although specific products that support S/MIME may make use of some of the others (e.g., a product that supports the Fortezza cryptographic "smart card" would use PKCS 11).

Unlike many of the standards described in this book, S/MIME currently is not defined by any RFC (although several drafts are in progress). In spite of that, quite a few companies are supporting it already (see www.rsa.com for a current list). In particular, Microsoft and Netscape both are supporting it heavily, and a number of products are available (which is unusual for a standard still in draft form).

Like MIME, S/MIME is an end-to-end protocol. It is important only for the sending and the receiving client to support it (intervening MTAs or e-mail servers need not know anything about it). Gateways, however, are another matter altogether. A gateway between a proprietary secure e-mail system (e.g., Microsoft Exchange) and S/MIME Internet e-mail would be a complex product. It also would need access to all user's public *and* private keys (not to mention any keys used by the proprietary scheme) to do its job. In a good secure messaging system, each user's private key *never* should leave a highly secure area of that user's computer (better yet, it should exist only in a physical smart card).

Like MIME, S/MIME involves new message headers in addition to new content in the message body. That makes it difficult or impossible to support S/MIME with a separate standalone program, as is possible with basic PGP (but not with PGP/MIME). Some plug-ins exist to add S/MIME capability to existing non-S/MIME compliant e-mail clients, but those must be able to emit and interpret the new S/MIME headers. However, the need for plug-ins or preprocessors is disappearing as more vendors support S/MIME directly in their e-mail clients (e.g., in Microsoft's Outlook Express).

There are two basic capabilities in S/MIME: digital signature and digital envelope (see Chapter 5 for detailed explanations of those concepts). However, there are some variants of those two basic capabilities. In addition, it supports multiple algorithms (e.g., both SHA-1 and MD5 for generating message digests).

S/MIME can be applied to any MIME subpart (including the entire message). A given part can be signed, enveloped, both, or neither. The digital signature can be done on the message in its original form or in a processed form (a so-called opaque signature). In an opaque signed message, the original message content is "canonicalized" (rendered into a standard form that is invariant as it goes through various e-mail components). The privacy is not particularly strong (no encryption is involved), but the message content is not plainly visible, as it is with a non-opaque signature. To read the message part of an opaque signed message, the receiver must have S/MIME support (or be willing to do some relatively complicated offline processing to the received message).

You can send a non-opaque signed message to anyone who has S/MIME support to read the message and verify the signature; without S/MIME support, at least the recipient can read the message part (and ignore the digital signature subpart as an extraneous attachment). You should send opaque signed messages *only* to people you know have S/MIME support, or they probably will not be able to make any sense of your message. The advantage of an opaque signature is that the digital signature still will work correctly even if gateways or MTAs modify the text of your message along the way (e.g., doing word wrap at column 72). A non-opaque signature will fail if the text of the message is modified in *any* way (that is one of the main things a digital signature is supposed to do). There is no way for it to distinguish between malicious, intentional tampering and innocent changes made by an e-mail system.

It also is possible to send digital certificates (see Chapter 5) in an S/MIME message, either together with other content or by themselves (using PKCS 10).

Following are some real examples of S/MIME messages (in a form you likely would never see; I had to send them to an MTA with an S/MIME client and extract them from the MTA's message store to get the complete message structure, MIME headers and all). You can see that some of the options lead to a dramatic increase in message size (the message body is the same in all the examples).

No signature, no encryption, no certificate (basic message)

```
Received: from lhughes by lhughes.worldnet.att.net Mon Nov 24 16:04:44 1997
From: "Lawrence Hughes" <lehughes@worldnet.att.net>
To: "Hughes, Lawrence" <lehughes@worldnet.att.net>
Subject: test
Date: Mon, 24 Nov 1997 16:04:44 -0500
Message-ID: <01bcf91c$96ade2f0$d2a50b0a@lhughes.worldnet.att.net>
MIME-Version: 1.0
Content-Type: text/plain;
          charset="iso-8859-1"
Content-Transfer-Encoding: 7bit
X-Priority: 3
X-MSMail-Priority: Normal
X-Mailer: Microsoft Outlook Express 4.71.1712.3
X-MimeOLE: Produced By Microsoft MimeOLE V4.71.1712.3

Now is the time for all good men to come to the aid of their neighbors
```

Normal signature, no encryption, no certificate

```
Received: from lhughes by lhughes.worldnet.att.net Mon Nov 24 16:08:18 1997
From: "Lawrence Hughes" <lehughes@worldnet.att.net>
To: "Hughes, Lawrence" <lehughes@worldnet.att.net>
Subject: test signed, don't include my certificate
Date: Mon, 24 Nov 1997 16:08:17 -0500
Message-ID: <01bcf91d$16385690$d2a50b0a@lhughes.worldnet.att.net>
MIME-Version: 1.0
Content-Type: multipart/signed;
      boundary="----=_NextPart_000_0014_01BCF8F3.2D5492F0";
      protocol="application/x-pkcs7-signature";
      micalg=SHA-1
X-Priority: 3
X-MSMail-Priority: Normal
X-Mailer: Microsoft Outlook Express 4.71.1712.3
X-MimeOLE: Produced By Microsoft MimeOLE V4.71.1712.3

This is a multi-part message in MIME format.

------=_NextPart_000_0014_01BCF8F3.2D5492F0
Content-Type: text/plain;
      charset="iso-8859-1"
Content-Transfer-Encoding: 7bit

Now is the time for all good men to come to the aid of their neighbors
```

```
------=_NextPart_000_0014_01BCF8F3.2D5492F0
Content-Type: application/x-pkcs7-signature;
        name="smime.p7s"
Content-Transfer-Encoding: base64
Content-Disposition: attachment;
        filename="smime.p7s"
```

MIAGCSqGSIb3DQEHAqCAMIACAQExCzAJBgUrDgMCGgUAMIAGCSqGSIb3DQEHAQAAMYIBZDCCAWAC
AQEwdjBiMREwDwYDVQQHEwhJbnRlcm51dDEXMBUGA1UEChMOVmVyaVNpZ24sIEluYy4xNDAyBgNV
BAsTK1ZlcmlTaWduIENsYXNzIDEgQ0EgLSBJbmRpdmlkdWFsIFsIFN1YnNjcmliZXICEGmtA2REwZvH
boy9pyMAoS4wCQYFKw4DAhoFAKCBhjAYBgkqhkiG9w0BCQMxCwYJKoZIhvcNAQcBMBwGCSqGSIb3
DQEJBTEPFw05NzExMjQxNjA4MTdaMCMGCSqGSIb3DQEJBDEWBBQ9nlCcnyLXvfJ1W5Ry7oWkp9kW
YjAnBgkqhkiG9w0BCQ8xGjAYMA0GCCqGSIb3DQMCAgEoMAcGBSsOAwIdMA0GCSqGSIb3DQEBAQUA
BECQGZEjs06nc2B6HHtlaFEVncp8XgNuqOx94V9YXuRIKMTMB4E7hRI6GnPxI0us73Bl3UrAH8en
0mQouj0HxVP9AAAAAAA

```
------=_NextPart_000_0014_01BCF8F3.2D5492F0--
```

Normal signature, no encryption, with certificate

```
Received: from lhughes by lhughes.worldnet.att.net Mon Nov 24 15:59:17 1997
From: "Lawrence Hughes" <lehughes@worldnet.att.net>
To: "Hughes, Lawrence" <lehughes@worldnet.att.net>
Subject: test signed
Date: Mon, 24 Nov 1997 15:59:16 -0500
Message-ID: <01bcf91b$d3e429f0$d2a50b0a@lhughes.worldnet.att.net>
MIME-Version: 1.0
Content-Type: multipart/signed;
        boundary="----=_NextPart_000_0005_01BCF8F1.EAF8C530";
        protocol="application/x-pkcs7-signature";
        micalg=SHA-1
X-Priority: 3
X-MSMail-Priority: Normal
X-Mailer: Microsoft Outlook Express 4.71.1712.3
X-MimeOLE: Produced By Microsoft MimeOLE V4.71.1712.3

This is a multi-part message in MIME format.

------=_NextPart_000_0005_01BCF8F1.EAF8C530
Content-Type: text/plain;
        charset="iso-8859-1"
Content-Transfer-Encoding: 7bit

Now is the time for all good men to come to the aid of their neighbors
------=_NextPart_000_0005_01BCF8F1.EAF8C530
```

```
Content-Type: application/x-pkcs7-signature;
      name="smime.p7s"
Content-Transfer-Encoding: base64
Content-Disposition: attachment;
      filename="smime.p7s"
```

MIAGCSqGSIb3DQEHAqCAMIACAQExCzAJBgUrDgMCGgUAMIAGCSqGSIb3DQEHAQAAoIIJHTCCAjww
ggGlAhAyUDPPUNFW81yBrWVcT8glMA0GCSqGSIb3DQEBAgUAMF8xCzAJBgNVBAYTAlVTMRcwFQYD
VQQKEw5WZXJpU2lnbiwgSW5jLjE3MDUGA1UECxMuQ2xhc3MgMSBQdWJsaWMgUHJpbWFyeSBDZXJ0
aWZpY2F0aW9uIEF1dGhvcml0eTAeFw05NjAxMjkwMDAwMDBaFw0yMDAxMDcyMzU5NTlaMF8xCzAJ
BgNVBAYTAlVTMRcwFQYDVQQKEw5WZXJpU2lnbiwgSW5jLjE3MDUGA1UECxMuQ2xhc3MgMSBQdWJs
aWMgUHJpbWFyeSBDZXJ0aWZpY2F0aW9uIEF1dGhvcml0eTCBnzANBgkqhkiG9w0BAQEFAAOBjQAw
gYkCgYEA5Rm/baNWYS2ZSHH2Z965jeu3noaACpEO+jglr0aIguVzqKCbJF0NH8xlbgyw0FaEGIea
BpsQoXPftFg5a27B9hXVqKg/qhIGjTGsf7A01480Z4gJzRQR4k5FVmkfeAKA2txHkSm7NsljXMXg
1y2He6G3MrB7MLoqLzGq7qNn2tsCAwEAATANBgkqhkiG9w0BAQIFAAOBgQBLRGZgaGTkmBvzsHLm
lYl83XuzlcAdLtjYGdAtND3GUJoQhoyqPzuoBPw3UpXD2cnbzfKGBsSxG/CCiDBCjhdQHGR6uD6Z
SXSX/KwCQ/uWDFYEJQx8fIedJKfY8DIptaTfXaJMxRYyqEL2Raa2Nrngv2U2k8LS12vc3lnWojX4
RTCCAnkwggHioAMCAQICEFifNR3ycH4AK77KWYcE1TkwDQYJKoZIhvcNAQECBQAwXzELMAkGA1UE
BhMCVVMxFzAVBgNVBAoTDlZlcmlTaWduLCBJbmMuMTcwNQYDVQQLEy5DbGFzcyAxIFB1YmxpYyBQ
cmltYXJ5IENlcnRpZmljYXRpb24gQXV0aG9yaXR5MB4XDTk2MDYyNzAwMDAwMFoXDTk5MDYyNzIz
NTk1OVowYjERMA8GA1UEBxMISW50ZXJuZXQxFzAVBgNVBAoTDlZlcmlTaWduLCBJbmMuMTQwMgYD
VQQLEytWZXJpU2lnbiBDbGFzcyAxIENBIC0gSW5kaXZpZHVhbCBTdWJzY3JpYmVyMIGfMA0GCSqG
SIb3DQEBAQUAA4GNADCBiQKBgQC2FKbPTdAFDdjKI9BvqrQpkmOOLPhvltcunXZLEbE2jVfJw/0c
xrr+Hgi6M8qV6r7jW80GqLd5HUQq7XPysVKDaBBwZJHXPmv5912dFEObbpdFmIFH0S3L3bty10w/
cariQPJUObwW7s987LrbP2wqsxaxhhKdrpM01bjV0Pc+qQIDAQABozMwMTAPBgNVHRMECDAGAQH/
AgEBMAsGA1UdDwQEAwIBBjARBglghkgBhvhCAQEEBAMCAQYwDQYJKoZIhvcNAQECBQADgYEAwfr3
AudXyhF1xpwM+it3T4dFFzvj0sHaD1g5jq6VmQOhqKE4/nmakxcLl4Y5x8poNGa7x4hF9sgMBe6+
lyXv4NRu5H+ddlzOfboUoq4Ln/tnW0ilZyWvGWSI9nLYKSeqNxJqsSivJ4MYZWyN7UCeTcR4qIbs
6SxQv6b5DduwpkowggRcMIIDxaADAgECAhBprQNkRMGbx26MvacjAKEuMA0GCSqGSIb3DQEBBAUA
MGIxETAPBgNVBAcTCEludGVybmV0MRcwFQYDVQQKEw5WZXJpU2lnbiwgSW5jLjE0MDIGA1UECxMr
VmVyaVNpZ24gQ2xhc3MgMSBDQSAtIEluZGl2aWR1YWwgU3Vic2NyaWJlcjAeFw05NzEwMTUwMDAw
MDBaFw05ODEwMTUyMzU5NTlaMIIBJjERMA8GA1UEBxMISW50ZXJuZXQxFzAVBgNVBAoTDlZlcmlT
aWduLCBJbmMuMTQwMgYDVQQLEytWZXJpU2lnbiBDbGFzcyAxIENBIC0gSW5kaXZpZHVhbCBTdWJz
Y3JpYmVyMUYwRAYDVQQLEz13d3cudmVyaXNpZ24uY29tL3JlcG9zaXRvcnkvUlBBIEluY29ycC4g
YnkgUmVmLixMSUFCLkxURChKTk2MTQwMgYDVQQLEytEaWdpdGFsIElEIENsYXNzIDEgLSBNaWNy
b3NvZnRQdGRnVsbCBTZXJ2aWNlMRowGAYDVQQExFMQVdSRU5DRSBFIEhVR0hFUzEoMCYGCSqGSIb3
DQEJARYbZGVodWdoZXNAZ29ybGRuZXQuQXlXR0Lm5lDBcMA0GCSqGSIb3DQEBAQUAA0sAMEgCQQCi
PBi2fC5gLo5cdrC0wdI09VENDKAb9Q6MPHrWgRdbPoJGltjxSRCMwpnSnmhIeCGv/IR3zvrWB5Ku
kuwdRw0fAgMBAAGjggGPMIIBizAJBgNVHRMEAjAAMIGvBgNVHSAEgacwgDCABgtghkgBhvhFAQcB
ATCAMCgGCCsGAQUFBwIBFhxodHRwczovL3d3dy52ZXJpc2lnbi5jb20vQ1BTMCGIGCCsGAQUFBwIC
MFYwFRYOVmVyaVNpZ24sIEluYy4wAwIBARo9VmVyaVNpZ24ncyBDUFMgaW5jb3JwLiBieSBSYZWl
cmVuY2UgbGlhYi4gbHRkLiAoYyk5NyBWZXJpU2lnbgAAAAAAADARBglghkgBhvhCAQEEBAMCB4Aw
gYYGCmCGSAGG+EUBBgMEeBZ2ZDQ2NTJiZDYzZjIwNDcwMjkyOTg3N3g3MjNTA2OWM3MzU5
YmVkMWIwNTlkT1kYTc1YmM0YmM5ZnAzxNzQ3ZGE0NDNmMjE0MWJ1YWriMmJkMmU4OTTxNWFkNjlmMWQl
MTE0ODllYTJiODQzZjRlNjk2NjU0MTAwBgpghkgBhvhFAQYHBCIWIDRlNTJhMmE2ZmZmZDjYzYw
ODkyYTUyODwgwMA0GCSqGSIb3DQEBBAUAA4GBAJigP2LY1+XlHocBwAezrnDSK9XJcsTL
9rIhLe48Mmy62qQdl3Fk4erjevEkOYUYpaIlzQFzCbz8O+tnVQl4nVmVQC5RyY5dppymqQEGh7Ii
shwN9ChwuDkHYwBn83d6OPp66tMU3rEpSyBT6jn8tn1La4XiDag5ajfVhJ+E3MeXMYIBVDCCAVAC
AQEwdjBiMREwDwYDVQQHEwhJbnRlcm5ldDEXMBUGA1UEChMOVmVyaVNpZ24sIEluYy4xNDAyBgNV
BAsTK1ZlcmlTaWduIENsYXNzIDEgQ0EgLSBJbmRpdmlkdWFsIFN1YnNjcmliZXIICEGmtA2REwZvH
boy9pyMAoS4wCQYFKw4DAhoFAKB3MBgGCSqGSIb3DQEJAzELBgkqhkiG9w0BBwEwGAYJKoZIhvcN
AQkPMQswCTAHBgUrDgMCHTAcBgkqhkiG9w0BCQUxDxcNOTcxMTI0MTU1OTE2WjAjBgkqhkiG9w0B
CQQxFgQUPZ5QnJ8i173ydVuUcu6FpKfZFmIwDQYJKoZIhvcNAQEBBQAEQD3EW9l2yLUvUsZMojS1
adXYVthV3dA0Gw/KpVyIbQAldea6d+cPW2zsSK2+gYhvrjWwA9+tRr0QYmj2DC3692UAAAAAAA=

```
------=_NextPart_000_0005_01BCF8F1.EAF8C530--
```

Opaque signature, no encryption, no certificate

```
Received: from lhughes by lhughes.worldnet.att.net Mon Nov 24 16:12:48 1997
From: "Lawrence Hughes" <lehughes@worldnet.att.net>
To: "Hughes, Lawrence" <lehughes@worldnet.att.net>
Subject: test, signed opaque, don't include my certificate
Date: Mon, 24 Nov 1997 16:12:47 -0500
Message-ID: bcf91d$b70fe010$d2a50b0a@lhughes.worldnet.att.net
MIME-Version: 1.0
Content-Type: application/x-pkcs7-mime;
        name="smime.p7m"
Content-Transfer-Encoding: base64
Content-Disposition: attachment;
        filename="smime.p7m"
X-Priority: 3
X-MSMail-Priority: NormalX
X-Mailer: Microsoft Outlook Express 4.71.1712.3
X-MimeOLE: Produced By Microsoft MimeOLE V4.71.1712.3
```

```
MIAGCSqGSIb3DQEHAqCAMIACAQExCzAJBgUrDgMCGgUAMIAGCSqGSIb3DQEHAaCAJIAEBEZyb20E
AjogBAEiBA9MYXdyZW5jZSBIdWdoZXMEASIEAiA8BBlsZWh1Z2hlc0B3b3JsZG5ldC5hdHQubmV0
BAE+BAINCgQCVG8EAjogBAEiBBBIdWdoZXMsIExhd3JlbmNlBAEiBAIgPAQZbGVodWdoZXNAd29y
bGRuZXQuYXR0Lm51dAQBPgQCDQoEB1N1YmplY3QEAjogBDF0ZXN0LCBzaWduZWQgb3BhcXVlLCBk
b24ndCBpbmNsdWRlIG15IGNlcnRpZmljYXR1BAINCgQERGF0ZQQCOiAEH01vbiwgMjQgTm92IDE5
OTcgMTY6MTI6NDcgLTA1MDAEAg0KBApNZXNzYWdlLUlEBAI6IAQ1PDAxYmNmOTFkJGI2ZmYxNzMw
JGQyYTUwYjBhQGxodWdoZXMud29ybGRuZXQuYXR0Lm51dD4EAg0KBAxNSU1FLVZ1cnNpb24EAjog
BAMxLjAEAg0KBAxb250ZW50LVR5cGUEAjogBAp0ZXh0L3BsYWluBAQ7DQoJBAdjaGFyc2V0BAE9
BAEiBAppc28tODg1OS0xBAEiBAINCgQCZQ29udGVudC1UcmFuc2Zlci1FbmNvZGluZwQCOiAEBDdi
aXQEAg0KBApYLVByaW9yaXR5BAI6IAQBMwQCDQoEEVgtTVNNYWlsLVByaW9yaXR5BAI6IAQGTm9y
bWFsBAINCgQIWC1NYWlsZXIEAjogBCVNaWNyb3NvZnQgT3V0bG9vayBFeHByZXNzIDQuNzEuMTcx
Mi4zBAINCgQJWC1NaW1lT0xFBAI6IAQqUHJvZHVjZWQgQnkgTWljcm9zb2Z0IE1pbWVPTEUgVjQu
NzEuMTcxMi4zBAINCgQCDQoESk5vdyBpcyB0aGUgdGltZSBmb3IgYWxsIGdvb2QgbWVuIHRvIGNv
bWUgdG8gdGhlIGFpZCBvZiB0aGVpciBjb3VudHJ5Lg0KBAAAAAAAAAAAxggFkMIIBYAIBATB2
MGIxETAPBgNVBAcTCEludGVybmV0MRcwFQYDVQQKEw5WZXXJpU2lnbiwgSW5jLjE0MDIGA1UECxMr
VmVyaVNpZ24gQ2xhc3MgMSBDQSAtIEluZGl2aWR1YWwgU3Vic2NyaWJlciBIQaa0DZETBm8dujL2n
IwChLjAJBgUrDgMCGgUAoIGGMBgGCSqGSIb3DQEJAzELBgkqhkiG9w0BBwEwHAYJKoZIhvcNAQkF
MQ8XDTk3MTEyNDE2MTI0N1owIwYJKoZIhvcNAQkEMRYEFLCnY1wFeX0yufhVZNLwfZXzyI7hMCcG
CSqGSIb3DQEJDzEaMBgwDQYIKoZIhvcNAwICASgwBwYFKw4DAh0wDQYJKoZIhvcNAQEBBQAEQFnY
OCZPsghBg4VPo28qIlx7j3rsjjiqZYXZqTAWgrnENpVrKgnNonR3qoaAqCWn+h0wYwzNR5N5MrWs
zj996ZoAAAAAAA=
```

No signature, encrypted,
no certificate

```
Received: from lhughes by lhughes.worldnet.att.net Mon Nov 24 16:01:48 1997
From: "Lawrence Hughes" <lehughes@worldnet.att.net>
To: "Hughes, Lawrence" <lehughes@worldnet.att.net>
Subject: test encrypted
Date: Mon, 24 Nov 1997 16:01:47 -0500
Message-ID: <01bcf91c$2d57f1b0$d2a50b0a@lhughes.worldnet.att.net>
MIME-Version: 1.0
Content-Type: application/x-pkcs7-mime;
        name="smime.p7m"
Content-Transfer-Encoding: base64
Content-Disposition: attachment;
        filename="smime.p7m"
X-Priority: 3
X-MSMail-Priority: Normal
X-Mailer: Microsoft Outlook Express 4.71.1712.3
X-MimeOLE: Produced By Microsoft MimeOLE V4.71.1712.3
```

MIAGCSqGSIb3DQEHA6CAMIACAQAxggGeMIHMAgEAMHYwYjERMA8GA1UEBxMISW50ZXJuZXQxFzAV
BgNVBAoTDlZlcmlTaWduLCBJbmMuMTQwMgYDVQQLEytWZXJpU2lnbiBDbGFzcyAxIENBIC0gSW5k
aXZpZHVhbCBTdWJzY3JpYmVyAhBprQNkRMGbx26MvacjAKEuMA0GCSqGSIb3DQEBAQUABEAqTzgD
4WVNEYXzesdrF1ttaA1kInfHE4ljmMl4R6vQHHnQJpecVFGr6PYBgAJWcCAUXgRKu7N2mmSWxFkQ
te9IMIHMAgEAMHYwYjERMA8GA1UEBxMISW50ZXJuZXQxFzAVBgNVBAoTDlZlcmlTaWduLCBJbmMu
MTQwMgYDVQQLEytWZXJpU2lnbiBDbGFzcyAxIENBIC0gSW5kaXZpZHVhbCBTdWJzY3JpYmVyAhBp
rQNkRMGbx26MvacjAKEuMA0GCSqGSIb3DQEBAQUABECOYjs0ImfKqD2bZoRiwwMxvVpbjSWEIX1W
9tKT09spDgygWm1h4vv4kU2sKIrm9Sn0WX5kQu2E2FS4uDEF22y3MIAGCSqGSIb3DQEHATAaBggq
hkiG9w0DAjAOAgIAoAQIecHKdqXRywCggASCAjhiFC7FGgd9Tqlq9JbWvuy2K/cF4LzepGTtoRA+
G2FXWUBf5PAzbGbauNDio5DK8X8j/kFrA5R4lLBq0fiexouMH3IEEo/XNTzYpNAHYnyQfXY4h6bh
C8256s31EzjZQjHmxk9wQBbwqxmAI4YzMRz450O80C8gGuCj0EVsUUR/yCap/j/c12Xi5wdDG188
tQU9QkX6mDTHCQnpsr4FQ4TXKgTaTiWaJe0GDBlIjoOBFtCn29EjwupfqMdvSKZNlxyri2HnjXZh
lNth93PqCuR4tvpgoxWnuPtVqZwvJqBnQPW9YsofTF5hAa0qoA9XNIblhcOKCpwrh0dNqLD6cGVw
JBwtQkeufZHsfn6upiAbufHEZEcmJn9cTHXqG2rFHXC0n0yTE23SU+mPVUod3OjsgwX6U/x0sAq6
q98Ifa2Ls6JXf5e8Kmrm/BOHXrjkKcZGzaLrvYZLJ8F+AttP8efhlEVMaF+URAwlUBhr82X5IASN
l7xPWdZnFF590riFxMY2Vl7CX51uNsjdtzzi1XYA/+eo45whRxQyY8uQKyJp02/2Xt3k3JgMFc6g
bxLVLN794OvGQ+FZ5wLckTBIWdP+QnADbqPIewnKmdrtqKpxAEA2kGFCaTvvMQz2+jb7iqd0fMPE
i8rFecWjc1iKxzOpVMOrkIzxOqGrBv4jnwrp6BFZE4Ugzs/arzA0P8GjxQpPfyJeBZMxTtf4UdJH
RuRuoMomdZ6InGMTmtJjRDzRfUMLOts8p5cf6HAAAAAAAAAAAAAA==

Normal signature, encrypted, no certificate

```
Received: from lhughes by lhughes.worldnet.att.net Mon Nov 24 16:10:26 1997
From: "Lawrence Hughes" <lehughes@worldnet.att.net>
To: "Hughes, Lawrence" <lehughes@worldnet.att.net>
Subject: test, signed, encrypted, don't include my certificate
Date: Mon, 24 Nov 1997 16:10:25 -0500
Message-ID: <01bcf91d$627f9ef0$d2a50b0a@lhughes.worldnet.att.net>
MIME-Version: 1.0
Content-Type: application/x-pkcs7-mime;
        name="smime.p7m"
Content-Transfer-Encoding: base64
Content-Disposition: attachment;
        filename="smime.p7m"
X-Priority: 3
X-MSMail-Priority: Normal
X-Mailer: Microsoft Outlook Express 4.71.1712.3
X-MimeOLE: Produced By Microsoft MimeOLE V4.71.1712.3
```

```
MIAGCSqGSIb3DQEHA6CAMIACAQAxggGeMIHMAgEAMHYwYjERMA8GA1UEBxMISW50ZXJuZXQxFzAV
BgNVBAoTDlZlcmlTaWduLCBJbmMuMTQwMgYDVQQLEytWZXJpU2lnbiBDbGFzcyAxIENBIC0gSW5k
aXZpZHVhbCBTdWJzY3JpYmVyAhBprQNkRMGbx26MvacjAKEuMA0GCSqGSIb3DQEBAQUABEAnBtDi
LYcRLgpdNxMpIGiahW/nGhZTP8S4Jclzfbo4ws1JJq/unw5wWaFxEPlDypqhhbFOlI6SeOqpgoVn
UhNgMIHMAgEAMHYwYjERMA8GA1UEBxMISW50ZXJuZXQxFzAVBgNVBAoTDlZlcmlTaWduLCBJbmMu
MTQwMgYDVQQLEytWZXJpU2lnbiBDbGFzcyAxIENBIC0gSW5kaXZpZHVhbCBTdWJzY3JpYmVyAhBp
rQNkRMGbx26MvacjAKEuMA0GCSqGSIb3DQEBAQUABEBeynMleG3wZUDUdFap6NrRxCisAy8jTXFg
htHs7/pC9mMw8equoq1a/pzretaXZ5DikF0DD1W5V/wTyxkVWmeEMIAGCSqGSIb3DQEHATAaBggq
hkiG9w0DAjAOAgIAoAQIphOCXhobN9uggASCBAA3/gFEQCVNtRhV6pSBYfmeKZGD2S7xZNxViGql
gte2ZRi8b2Z4LtzFr4ALR2TKKcgK2ym1r1PL2qIdknUXMrFGBR0SxxCf0qTrKpBmWtnIwS+8NPt+
k4kT/GKFbQUin9oyJb9SFU/qZTl/FFjAzdJcaisfTGcQYcCKHsLGPQp8PO1d5EVnXVAaGg3PKdLb
KpswxhIV+z8Yu2OqmOkdOCGArJraN3oNvL3QArtBf0vNrVgRVJVdnaBiAQdDf1H33CQVPRtE0eYo
gJyw+ayDc1Qhc846izB08PAGLmiQfyY9HtlIB4e0Og7aKYa8wtEefNUehOObT/uAM8VDTcBP0w1c
xakfruDShENhRBXszwnn3vMrHN0bcwktp/0pKdKyCSInnDmrFIRNpa1Dp9kfjuvuMrYp2/hQTsbu
VS7i+nspMkydBxm5XQDyS+LMZiabY/aidBfYTH9X5eK6AuhDpACfMZwYBoKhG7+GbIrnQgTLCyDu
FYKVD4kZzuyrp+uw0NSQ6PQMZdvZiNJjaxptAXhYG0iNmduKZl5KaMgLnLQlj5XvkY3cPqN0XwMU
ypxFrSHVLWyuNWamGN+T40oHMeP8ujE/YNsVvHAM6wyyx1Asoti3Fnfn9AvxhseTmKeKPJj/RaeU
zowYGrmZS4PHelSS5KwdNaIpx7Y+ybvdHTyInrcfRbR2cjnx3sYgEN4V/HMkOqqoScdAQknesCpR
0uSabAL7HMSQOUDrbESNpzUWk0y/C3C6rNMxRqQlyuzu0DA4h7nftpb8YCMBJLvzkzD3c75EH8+z
wtFUbAWcZ+OrEIE8Sw/Eu7X8iu5TETuS1Lp5UIIPYWCsREsl/t8B0TIJ/7zhTka3t088s9DGOp3c
S2a4KA+cuxwsCJzIMVX6SncrkH4qBJd/OhYhrc/WuC8djR8DW/OrDJ5OUgQ2PfZP3PNJkRUyt/3A
eb17bIcNeRoDNrTU6if9qzQ7ojagCE9veOJyiyIi80gZ0pyd75w78QPUgKzh2MQvpORSlDrSx+nZ
Yjf83m+V5jH+LdwLH8pBB6oD7JoXXmVgh+mDJEFY1284U1vDHeRrA8aSow9a9SLM3X7v3R1gBIog
3sslEZYVstep5OreCQorqEynQ8xT6DgWagON6KdAxafJD9OOBxhyiB5flG0WLw85qn0TCihJ/Hqx
5F/S/6JaGGmCeNOpfkd4cefKEwe3ywhv4pzhPhVPD4xJhYMwU/3B6qKLGx7fxkDQWf1ft1uLrhrk
UHbZLEgTkj8Vb8Sy+0wd3XR0J50g5VP/u9ENeKv6rvKTwgJjZZmcuhB1wxeqa0zkbWxEMXOojR4b
/C11SZ2LP3/mefXqRC3qytbCvBKPUPzmUg37BIICgPa1BeCXFnjuix5uhNM2BRRghFxiaF5q1tGo
fQxG7bkfft8OcdO1g+jOhbIkpORSlDvL96S0gWGGA/luuXg4kWSzbcdhNBrel1aX/nkEqSB7/TTk
EOZgHwkqYXc0mgHToQWQ1xMSquerWuqfxDJsaeo/9oUF7FRego1HkBaGFIZLbj+eo36zUiEjkyM/
WcCzrLqD8SPKZvyI5kqbtDi41jzJt/AH+Hhlik5ZJz6fk21o4+axmV9FTMbpKrCYAr4IfbxzHxl4
YdMEelAP1CIIsDE8OVeuZQoKNXlRcJnYLqXSZ9iSF36qNQL1GymlmR2RskHvij+bgwYsEzbP67tU
CHaF41wvkLXXkeyu42ivXyxdP0xhbz2bsE5SAsUENMlpf+9cxU+xqOVi1ID18cMc+zrGZ/iwHWR4
5kGLVWXMe53+/SATefjEextPwJqX2r0LnmF4Z183/sU2m0ma5TUro4rRFRV4xtiJAWuSzLCcEFVC
Ns6dDqusGSxtYOkTcRjDB5D6CNbYBPyxYjirXorqYJ2a2rOebHwjJ0g1G9tkhNozBhjrOyy/xjDX
```

bIXnNAHz5bzV6AMajhWVkLLQHrv34OZqmpItCsKLtintM2geuzi87utQSg8V8ZRK0IW9UECJ9kfR
MVVY7TSSgk6TGJWUjvH9CfjncauSYTEsogRh6r2jqCT5dakGCOr/L9L3h2OGTvosItVoj6zxnhEM
G9Swbm8RbCBPnvMAlkiVIfK+aNT0ODgF3C7AjgJTdn158sliFkHjcOQSGQC1++s2KIQOJd6ou6IR
YyDuXgUui3bzjOEMDN6tIH3mPP1nyXG6TAH1vV11+S76WN7y6DQO4WVRTSUAAAAAAAAAAAAA

Opaque signature, encrypted, no certificate

Received: from lhughes by lhughes.worldnet.att.net Mon Nov 24 16:13:56 1997
From: "Lawrence Hughes" <lehughes@worldnet.att.net>
To: "Hughes, Lawrence" <lehughes@worldnet.att.net>
Subject: test, signed opaque, encrypted, don't include my certificate
Date: Mon, 24 Nov 1997 16:13:55 -0500
Message-ID: <01bcf91d$df699e20$d2a50b0a@lhughes.worldnet.att.net>
MIME-Version: 1.0
Content-Type: application/x-pkcs7-mime;
        name="smime.p7m"
Content-Transfer-Encoding: base64
Content-Disposition: attachment;
        filename="smime.p7m"
X-Priority: 3
X-MSMail-Priority: Normal
X-Mailer: Microsoft Outlook Express 4.71.1712.3
X-MimeOLE: Produced By Microsoft MimeOLE V4.71.1712.3

MIAGCSqGSIb3DQEHA6CAMIACAQAxggGeMIHMAgEAMHYwYjERMA8GA1UEBxMISW50ZXJuZXQxFzAV
BgNVBAoTD1Z1cmlTaWduLCBJbmMuMTQwMgYDVQQLEytWZXJpU2lnbiBDbGFzcyAxIENBIC0gSW5k
aXZpZHVhbCBTdWJzY3JpYmVyAhBprQNkRMGbx26MvacjAKEuMA0GCSqGSIb3DQEBAQUABEAuPs4T
3fYFess7/j11aUd7oeURjLjgJXwINbUcu1LH9CMVC9mknffa1EX57uY6XMXf+78vQexUQby4ruqx
LgYSMIHMAgEAMHYwYjERMA8GA1UEBxMISW50ZXJuZXQxFzAVBgNVBAoTD1Z1cmlTaWduLCBJbmMu
MTQwMgYDVQQLEytWZXJpU2lnbiBDbGFzcyAxIENBIC0gSW5kaXZpZHVhbCBTdWJzY3JpYmVyAhBp
rQNkRMGbx26MvacjAKEuMA0GCSqGSIb3DQEBAQUABEBRJuPPYx38daGiGdAXJbrksJL8bVWQg4I1
MNNQIy+wcqlQqbTRjcY67TLLBFGWuNGY/nY301OO3Twq1qKb02ETMIAGCSqGSIb3DQEHATAaBggq
hkiG9w0DAjAOAgIAoAQIrwmMn6/imM6ggASCBADNdSdcBAe0HVpGfdpGInDzdDdEl2xmwx9mwMzJ
fX2GzDpySNiOwLOFIvH/LmFPs2ZU63NoiAzx5Xtly4urqshccNOxN2mh+Rd/+MC3iI2jOA33QXvg
JYFAPiVQ98nW4ATPvTy6nJtVBIhY1Cd96iyL4NnQw6Qp9yc7DC1gauMhbNo086yq3loXEMTJldP
3tAo/x29LiQRBnWQmBl8KqyRPeOM//5gl6nG+FEuRM55YMqIyzOx4ydA3HxzCTUjbNiib1/Dk1MC
c1YZjifubcoHTFdm1S6jipEj+5gbysVO0xzYpqWnO5A+BlHB/HgwU3F4qi/K9hout4KPXILwNFbG
EX18wNOpIz0TDP2jPmxNRQl82bng6UYSE13bAakXh2lafZrGXwuMu5XCs5hsN3Et59mazWYvSNQ4
9Hj4BIpAnlDQqzfMFCzwZWjWH468oBPjV6ILjpItYuRgYRJLDt/tYA5zXCvJGOJEtaoHA83kvFCy
vYSn5XQyWBqSXqWbWmGvwkzxoyu26U8SJ6Tij1SZZ2bZL7sX1OV8mihAVM57cbHAr7NQVNSauJ+s
+eE/am6iATaiR2oKxyGf4r2T9D4a9sS0ikq8i4hqt3lO4JbdDO0qQfnC4DX1OHGTYLnP6H5kI/Fo
lL3Ec47qiEwt2nIIeKNf3tKto9mA+OEpmU/cNcEg6y2V6tdrA/lIo8WNSYb3QnXXsra/iRVL+fFL
Al74OhaG4LbShXVs/xNanhLUaHBE9XJYLrDUFst6TUcahhF5dqM4kCa09fwxHU1Nm2tQKK/j6foO
08KwEJBv1TTVQQkRz2WVE3adXCA+zAfzedQvOtagUaOjjZdM3/T8GkX/7GjdhuLdEu0dNostdVpJ
2xD9NEm81iWdR6RcHeHni3V0k7x66MazQZSEpXpjvjqegtvz2UlyOXYpGWk5PuG5Y0npoM3qGZbB
Dn1/6CpckMwgx1DpIVUFS1c0UU3TxMiayDCkxLJDOA07ykEGpSCvEkjWvX6vb0ChNudGp7RwObYT
O6eIBu8mGok9jzUO2KBPfrRTcw9K1CMs7pNQwKn9c08Qj9YB9mDWk8bt9PudjMvuCXkFf3rYWPYt
iPqrkMtqL651elWXhCIGYoJ6Zdp40olhKbz91/KihAJZpVD2eEdcI6cnMfy3V8X5wbsYa2EBM0tL
ynWuV13TBpORydVCxvfq+iCHi7vw9jRTmbb77DC9punVHNKnJZ2rGOZlSMc1oM4IRGxXIYYid0zD
hwa73CCLIUEwP5vvM1TRGPddFCFIBr88cQ/dtFxrGqbiuyvX8U9HnGsPVd1YsJHBxJr/8jtECwy5
/1WvNOertZW5nA4cAEEB+m6f+AUOQEg76nGrBIICiCkb3VSgarC9LEPwdgYlHR92z+u5MPUUZLS0

408   Internet E-mail: Protocols, Standards, and Implementation

HTcDmnn9Gq0HZawQOxc/43dP3+Qm+A6WjPc7cre01YsSKc/Os04fxVw9zvdyUqqmSr4BN4pP5uls
9jNGghLMPrX4+YyfUSI7zKcLAwzzVOCD0CYPn1Afe178+gDqN/fya/LDo9pPto6qFjg/yU+mirp5
7p7Pq+sOPo1lRP2hcw2mSGYdVJZSI9JOR3E3mdVlhKn48Mf4PXBasiIIcqZHrWIrqwIPk6N0j6WG
vPuAfOuksWRqTVpEFVqzmCbxDIPg1FYFdnOooHELpuC/9MUmvePjc5dZfirVtYABZv3YECy9gLxW
8XEa3HBs0Vtn7gZdpwanY3WFlDL2yDlQRGOorpHX8PGbD6dw5vXZK7jVvo3RiiF+BL3qby+57ZII
2QxYFg7GFwZFY9B2Jq9gozXPI9/CHNAnZ84YCfIw/e8P8ZLPG4Y7Ih9PzKW/chkhXyvQYsLCalSe
77guKosStyeDHdQuTRYa7HYIpKEL6Yy4VCdKUiYCgsWBq5/LzArYpAsmSZy3gjXrNVPnHD3fl/ga
B0kGe/F37VYvPBx+BM4rFyJ6xfCtOp64oCT2vxPdLiFT2aUlyOIaHps+YI/hG0ov5kDlijTp/T/u
lWE95e+lrqxxizjQAKL9pVErADNvwD5VyQgOMr0T+JveLShLDnwE1x/wKX31pgJxI9ETOBgqtctK
lCWkDiARZRr24YzYLKgNu3hnQJ6WAIcgz1F1Tve1C2h2lMo7RvnQ6oevYM9rBvucmE9foOupSKSG
DQD+7+/V1ABFb8RFFdX/g2aHkTRy3QS5zyu4f5IVf0ILhtjxUK7aa5weXhaEmvaRe3i+3AAAAAAA
AAAAA=

Normal signature, encrypted, with certificate

Received: from lhughes by lhughes.worldnet.att.net Mon Nov 24 16:03:12 1997
From: "Lawrence Hughes" <lehughes@worldnet.att.net>
To: "Hughes, Lawrence" <lehughes@worldnet.att.net>
Subject: test signed, encrypted
Date: Mon, 24 Nov 1997 16:03:11 -0500
Message-ID: <01bcf91c$5f9ee2a0$d2a50b0a@lhughes.worldnet.att.net>
MIME-Version: 1.0
Content-Type: application/x-pkcs7-mime;
       name="smime.p7m"
Content-Transfer-Encoding: base64
Content-Disposition: attachment;
       filename="smime.p7m"
X-Priority: 3
X-MSMail-Priority: Normal
X-Mailer: Microsoft Outlook Express 4.71.1712.3
X-MimeOLE: Produced By Microsoft MimeOLE V4.71.1712.3

MIAGCSqGSIb3DQEHA6CAMIACAQAxggGeMIHMAgEAMHYwYjERMA8GA1UEBxMISW50ZXJuZXQxFzAV
BgNVBAoTDlZlcmlTaWduLCBJbmMuMTQwMgYDVQQLEytWZXJpU2lnbiBDbGFzcyAxIENBIC0gSW5k
aXZpZHVhbCBTdWJzY3JpYmVyMVyAQNkRMGbx26MvacjAKEuMA0GCSqGSIb3DQEBAQUABEAqALVS
5b7r1VBLpRArlhSNNtlTmVixuVrmfjysTjIvsOU7I+x/MZB0axMxbEQEaaoHJ96FzyoFNcb3ZEJh
2YW4MIHMAgEAMHYwYjERMA8GA1UEBxMISW50ZXJuZXQxFzAVBgNVBAoTDlZlcmlTaWduLCBJbmMu
MTQwMgYDVQQLEytWZXJpU2lnbiBDbGFzcyAxIENBIC0gSW5kaXZpZHVhbCBTdWJzY3JpYmVyVyAhBp
rQNkRMGbx26MvacjAKEuMA0GCSqGSIb3DQEBAQUABEA2O+VcRBwoe6fHtP9m4z4NZgVPW4pJPCY9
hg4aK6wZVxyfIg/lOSOzpNB12p/pn4IQGq4Ly/iI/0d3Y1hgUieGMIAGCSqGSIb3DQEHATAaBggq
hkiG9w0DAjAOAgIAoAQIq7jMrnzWhpSggASCBAAU9F7z7WiJOEL07YX21aqUAwklplTITi5axnWu
icQ5g01GmEBcdbg+OwOD0FhpnhXzx9quQ60gLoQlCha6ez5vjJdTwO6z1252jXubQrfnEioIhXGO
kG7kXBpNLK6BSYmQQzNfVHxZuD4/sFf5JnvP3bLtuSU0YMxapoBXl5c6DGw0P0eqjOwMSldKbkza
Ha9jjEnQB/RymFBcQ5OrhtJOuzqKf9YyYbXJbTZ8LB5k2L0ShR3UVNJib/l/zIBwxXc3Ke2csEsK
lihqyXiKtaiaLm0kDkj4nXXyB3HEBe5SyG+Mw0t8JJ4/S4zlPokzVTe1roWApgXFnLoJ3yH7bAiI
Zxp5MVWIqdh6kmC70KkaLnOfHiw/Bo8MFCZqfveCDiLdFg3Dn01AVkpVb5f6Zu+jEz+TXalHMZ8b
e0e1TVbImCl3irtKE+ykauaXJP4JrtDN0ir3W/QpUHIVJlLVLBIeWtn2SCXFHdUFqQxNgJ552JNf
4AsKpvGUevu0DwdHKfbOAoTMgqAwghswDHSqRyr5CgFtggfxpTzvOY2DSYUIStb6wyO3YWE6nQZo

kJLvriphrwGEIyaK3Yfh9svjo6xJaudWhfl5SMJPJ0J0oJKGdnDiSYq6wbxyK/R8xeCMxqcMfaOH
tye+TzFyGt3JGVKVZe3Q8csaSdZfEtZ3mJaEE8oIJxEX4DWyrvvKe53bC7TMgnojDFfYlwg6FN44
Q7wQ2iifX7yemxSSYeh2ErrXsNNIRo+5FoE1096TSARegNd+qVRN36mF+aGrjRNb9cH10mee14ol
rRkQWEfL0ik0dc7bXk/TULNe/HFbc7mT52mOull1p/ZQbYZ8fJIGlNdQy+y11WygzpMwmMqYWkgJf
sT3phDM7UmSFVeODIVDpS7/ZaHSI+GzkSPzy9glZINaqhYnd3RaSxgPrqNwRRIY+p0ZZbD7J+t2H
DfgG3H16X3AGr2Ul/3YKGWkKNp8sns8V4XJasymHYOFgjkvinLP+ehE1nFL7wjdSdwZ5q9MXUhCd
daAyjqJWOdOfefHL7bKx/Ado02YH2pjsyrOsHq0VsMTnkl+oFeFLE9e6hANtP3RJ1jR44tnPJM8x
LV//X4+MC/Fn8Ko3w7cIIzoTVrHihSW1F1ELKothZ3h53isiqCXwGwnvhde/hZWEIunLCrsUT5SY
FcA+Y4jihe8uiMvmvcDIQFArvfSXKHX/d+2qj+t4//PBgiWgTNvFokSeL/yJG5PPdXLXy6uvmaAA
gZS0+QVQLU2vJcZHOT8DxEb5cLTLUjRFx8cbC7Rfq8N6mnyNemJgsynRwMJ7CPBo/nn12/K8GVoM
EVcgYoYotLpqwuQHImDVyxdLVdkj9Xg59BHWBIIEAMP3CgSD0926AGxKaQjpiVCdLJI6VaZCLixO
U1dv5fxGf7HACJnoa+XNzQOEZh7iXPEpfzKrBFZlHrNSqQWgwN/RSLqwJE9rY79CH9La/9x8SLQj
SH7k+MugO7C23uQFFL0n19xgEriZhT438Yg77H+HG+KGoLrZ0NAwaZc0gfyhsxSApiLtMuQvJkvP
8fKaF2/plDd0XHrQkHVN6qsiZAdt0gnvcBKvyZItcBExGv2N3pd70nEcRTgEB0Kr0yGpxYDPNczq
8yC21rC0sGK0IBt+O1Owza3fk5ENW+lQAOZeQ4NFWpt1vSbH5XNeqCWr7mp1AH51VJIJCPH6UCLC
mUQCy4p00MC263HpsgefF7ppOi5E3uxBlyLCNhQOgIKYNPe61DllIiiZAMXTTm/11rFgKi8mQwu+
6yU/1pERdWy7h1GmSU3jTn/aN9hdhPvKaYlyKkJKW4B/HZGHX0y+feRRMCWj6bzfQi70Evpt6iZS
LTGuAHf/i9j30eFF+1IGS0kwkCN8FhczOcgGNRf/eKLQBWd/2nqOcRxZVnGopkemF66yjsTICK94
Vd85tJrBEPq2Wg3mUFEb16DQo+S7c+qhV808ugcwMFbA7CDGeAu3Dw2kySMuFghXFZ9LWfgLMlhg
idGRcV5ih70NxyzJOGtGzWEnX/W98nlMTL15Ul5PdKPJ9mx5W09hN33smQLfTaMAD1/SLnNKB0Zz
Ut24QioDPcAYT29RH+q66c5AaZDUX7bKs3sAVy1JnSthkWQsIwHaKSmn6J1VE+dN2QteLvXPoOYk
t2/OsnDF+kfFSSmseUWOpJdCh151ndQpDSv6ujJrhihqRGgmLLKSa+OcmHSke5g5m3DtOa/dLmN5
cIxkUcFuF+sa7k/WkGUJAmPKLsQ5KIOjDK9OQnVGK4m2WeK3mog4+GZlCNaH+XqGOhH3bMoTKWgf
bahHmKG2kbBwiS60iHqx9IhukMaqEOCS+J1x4dKGYP1FYXHFytosiNuxJVS1XE++l4ZFHI0j2w0H
TaFmAX5Sx0OwzyqUw/KbHA3etWAzXP/miXn0ivPkA7cnI1HrOVpTe0tEYwWbePdDgagjlwmeThPD
wFqNwlYo2rXpgq5kEBFhZF5dK5QBuT0GjSfxPNuWRAODqYBQMdlDsqoU5EFRDQWGUWgRsONhFXKH
Fn66eftfhYX1iv/EKXbnMUMCUC0BHyLicN0VRB4oiYw8eEhxAna5FkLqL3phudbeKMpmfk9NzEDr
u/6jvwlZgiaqjSDdWqxEHFSlZJiq32Nw7tySi7WcNNegSg36/FXbm0l6Czr1940un+Q+PnawMeZV
Y20i7h2Y7V1P0LrYItQYPK08kqSFGWei5MLXnbMEggQAFWnu9WRTlsuva2cPFmBkzriTJ9DNZTn3
koGdFdJHH+dtGyAuptrzG3Hskq78Ac1ZE5tpFJOz5cvNpdYfdZ01z+3KXmlngFMB7fFFnVrrokuIU
xccetaF4mdwZ38XQ84h342rydkzHBDKCfg66e7ttHptMR7cZaoqO9yfuVciBrV8pws5/Z6kqmR5n
Yx+DiLwPu82ivAGVb0O3NzQYcSL14Qr7Tr2CnEPZspTNzwZ/fjNuPps3cwbJHhMifa7PB/W+Ragz
liEiptAbOHk5wtA5uFvegsQSiMi5zxNXY5YbdAKM49hN6KFqCFjwhTAG/dxHz2u4j1Ipc1u62qmK
gtpXua43T+3Qfeihr7/MBJmxB0YMWPlzn5kzmlZ2K9ZGW8q0zg90ROR7VZXu0RFMFr8f0IadTZ9K
8SXVP3vL62nBHtqV9VrH2zJTPc1OQUO+MGxnNRSzgt0uX0zzyJfd1RfgJhyTkT/eWtU66qUqQBif6
ZaYH3M2fXWwyYo14k4idnXWkV4H2yJA2cf7ZlYMtkmdi7wAkmYpQUNQnvdH6MDbzRzMggg5sJbgZ
4WX8t2Y6jlZ+DKS/0dq0zRdgVVUv+mHh+HTqkILCWkudD9uYYHPYMHE/wnGygh/9B19NK4MDg/hw
7k8MU1pFFKSPfxoL/YOggCDvR9gS7CqGAjSOdHRZQG31SSKHGEzMjEY6X02QpztHCNgLKJDmUeSCU
b/sL4rCOg3DFEvvaHip171RhImjVDg+sKX2lcKG6IvFoDdcnza/IjrWiJIXcwlt2fjGWlK9i1+Af
QSjqwtjJMB32advu0YHrScTaANWT8JS+DMYjZoTkaVF3uE1mJvIKdIi0h6HN26Ff7wax0ig1ruFJ
u0KW2uUMvHMIxWeZBZyl/8MJyVHzWRaEcORdGD97A3IHjeLFVQrSDGL7ll7bqF1osQ3S9RFLWX+o
YxP0rR6baEoi7EGFRd7rIzwqrhD7IALIV0ZgeOsvdR3/IuExDOOL58FtXcomoL8wvwkZ6B0YnfQu
bg0VTznTBq6wZ2xAqe0IW+KuMet/VUFcOcVJeuyjSdeT7xzOTBzpw1pWuF61aEb3Ej+feL7Sb85T
NzRoV1TX/e4p9JIfzX3Bjy0vc2pF6bEIq27/zPBdXHSYa9kTVIkGrMXPKifNvpD/sQAdatQyuCik
0KfsLx/dn6QKQxZDeMsqdBxPuA8Jz3UjvtVC27swrcxubyT7hptxMPBj3sY2nQzU83/z8xnwJLWt
LZUqBtBw/D4EilWe+0aOgi19UhP+OERlnteM8OyVzL15l1jq8+eMcd9n96g3rYBp6X6GL99V6u8y0
1UMq8KEkrbz5BbiJra9khYvNpZUr8C+mrko6fxEaHASCBACCjxlFoBsT7m10j08ASIn7TT2ovefI
a+ieCWXk3liqa6EdS0SdWM6MyZ6f3R6bgM5uC8UlqHhL8ubKZcky36QrMJ+lynvsiYOCVexoiQd3
1GdBGhjZoQoal/fUZV5MaqRrUOmZnlF8K75Tw5FwbUMiTZkxqphIjN2rOoW4RIeokCWBjJNDnKWMU
2U0GeHDDgoCx3oUbrmIs5xfmdjqqCbYMMNmRkeF1Xxnr1nHej2UQk5VnkTn5O63f5i3t7XKQiU6s
MXFYvHjb8PUO661T3Bdcpvsrs+EQwJXG+nMYk7rfCCt7zGcjxtd4gxjfmJ44V4R0G56ice3PE5yh
+pKUhA67UPk8aXXzog/tLClGl2f1oAS5yXNgI/ntwj3VtgV0WTkoW7c5gdcVHTH5JJov1sNzQ2Ss
t9/+R3aXoQEjaP6Mkue95qoYzQD0sNaYA8nlN6LI2t2I6z0TpvfH49t7EBwQpreU9NAOLLL51ZW0
z7NhM4tc6+MIwslkGCO+RTQhCm6EkBQUAkwqBf6YItHfBG4RJmi5qtZMmLABMKu4oeQJjF7sqt/3
MKiHr/oYCXziuKpby/OsLLf66U0xik4LjEtW51Dg9166yDdc0ZzyeyNWEWK9PpARFsS4ji3YgKpD6
ltAqgvbEzpt8JPBj5/TD997IzqzqRHjmCKI7AUxrP+c/0mvnw/WjVT7h7usuDD4ICieRQcOqdVUu
qq+uJDj70LxoUHODXgZ+nNrjHyufJ/bzn+9e/FBsZKha9fTt0k6XlrXxB0s2sUsfeuFh61pZgREC

```
mewZxtkOIvGL9ekHS79fCKOwfcPJHJVJlPU2/E+QUQV/QiN79WbXZpvKd6XaW5NWGq/GpOfV1MWp
eZknp14K/UfXqL1F4s9MGmBewIiO16dVXzoktDPv7uuo6REdoereBjwVh+1pZ36soUn9B9TSnfG9
Cp5pW9ouu4tsqvEMT7F3OXscfYUwJe91P1/d1vp3rufGG3a2BsMb9PqYJIR9vMEVnCwJffOXWK+r
9Q09WIOuA6kRaLvqSwC9De2Tk2oD3rtaFKbVwWKCs8AokiUQmwMVxhT3VHwqK94pEIY5rKbNuJSy
qwrV3S6jlUVzR7IsUnai2ggcnbKK7j22RPhik6JTdkIAB3kRVuIX5As9AomeP0UFJb/m6CpBlJP5
hZaS86HQunHcunjVzh5Q1O46SDDcv8XP11p0sQlWprsHsSuYfF245X6UNQ9mTGH8Av5oCqmLR0P1
shZWUnNc6L5o2RPreI36v1ENkFumwf+qebTsou7tzVWIapZ1LZUVuPutBCWnD/x/c9XN7zxfjX8M
1v6vSWi0h9D5zOJl3FrDBXkkSK3LtDOfGFc8OCX32CleBIICwHNM1hJeVMd498L2GHBry7aOdWsp
sCp6DPMwM2Tj1IiO8tPExhDyYBuumGV+2UVxNjFDw1NorwaoXBeARaO1gSL5ymtxUYZxlFpRyV37
vpvLEBXolE2knJIXH7MsZjtBr3dhIU5mviieLvWAhVh+ymzdghz89OHZgPl2NURGtO8eY6PfsGiU
AjuDxx1Pa3luk5ER3BF4qSNYJ5WswpmQz8zUoS0DcGd4ezadRvTP/zU25gTvS3R7rRAp8TVnwzKm
keS0oVhnFqwwgH0Sf0jvGDnMncBpSWmVfUBnXvdLHuJU376Giz8cs/XEhBstBcvShDuQL6jV+t5D
IBTgPiSL2v0r4LSb/zl1PBONoDgLco8JW9uECIrsrU2K0MC2+DKL8UpjuaZJ/1nZoUDjMwHd9sTf
MCGJTGHk5YFj57zG1mVHqCHoIuzTunnHPPl+1Dk/VHU1QScpJUyeqYAdFwS28D2PYLS9e08mqRjk
OorW1yu03pkf7pk4MoIjTDS8RcqG5FcKj0yGoE2wVCh/kJX3f3KjqDjqO9XwCCx3rusZoQDGpfRk
B220PgehiTtBxvcyX7RieOsqK7L3Mznu8fh/0S/9UZhKbYbTqvXGyB3rF9+1hTvmOqOPb8yxOO0
ZmzzGFDNnmqpBUwT8F1D1RgCE613uAo3YEhHDBptSV49XhdqemEpbA1HqU35ERzMQ9a5GnavJKzU
sNRg29OLa5Z/dw6p429ReXW7ps7TcndmZJDSK5/dv5IX3dItoVFvC/LioMyvjlniwpiSkxMesYYn
RcH12p2ecUIr4/FpH+lUoW31GWuuvsJpnNux3u73XgW+Cwu4bLCXenfNdjYCvs2fQ/DjNHarjzkT
q0FfL9a4vBJ773LD2nkq6EUjXkD/b61fUM+QHHtruVguV+z/FzwsiLTv8aNO9WPcbjZv5jWxRnIU
AAAAAAAAAAAAA==
```

Notes

A non-S/MIME mail client would receive a normal digital signature as an attachment with a file type of p7s. An opaquely signed message or an encrypted one would receive the entire message as a single attachment with a file type of p7m.

The internal structure of the .p7s and .p7m subparts is beyond the scope of this book, but it involves asn-1 (an abstract syntax notation for representing cryptographic keys, algorithms used, signatures, encrypted parts, etc.). The result is further encoded into base64 ASCII, as with basic MIME. If you are interested in the details of these representations, see PKCS 7 and 10 (included on the accompanying CD-ROM).

The size of the basic message (no signature, no encryption, no certificate) was 632 bytes (including message headers). The size of the same message with the whole S/MIME works (normal signature, encrypted, with certificate) was 7,919 bytes which is a factor of 12.5 expansion. Most of that increase is due to the inclusion of my public key certificate (never include your public key certificate if you know the recipient already has it or can obtain it easily).

A good e-mail program will store a received public key certificate in the personal address book (Outlook Express can do that). In this case, the certificate need only be sent with the first message to a given person. In time, global LDAP directory servers will include everyone's public key certificate, in addition to their e-mail address. Once that public key infrastructure (PKI) is in place, you need never send any certificate, even the first time. The first time you receive a message, you will retrieve the sender's public key certificate from the PKI and store it for future reference. Ideally, your client (whether it receives a certificate as part of the message or retrieves it from a PKI) will verify the certificate and check a CRL before it is willing to accept the certificate as valid. A good PKI also will support that kind of lookup. While you can retrieve a user's e-mail address and public key certificate from some LDAP servers (e.g., the one at Verisign), we are still years away from having a fully functional PKI in place. That will happen eventually. I just hope that the federal government does not hold it hostage to force its key recovery scheme or key escrowing scheme on us.

Late breaking news:

S/MIME v3 is being specified as this book is going to press. The big breakthrough is the creation of a public domain (royalty free) API of the necessary asn-1 and RSA algorithms for anyone to use in implementing S/MIME software. This virtually assures its adoption as the secure e-mail standard.

APPENDIX

A

Contents

Character sets

Character sets are the basic building blocks of e-mail messages. The lingua franca of e-mail is US-ASCII. However, the US-ASCII character set does not include a number of the characters common in European languages (e.g., the German umlaut). Some of those characters do not convey critical information, and a message can be understood with only the US-ASCII equivalents. Many of them convey critical distinctions from their nearest US-ASCII equivalent (e.g., many German words are pluralized by changing a normal vowel to the umlaut form). Therefore, an e-mail system must support characters beyond simple ASCII to allow communication even in other Western European languages. The situation gets even more complex with languages that use Cyrillic characters (Russian, Bulgarian, etc.). Some languages do not even read from left to right (Hebrew). Farsi (an Arabic language) has characters that change shape depending on context. Most Far Eastern languages (Japanese, Chinese) have far more than even 256 distinct characters.

A truly worldwide e-mail system (one that is able to support communication in all the above-mentioned languages) could be designed using the Unicode character set (which is beyond the scope of this book). Fortunately, it is still possible to support quite a few (Western) languages by the use of various 256-member character sets, as defined in ISO-8859.

US-ASCII

The most widely used computer character set in the world is known by various names. The most common name is ASCII, which stands for American Standard Code for Information Interchange (first defined by ANSI). A character set standard defines not only which characters are included but also the binary representation for each character (which indirectly defines a default sort order). ASCII defines 128 distinct characters, some of which are printable ones, while some are control codes, like carriage return. It takes 7 bits to represent any one of 128 possible characters ($2^7 = 128$). For example, the character uppercase A has the binary representation 0100001 (which also can be represented as 0x41 in hexadecimal or 65 in decimal).

Recently, it has become common to see ASCII referred to as US-ASCII (probably to remove all doubt about its not being international).

Table A.1 lists all 128 of the 7-bit ASCII characters. The hexadecimal code for a given character is the value at the top (first hex digit) plus the offset at the start of the row (second hex digit). For example, the hexadecimal code for A is 0x40 + 1 = 0x41, and the code for ESC is 0x10 + B = 0x1B. Correspondingly, to convert from a hex value to the character, go to the column based on the first hex digit, then down to the row based on the second digit (e.g., 0x43 would be the column headed 0x40, then row +3, or uppercase C).

The 33 special characters (i.e., characters that are not control codes, digits, or alphabetic characters) are scattered throughout the table (in the leftover positions). Table A.2 lists the special characters, along with some of the names by which they are known.

Table A.1
ASCII Characters

	0x00	0x10	0x20	0x30	0x40	0x50	0x60,	0x70
+0	NUL	DLE		0	@	P	`	p
+1	SOH	DC1	!	1	A	Q	a	q
+2	STX	DC2	"	2	B	R	b	r
+3	ETX	DC3	#	3	C	S	c	s
+4	EOT	DC4	$	4	D	T	d	t
+5	ENQ	NAK	%	5	E	U	e	u
+6	ACK	SYN	&	6	F	V	f	v
+7	BEL	ETB	'	7	G	W	g	w
+8	BS	CAN	(	8	H	X	h	x
+9	HT	EM	)	9	I	Y	I	y
+A	LF	SUB	*	:	J	Z	j	z
+B	VT	ESC	+	;	K	[	k	{
+C	FF	FS	,	<	L	\	l	|
+D	CR	GS	-	=	M	]	m	}
+E	SO	RS	.	>	N	^	n	~
+F	SI	US	/	?	O	_	o	DEL

Table A.2
ASCII Special Characters

0x20		Space/blank
0x21	!	Exclamation mark, factorial, "bang"
0x22	"	Quotation mark, double quote
0x23	#	Number sign / pound sign
0x24	$	Dollar sign
0x25	%	Percent sign
0x26	&	Ampersand
0x27	'	Apostrophe, single quote
0x28	(	Left (open) parenthesis

Table A.2 (continued)

0x29	)	Right (close) parenthesis
0x2A	*	Asterisk
0x2B	+	Plus sign, positive
0x2C	,	Comma
0x2D	-	Hyphen, minus sign, negative
0x2E	.	Period, decimal point, full stop
0x2F	/	Solidus, slash, virgule
0x3A	:	Colon
0x3B	;	Semicolon
0x3C	<	Less than sign, left angle bracket
0x3D	=	Equals sign
0x3E	>	Greater than sign, right angle bracket
0x3F	?	Question mark
0x40	@	Commercial at, at sign
0x5B	[	Left (open) square bracket
0x5C	\	Reverse solidus, reverse slash, backslash
0x5D	]	Right (close) square bracket
0x5E	^	Circumflex, caret
0x5F	_	Underscore, underbar, low line
0x60	`	Grave accent, reverse single quote
0x7B	{	Left (open) curly bracket, left (open) brace
0x7C	|	Vertical line, vertical bar
0x7D	}	Right (close) curly bracket, right (close) brace
0x7E	~	Tilde

ASCII is laid out in a logical manner. The following rules make certain conversions much easier than with codes such as EBCDIC (the proprietary IBM character code).

▸ All the control codes are in the first 32 characters (except DEL, at 0x7F). Control codes are all characters outside the range [0x20,0x7E].

▶ The control codes are exactly 0x40 (64 decimal) less than the character you use together with the CTRL key to generate the character on a keyboard. For example, CR (0x0D) is generated with CTRL+M, and M = 0x4D. Therefore, to convert from character to CTRL+<character>, subtract 0x40 from the hex value for that character. To convert from CTRL+<character> to character, add 0x40. That also can be accomplished by inverting a single bit (specifically, the bit for 2^6, or 0100 0000).

▶ The uppercase alphabetic characters are contiguous ascending in alphabetic order, so the nth uppercase character in the alphabet has a code $n-1$ greater than the code for A (0x41). Likewise, the lowercase alphabetic characters are contiguous ascending in alphabetic order, so the nth lowercase character in the alphabet has a code $n-1$ greater than the code for a (0x61). Uppercase alphabetic characters are in the range [0x41, 0x5A]. Lowercase alphabetic characters are in the range [0x61, 0x7A].

▶ Each uppercase alphabetic character has a value exactly 0x20 (32 decimal) less than the corresponding lowercase alphabetic character. Therefore, to convert an uppercase alphabetic character to lowercase, just add 0x20. To convert a lowercase alphabetic character to uppercase, subtract 0x20. This also can be accomplished by inverting a single bit (specifically, the bit for 2^5, or 0010 0000).

▶ The digits are in contiguous ascending numeric order, so you can convert the ASCII character for n to the numeric value n by subtracting 0x30 (the code for '0'). To convert from a numeric value (0 to 9) to ASCII ('0' to '9'), add 0x30. Decimal digits are in the range [0x30, 0x39].

The control codes are used extensively in data communications protocols, especially in very low level protocols (but not so much in Internet protocols, other than CR and LF). They also typically are used in dumb video terminal (e.g., VT100) and printer control sequences. Table A.3 lists the control codes and their meanings.

Table A.3
Control Codes

Code	Ctrl	Meaning
NUL	^@	Null; padding or timing; end of string in C ('\0')
SOH	^A	Start of heading; beginning of the header field
STX	^B	Start of text; start of data part of packet
ETX	^C	End of text; end of data, often checksum follows
EOT	^D	End of transmission
ENQ	^E	Enquiry; query for status
ACK	^F	(Positive) Acknowledge; Yes/OK/go ahead
BEL	^G	Bell; ding!
BS	^H	Backspace; move cursor back over previous character
HT	^I	Horizontal tab; skip forward to next tab stop ('\t')
LF	^J	Line feed; advance cursor to next line ('\n')
VT	^K	Vertical tab
FF	^L	Form feed; erase screen (eject page)
CR	^M	Carriage return; return cursor to first column ('\r')
SO	^N	Shift out; shift out of one mode
SI	^O	Shift I; shift back into previous mode
DLE	^P	Data link escape; interpret following char literally
DC1	^Q	Device control 1; device specific; also XON (continue)
DC2	^R	Device control 2; device specific
DC3	^S	Device control 3; device specific; also XOFF (pause)
DC4	^T	Device control 4; device specific
NAK	^U	Negative acknowledge: error, no, not right now
SYN	^V	Synchronous idle; lead-in character in synch data stream
ETB	^W	End of transmission block
CAN	^X	Cancel; abort
EM	^Y	End of medium
SUB	^Z	Substitute
ESC	^[	Escape; modify meaning of following character(s)
FS	^\	File separator

Code	Ctrl	Meaning
GS	^]	Group separator
RS	^^	Record separator
US	^_	Unit separator

ISO 8859

US-ASCII is sufficient for English, but it is missing many of the characters necessary even for Western European languages (German, French, etc.). Those languages can be handled by defining characters for binary representations in the range 128 to 255 decimal (which together with the basic ASCII characters from 0 to 127, are sometimes called 8-bit codes). Various sets of definitions for those characters are in use for different language groups. ISO 8859 is an international standard that defines 10 extended character sets, each with US-ASCII as the first 128 and various extended characters in the upper 128.

The character sets defined in ISO 8859 follow. In each case, the printable extended characters in the range from 0xA0 to 0xFF are shown. The characters from 0x00 to 0x7F are the same as US-ASCII in each case. An e-mail package that supports US-ASCII and 8859-1, -2, -3, -5, -7, and –9 character sets would cover most Western languages. Messages using any of the 8859-x character sets would need to use one of the MIME schemes for 8-bit content (8BITMIME, quoted printable, etc.).

The information on ISO 8859 and the drawings of the characters came from a Web page by Roman Czyborra, in Berlin. For the original Web page and much more information on 8859 (including BMP images of the 8859 character sets), see the following:

http://wwwwbs.cs.tu-berlin.de/~czyborra/charsets/

ISO 8859-1

This character set, also known as Latin-1, covers most Western European languages such as Albanian, Catalan, Danish, Dutch, English, Faeroese, Finnish, French, German, Galician, Irish, Icelandic, Italian, Norwegian, Portuguese, Spanish, and Swedish.

	¡	¢	£	¤	¥	¦	§	¨	©	ª	«	¬	–	®	¯
°	±	²	³	´	µ	¶	·	¸	¹	º	»	¼	½	¾	¿
À	Á	Â	Ã	Ä	Å	Æ	Ç	È	É	Ê	Ë	Ì	Í	Î	Ï
Ð	Ñ	Ò	Ó	Ô	Õ	Ö	×	Ø	Ù	Ú	Û	Ü	Ý	Þ	ß
à	á	â	ã	ä	å	æ	ç	è	é	ê	ë	ì	í	î	ï
ð	ñ	ò	ó	ô	õ	ö	÷	ø	ù	ú	û	ü	ý	þ	ÿ

ISO 8859-2

This character set, also known as Latin-2, covers most Latin-written Slavic and Central European languages, such as Czech, German, Hungarian, Polish, Rumanian, Croatian, Slovak, and Slovene.

	Ą	˘	Ł	¤	Ľ	Ś	§	¨	Š	Ş	Ť	Ź	–	Ž	Ż
°	ą	˛	ł	´	ľ	ś	ˇ	¸	š	ş	ť	ź	˝	ž	ż
Ŕ	Á	Â	Ă	Ä	Ĺ	Ć	Ç	Č	É	Ę	Ë	Ě	Í	Î	Ď
Đ	Ń	Ň	Ó	Ô	Ő	Ö	×	Ř	Ů	Ú	Ű	Ü	Ý	Ţ	ß
ŕ	á	â	ă	ä	ĺ	ć	ç	č	é	ę	ë	ě	í	î	ď
đ	ń	ň	ó	ô	ő	ö	÷	ř	ů	ú	ű	ü	ý	ţ	˙

ISO 8859-3

This character set is suitable for Esperanto, Galician, Maltese, and Turkish.

	Ħ	˘	£	¤		Ĥ	§	¨	İ	Ş	Ğ	Ĵ			Ż
°	ħ	²	³	´	µ	ĥ	·	¸	ı	ş	ğ	ĵ	½		ż
À	Á	Â		Ä	Ċ	Ĉ	Ç	È	É	Ê	Ë	Ì	Í	Î	Ï
	Ñ	Ò	Ó	Ô	Ġ	Ö	×	Ĝ	Ù	Ú	Û	Ü	Ŭ	Ŝ	ß
à	á	â		ä	ċ	ĉ	ç	è	é	ê	ë	ì	í	î	ï
	ñ	ò	ó	ô	ġ	ö	÷	ĝ	ù	ú	û	ü	ŭ	ŝ	˙

ISO 8859-4

This character set is suitable for Estonian, Latvian, and Lithuanian (since replaced by ISO 8859-10).

	Ą	ĸ	Ŗ	¤	Ĩ	Ļ	§	¨	Š	Ē	Ģ	Ŧ		Ž	¯
°	ą	˛	ŗ	´	ĩ	ļ	ˇ	¸	š	ē	ģ	ŧ	Ŋ	ž	ŋ
Ā	Á	Â	Ã	Ä	Å	Æ	Į	Č	É	Ę	Ë	Ė	Í	Î	Ī
Đ	Ņ	Ō	Ķ	Ô	Õ	Ö	×	Ø	Ų	Ú	Û	Ü	Ũ	Ū	ß
ā	á	â	ã	ä	å	æ	į	č	é	ę	ë	ė	í	î	ī
đ	ņ	ō	ķ	ô	õ	ö	÷	ø	ų	ú	û	ü	ũ	ū	˙

ISO 8859-5

This character set contains Cyrillic characters for Bulgarian, Byelorussian, Macedonian, Russian, Serbian, and Ukrainian (although another character set called KOI8-R, not part of ISO 8859, is preferred for Russian).

	Ё	Ђ	Ѓ	Є	Ѕ	І	Ї	Ј	Љ	Њ	Ћ	Ќ	–	Ў	Џ
А	Б	В	Г	Д	Е	Ж	З	И	Й	К	Л	М	Н	О	П
Р	С	Т	У	Ф	Х	Ц	Ч	Ш	Щ	Ъ	Ы	Ь	Э	Ю	Я
а	б	в	г	д	е	ж	з	и	й	к	л	м	н	о	п
р	с	т	у	ф	х	ц	ч	ш	щ	ъ	ы	ь	э	ю	я
Ñ	ё	ђ	ѓ	є	ѕ	і	ї	ј	љ	њ	ћ	ќ	§	ў	џ

ISO 8859-6

This character set contains Arabic characters. Due to the fact that these characters change shape significantly depending on where they occur in a word, a much more complex scheme for character rendering is required than with other languages. To complicate things even further, it is written from right to left.

ISO 8859-7

This character set contains modern Greek.

	'	'	£			¦	§	¨	©		«	¬			—
°	±	²	³	΄	΅	Ά	·	Έ	Ή	Ί	»	Ό	½	Ύ	Ώ
ΐ	Α	Β	Γ	Δ	Ε	Ζ	Η	Θ	Ι	Κ	Λ	Μ	Ν	Ξ	Ο
Π	Ρ		Σ	Τ	Υ	Φ	Χ	Ψ	Ω	Ϊ	Ϋ	ά	έ	ή	ί
ΰ	α	β	γ	δ	ε	ζ	η	θ	ι	κ	λ	μ	ν	ξ	ο
π	ρ	ς	σ	τ	υ	φ	χ	ψ	ω	ϊ	ϋ	ό	ύ	ώ	

ISO 8859-8

This character set contains Hebrew, which is written from right to left.

		¢	£	¤	¥	¦	§	¨	©	×	«	¬		®	¯
°	±	²	³	´	µ	¶	·	¸	¹	÷	»	¼	½	¾	
															=
א	ב	ג	ד	ה	ו	ז	ח	ט	י	ך	כ	ל	ם	מ	ן
נ	ס	ע	ף	פ	ץ	צ	ק	ר	ש	ת					

ISO 8859-9

This character set, also known as Latin-5, is basically the Latin-1 set with the Icelandic characters replaced with Turkish ones.

	¡	¢	£	¤	¥	¦	§	¨	©	ª	«	¬	-	®	¯
°	±	²	³	´	µ	¶	·	¸	¹	º	»	¼	½	¾	¿
À	Á	Â	Ã	Ä	Å	Æ	Ç	È	É	Ê	Ë	Ì	Í	Î	Ï
Ğ	Ñ	Ò	Ó	Ô	Õ	Ö	×	Ø	Ù	Ú	Û	Ü	İ	Ş	ß
à	á	â	ã	ä	å	æ	ç	è	é	ê	ë	ì	í	î	ï
ğ	ñ	ò	ó	ô	õ	ö	÷	ø	ù	ú	û	ü	ı	ş	ÿ

ISO 8859-10

This character set, also known as Latin-6, replaces ISO 8859-4 (Latin-4) by adding Inuit (Greenlandic) and Sami (Lappish) characters.

	Ą	Ē	Ģ	Ī	Ĩ	Ķ	§	Ļ	Đ	Š	Ŧ	Ž	-	Ū	Ŋ
°	ą	ē	ģ	ī	ĩ	ķ	·	ļ	đ	š	ŧ	ž	-	ū	ŋ
Ā	Á	Â	Ã	Ä	Å	Æ	Į	Č	É	Ę	Ë	Ė	Í	Î	Ï
Đ	Ņ	Ō	Ó	Ô	Õ	Ö	Ũ	Ø	Ų	Ú	Û	Ü	Ý	Þ	ß
ā	á	â	ã	ä	å	æ	į	č	é	ę	ë	ė	í	î	ï
đ	ņ	ō	ó	ô	õ	ö	ũ	ø	ų	ú	û	ü	ý	þ	ĸ

Unicode

US-ASCII supplemented with ISO-8859 is sufficient for most European languages, but it falls far short for East Asian languages. A much larger character set, requiring 16 bits for each character, called Unicode, includes essentially all human languages (even ideographic ones like Chinese and syllabic ones like Korean Hangul). Unicode happens to be the native character set of Windows NT (to the best of my knowledge, it is the only such operating system in the world). The NTFS file system actually stores filenames (and information such as the owner of the file) in Unicode. For details on Unicode, see *The Unicode Standard, Version 2.0*, by the Unicode Consortium (Reading, MA: Addison-Wesley, 1996, ISBN 0-201-48345-9).

There are standards for how to represent full Unicode in either 7-bit US-ASCII (UTF-7) or in 8-bit codes (UTF-8). An e-mail client that supports entry, editing, transmission, and display of arbitrary Unicode content could truly be called world-aware software and be equally suitable for use in essentially any country (and even allow mixed languages in a single message, e.g., English and Chinese). To the best of my knowledge, there currently is no such program, but you can bet it will happen first in Windows NT.

Contents

Netography

All technical books must have a bibliography (as does this text). This book is one of the first, however, to incorporate a new section—a "netography"—which I hope will become standard. A list of references to relevant information on the Internet, this netography should serve as a starting point to help you go well beyond the information contained in this book or to further pursue specific topics.

Three types of references are included here:

▶ Websites, which are relevant pages on the World Wide Web, to be viewed with a Web browser, such as Microsoft's Internet Explorer;

▶ Usenet newsgroups, which are relevant news groups, to be viewed and responded to with a news reader program, such as the one in Microsoft's Outlook Express (part of Internet Explorer 4.0 and later);

▶ Mailing lists, which are public e-mail mailing lists (as managed by list managers such as Majordomo or L-Soft's List Manager), to be read and responded to with an Internet e-mail client, such as the one in Microsoft's Outlook Express.

Websites

Cryptography and privacy

Cryptography, PGP and your Privacy
http://world.std.com/~franl/crypto.html

Crypto-Log
http://www.enter.net/~chronos/cryptolog1.htmlElectronic
Frontier Foundationhttp://www.eff.org/

Encryption Policy Resource Page
http://www.crypto.com/

Electronic Privacy Information Center
http://www.epic.org/

Privacy International
http://www.privacy.org/pi/

The Privacy Forum
http://www.vortex.com/privacy.html

International Cryptography Software Page
http://www.cs.hut.fi/crypto/

PGP, Inc. Homepage (Phil Zimmerman's new company)
http://www.pgp.com

MIT distribution site for PGP
http://web.mit.edu/network/pgp.html

Popular Cryptography—The Journal of Internet Privacy
http://www.eskimo.com/~joelm/

Quadralay's Cryptography Archive
http://www.austinlinks.com/Crypto/

Ronald L. Rivest's homepage (the R of RSA)
http://theory.lcs.mit.edu/~rivest/

The National Security Agency Website
http://www.nsa.gov:8080/

Security Resource net
http://nsi.org

Stronghold (secure Web server)
https://stronghold.c2.net/

International Computer Security Association
http://www.icsa.com/

National Institute of Standards and Technology
Computer Security Resource Clearinghouse
http://csrc.ncsl.nist.gov/

Computer and Network Security Reference Index
http://www.telstra.com.au/info/security.html

Organization for Economic Co-operation and Development
http://www.oecd.org

Bruce Schneier's homepage
(author of leading text on computer cryptography)
http://www.counterpane.com

Michael Froomkin's homepage (University of Miami School
of Law)
http://www.law.miami.edu/~froomkin/

E-mail

E-Mail Web Resources
http://andrew2.andrew.cmu.edu/cyrus/email/email.html

IMAP homepage
http://www.imap.org

ACAP homepage
http://andrew2.andrew.cmu.edu/cyrus/acap/

X.500/LDAP

```
ISODE Consortium
http://www.isode.com/
```

Usenet newsgroups

E-mail

```
Info.ietf.smtp
Comp.mail.imap
Comp.mail.headers
Comp.mail.mime
Comp.mail.misc
Comp.mail.multi-media
Comp.mail.sendmail
lt.e-mail.lists
Alt.mail.com-priv
Alt.mail.firewalls
Microsoft.public.internet.mail
```

Usenet (NNTP)

```
News.software.nntp
Ucb.news.stats
```

Privacy

```
Alt.privacy
Alt.privacy.anon-server
Alt.privacy.clipper
Alt.security.pgp
Comp.security.privacy
```

Cryptography

```
Alt.sources.crypto
Sci.crypt.research
Talk.politics.crypto
Demon.security
```

Firewalls

```
Comp.security.firewalls
Comp.security.misc
```

Mailing Lists

To obtain information on the relevant mailing lists in Table B.1, including how to subscribe, send e-mail to the associated address with the one-line message body "info<list-name>", for example:

info imap

Table B.1
Relevant Mailing Lists

List	Address
Security-email	majordomo@apollo.it.hq.nasa.gov
IMAP	listproc@u.washington.edu
Firewalls	majordomo@greatcircle.com
Ietf-sectest	majordomo@toad.com
WINSEC	listserve@mailbag.intel.com

C

Contents

Contents of the CD-ROM

These days, any respectable technical book, especially one on computers, has an associated CD-ROM. This book is no exception. In the process of mastering this subject and writing this book, I ran across an amazing quantity of information related to Internet e-mail. I collected the useful public domain material and put it on the accompanying CD-ROM to save you pointless hours of finding and downloading things such as RFCs.

Drafts

Relevant IETF drafts (standards documents in progress).

MailRFCs

The RFCs referred to in this book, organized by protocol:

ACAP
DHCP
DNS
IMAP2
IMAP4
LDAP
MIME
NetBIOS
PGP
POP2
POP3
SMIME2
SMIME3
SMTP
X500

PKCS

RSA's Public Key Cryptography Standards:

PKCS #1
PKCS #3
PKCS #5
PKCS #7
PKCS #8
PKCS #9
PKCS #10

AllRFCs

All RFCs, grouped by number, current as of the writing of this book:

RFC00xx
RFC01xx

.

.

.

RFC22xx

SSLEAY

Information on public key development tools in the public domain

SSLEAY_FAQ.HTM

Bibliography

In such a rapidly changing field as the Internet, you should make liberal use of the search capabilities of online bookstores, such as:

www.bookpool.com
www.amazon.com

Cryptography, computer security

Warwick Ford and Michael S. Baum, *Secure Electronic Commerce*, Prentice Hall, ISBN 0-13-476342-4, $48.00.

Simson Garfincel, *PGP—Pretty Good Privacy*, O'Reilly & Associates, ISBN 1-56592-098-8, $29.95.

Bruce Schneier, *Applied Cryptography*, 2nd ed., John Wiley, ISBN 0471117099, $49.95.

William Stalling, *Practical Cryptography for Data Internetworks*, IEEE Computer Society Press, ISBN 0-8186-7140-8.

Rita Summers, *Secure Communications: Threats & Safeguards*, McGraw-Hill, ISBN 0070694192, $60.00.

Policy, legal, and social issues

Computer Science and Telecomm Board, National Research Council, *Cryptography's Role in Securing the Information Society (CRISIS)*, National Academy Press, ISBN 0-309-05475-3, $44.95.

Winn Schwartau, *Information Warfare*, 2nd ed., Thunder's Mouth Press, ISBN 1-56025-132-8, $16.95.

Windows NT

Rutstein, *Windows NT Security: A Practical Guide*, McGraw Hill, ISBN 0-070-57833-8, $34.95.

Tom Sheldon, *Windows NT Security: Everything You Need to Know to Protect Your Network*, McGraw Hill, 1997, ISBN 0-078-82240-8, $34.99.

Stephen Sutton (Trusted Systems Services, Inc.), *Windows NT Security Guide*, Addison-Wesley, ISBN 0-201-41969-6, $29.95.

Secure e-mail

Bruce Schneier, *E-mail Security—How to Keep Your Electronic Messages Private (PGP/PEM)*, John Wiley & Sons, ISBN 0-471-05318-X, $24.95.

Firewalls

Cheswick and Belloni, *Firewalls and Internet Security: Repelling the Wily Hacker*, Addison-Wesley, ISBN 0201633574, $35.45.

Authentication

Oppliger, *Authentication Systems for Secure Networks*, Artech House, ISBN 0890055101, $59.00.

About the author

Lawrence Hughes obtained his B.S. in pure math from Florida State University in 1973. While a student, and after graduation, he worked at the Computing Center where he pioneered a microcomputer applications group.

In 1982 he founded his own company, Mycroft Labs, which he ran until 1987. Mr. Hughes also spent five years with Intergraph Corporation, becoming a senior system engineer for their Asia/Pacific branch, and worked as a principal software engineer for an Internet startup company in California. He has written a full function secure SMTP and POP3 server for Windows NT, called KIMS (still available from Hayes Computer Systems, at www.hcsys.com).

Mr. Hughes is currently the Director of Security Products at SecureIT, Inc. (www.secureit.com), based in Atlanta, where he heads a development group creating security products for Windows NT and UNIX. He has several certifications, including a Microsoft Certified System Engineer (MCSE), and the McAfee Certified Security Specialist. He has also published several articles on messaging and security in *Windows NT* magazine. He can be reached at lawrence.hughes@mindspring.com.

Index

MX record. *See* Mail exchange record
Myrights command, 370

N

Nack command, 304–5
Named pipe, 12–13
Name resolution, 147–50
Name server record, 155
Namespace, 153
National Institute of Standards and
  Technology, 76, 112
National Security Agency, 70, 73, 76
NDIS. *See* Network Device Interface
  Specification
NDS. *See* NetWare Directory Services
NetBEUI, 144
NetBIOS, 26, 138, 144–45, 188
Netscape Navigator, 137, 204
NetWare Directory Services, 57
Network access layer, 126–32
Network architecture, 27–28
Network Device Interface
  Specification, 131
Network File System, 135, 138, 145
Network General, 218
Network Information Center, 176
Network interface card, 27, 127–31,
  200, 204, 219
Network layer, 200
Network Monitor, 212, 217–20
Network News Transfer Protocol,
  117–18, 136–37, 205, 337,
  377–80
  commands, 384–94
  sample session, 394–96
Network operating system, 9
Network protocol, 24
Newgroups command, 392
New network connection, 230–31
Newnews command, 393
News Client, 140

Newsgroup, 117–18, 137, 337,
  377–80
  See also Network News Transfer
  Protocol
NewsNet. *See* Network News
  Transfer Protocol
New technology file system,
  88, 424–25
Next command, 391
Next server state, 301
NFS. *See* Network File System
NIC. *See* Network Information
  Center; Network interface
  card
NICNAME, 176
NIST. *See* National Institute of
  Standards and Technology
NNTP. *See* Network News
  Transfer Protocol
Node, 26, 150
Nodename, 28, 138–39, 147–49
Nonauthenticated state, 339–40,
  343–53
Nonsynchronizing literals
  extension, 370–71
Noop command, 233, 237, 244–45,
  321, 342–43, 353
NOS. *See* Network operating system
No separate news reader
Notepad, 7
Notify keyword, 282–84
Novell, 10, 178, 188, 218
N-Plex Internet Mail, 217
NSA. *See* National Security Agency
Nslookup, 156
NS record. *See* Name server record
NTFS. *See* New technology file
  system
NT/LAN Manager, 191
NTLM. *See* NT/LAN Manager
NT service, 116–17
Number in data representation, 340

Setquota command, 373–74
7bit encoding, 271
SGML. *See* Standard Generalized
    Markup Language
SHA. *See* Secure Hash Algorithm
Shared address book, 55
Shared-file system, 8–9, 113
    MS Mail, 113, 115
    scalability, 11, 16–18
    versus client/server
        system, 10–15
Shared folder, 338
Shared mailbox, 337–38
Shared secret, 308–11, 314
Simple Mail Transport Protocol, 7,
        15, 19, 23–24, 116, 120,
        135, 139–40, 155, 183–84,
        192, 205, 215, 221–22,
        335, 339
    basic protocol, 229–37
    commands and syntax, 237–39
    Delivery Status
        Notification, 277–86
    extended, 239–42
    extensions, 252–53
    header definition, 223–29
    large/binary message, 247–50
    message syntax, 222–23
    MIMEtransport, 246–47
    pipelining, 244–46
    remote queue, 250–52
    requests for comment, 253–55
    size extension, 242–44
Simple Mail Transport Protocol
    proxy, 209
Single-part organization, 263–68
Single-server system, 17
Single-threaded system, 13, 28
Size extension, 242–44
S/KEY encryption, 102, 104–5,
        315, 344
SKIPJACK, 73–74

Slave command, 394
SMB. *See* Server Message Block
S/MIME. *See* Secure/Multipurpose
    Internet Mail Extension
SMP. *See* Symmetric multiprocessor
SMS. *See* System Management
    Server
SMTP. *See* Simple Mail Transport
    Protocol
Sniffer, 218–19
SNMP, 29, 212
Snooper, 218
SOA record. *See* Start of authority
    record
Socket, 12–13, 39, 205
Software, network, 131–32
Software layer, 27
Soml command, 236–38, 242, 244
Source Quench, 133
Spoofing, 201
SPX, 178, 202, 204, 206
SQL Server, 12
SSL. *See* Secure socket layer
Standard Generalized Markup
    Language, 264, 266
Standards. *See* Open standards
Standards bodies, 110–12
Star architecture, 127, 130
Start of authority record, 154
Stat command, 317, 390
Status codes, SMTP, 288–91
Status command, 351–52
Status field, 294
Steganography, 61
Store and forward system, 31
Store command, 356–57
Stream cipher, 68–69
StreetTalk, 57, 188
String, 341
Subdomain, 151
Subscribe command, 349–50
Symmetric key encryption, 65, 315

The Artech House Telecommunications Library

Vinton G. Cerf, Series Editor

X Window System User's Guide, Uday O. Pabrai

For further information on these and other Artech House titles, including previously considered out-of-print books now available through our In-Print-Forever™ (IPF™) program, contact:

Artech House   Artech House
685 Canton Street   Portland House, Stag Place
Norwood, MA 02062   London SW1E 5XA England
781-769-9750   +44 (0) 171-973-8077
Fax: 781-769-6334   Fax: +44 (0) 171-630-0166
Telex: 951-659   Telex: 951-659
email: artech@artech-house.com   email: artech-uk@artech-house.com

Find us on the World Wide Web at: www.artech-house.com